The Aging Brain
Neurological and
Mental Disturbances

ETTORE MAJORANA INTERNATIONAL SCIENCE SERIES

Series Editor:
Antonino Zichichi
European Physical Society
Geneva, Switzerland

(LIFE SCIENCES)

The Aging Brain
Neurological and Mental Disturbances

Edited by

G. Barbagallo-Sangiorgi
University of Palermo
Palermo, Italy

and

A.N. Exton-Smith
University College Hospital Medical School
London, England

Plenum Press · New York and London

Library of Congress Cataloging in Publication Data

International School of Physiopathology and Clinic of the Third Age, 2d, Ettore
Majorana Center for Scientific Culture, 1980. The Aging brain.

(Ettore Majorana international science series: Life sciences; v. 5)
"Proceedings of the second course of the International School of Physiopathology
and Clinic of the Third Age, held at the Ettore Majorana Center for Scientific Culture,
Erice, Sicily, Italy, March 3-8, 1980."
Includes index.
1. Brain—Diseases—Age factors—Congresses. 2. Brain—Aging—Congresses. 3. Senile
dementia—Congresses. 4. Geriatrics—Congresses. I. Barbagallo-Sangiorgi, G. II. Exton-
Smith, A. N. III. Title. IV. Series. [DNLM: 1. Aging—Congresses. 2. Brain diseases—
In old age—Congresses. 3. Mental disorders—In old age—Congresses. WT 104 A26741
1980]
RC386.2.I54 1980   616.97'68   80-24567
ISBN 0-306-40625-X

Proceedings of the Second Course of the International School of
Physiopathology and Clinic of the Third Age, held at the Ettore
Majorana Center for Scientific Culture, Erice, Sicily, Italy,
March 3-8, 1980.

© 1980 Plenum Press, New York
A Division of Plenum Publishing Corporation
227 West 17th Street, New York, N.Y. 10011

Printed in the United States of America

PREFACE

This symposium on the Aging Brain held at the Ettore Majorana Center, Erice is notable for two important reasons:

1) It was the first symposium to be held jointly between the Italian Society of Gerontology and Geriatrics and the British Geriatrics Society.

2) It provided the opportunity for a multi-disciplinary approach to the studies of the effects of aging and of disease processes on brain function by involving experts in the fields of geriatrics, neurology, biochemistry, pharmacology, psychiatry and psychology.

It became apparent during the course of the symposium that there are many similarities between the two countries in the ways in which these difficult studies are being undertaken. The sheer magnitude of the social and clinical problems imposed by dementia and other disabling neurological conditions,together with the fact that the patients suffering from these disorders are the greatest consumers of health and social services in all developed countries is only just beginning to provide the necessary stimulus for research. Thus our lack of understanding is in contrast to the importance of the subject in terms of the devastating nature of the illnesses and the enormous size of the problem. The scantiness in research can in part be attributed to the fact that patients are cared for in institutions remote from centres with research facilities and expertise. In recent years the opportunities for collaborative studies between clinicians and other scientific workers have increased in both Italy and the United Kingdom with the establish-ment of academic departments of gerontology and geriatrics. We trust that the proceedings of this symposium as reported in this book will indicate the number of promising pointers which have already emerged and the even greater opportunities that lie ahead with the strengthening of the links between clinical disciplines and centres of laboratory expertise.

G. Barbagallo-Sangiorgi
A. N. Exton-Smith

CONTENTS

FOREWORD AND WELCOME

G. Barbagallo Sangiorgi

Ladies and Gentlemen,

It is a great pleasure for me to greet and welcome all of you to the Second Course of the International School of Physiopathology and Clinic of the Third Age, of the E. Majorana Center. I particularly welcome Professor Exton-Smith, The President of the British Geriatrics Society and Professor Butturini, the President of the Italian Society of Gerontology and Geriatrics and the British and Italian workers in the fields of geriatrics, neurology, psychiatry and psychology taking part in this Course.

The Ettore Majorana Center, the town of Erice and Sicily are glad and proud to be able to have you as guests and to give you the opportunity of discussing the subject of the "Aging Brain" in the coming days.

The idea of organizing a joint Course under the auspices of the British Geriatrics Society and the Italian Society of Gerontology and Geriatrics arose two years ago here, in Erice, during the First Course. Professor Exton-Smith was invited with other Professors to take part in this course and with him we laid the foundation for the organization of the present Course. We immediately agreed on discussing the "Aging Brain", an interesting, fascinating and important subject with psychological, medical and social implications.

We agreed also on an interdisciplinary approach to the most important aspects of the subject chosen and I hope that the programme we have worked out will comply with this requirement both for the wide range of subjects to be discussed and for the

xi

high qualification of the invited speakers.

During the next four days, we shall hear talks by several important British and Italian speakers: geriatricians, neurologists, psychiatrists, neurochemists, neuropharmacologists and rehabilitation specialists, all particularly advanced in their fields of research.

As to the choice of the individual subjects, we have focussed attention on metabolic and biochemical alterations in the ageing brain, since we believe that cerebral failure in aged subjects is chiefly determined by biochemical, enzymatic and molecular changes rather than by vascular mechanisms, or, if you like, besides vascular ones.

I think we can expect the most decisive results for the better understanding of the aging brain from investigations concerning enzyme structure, the mechanisms of energy production, protein synthesis and the complete identification of the role and level of neurotransmitters. Once such knowledge has been acquired, we shall perhaps be able to devise more easily and successfully the correct therapeutic means for effective prevention and treatment of brain aging than has been possible to date.

From a pathological point of view we must now consider, besides the typical vascular syndromes, the existence of a wide range of conditions in which a disorder of metabolism may either play the primary role or represent the triggering event, whose importance - though superimposed - is not a secondary one.

In a number of cases the secondary, superimposed, metabolic disorder may be identified and corrected, and this is sometimes sufficient to improve the clinical symptoms, although the ischaemic disorder may remain unchanged.

Among such conditions iatrogenic damage plays an important role, and this is often due to the use of psycho-active drugs.

In the present Course, however, much attention is devoted to investigations on circulation pathology and ischaemic disorders, and particularly to the reversible ones, which have been the object of the most recent studies.

We shall also devote one morning to psychometric techniques for the evaluation of mental disorders. It is a particularly difficult field and everyone of us is aware of how arduous it often is to come to reasonable conclusions, i.e. conclusions of general validity and controllable in time. For this reason we want to offer the opportunity of comparing the various methods and we hope that both the negative and the positive aspects may emerge in order that they may be tested and evaluated.

We shall also discuss psychogeriatric problems both from the clinical and the service point of view and the clinical and therapeutic aspects of Parkinson's syndrome.   Rehabilitation problems and techniques will also be discussed fully.

I hope that the fine weather will anticipate the coming spring so that you may better enjoy our country and your stay here, and I wish you success in your work.

I now have the honour of declaring the Symposium open.

AGING AND BRAIN ENZYMES

G. Benzi, E. Arrigoni, F. Dagani, F. Marzatico,
D. Curti, M. Polgatti and R. F. Villa

Department of Science
Institute of Pharmacology, University of Pavia
Italy

ABSTRACT

Age-dependent changes of some cerebral enzymatic activities
(lactate dehydrogenase; citrate synthase and malate dehydrogenase;
total NADH-cytochrome c reductase and cytochrome oxidase) were
studied in the homogenate in toto and/or in the crude mitochondrial
fraction of the brain in rats aged 20, 60, 100 and 140 weeks. With
age, from youth to senescence, all the activities studied exhibited
a natural decrease to low values. The drugs tested (trimetazidine,
papaverine, vincamine, theophylline, nicergoline, CDP-choline) were
administered daily for periods of 4 weeks each (16-20, 56-60,
96-100 and 136-140 weeks of life) by intraperitoneal route and at
one dose level (1 or 5 mg/kg). The drugs tested exerted typical
effects on the various enzymatic activities of the brain. At any
rate, the range of drug-interference with these enzymatic activities narrowed remarkably during maturity and even more during
senescence. The possible mechanisms of such behaviour are
discussed.

INTRODUCTION

In a typical post-mitotic tissue such as the brain, the
evaluation of drug action in both maturity and senescence can be
conveniently tackled through the behaviour of enzymatic activities,
as an expression of genomic[1] or hormonal and metabolic functions[2].
However, the biochemical data are often in disagreement[3] because of
the relationships between enzymatic activity, protein synthesis and
age[4-6]. On the other hand, cerebral disorders can be classified[7]

1

into cerebrovascular insufficiency and psycho-organic disease, the latter including metabolic homeostasis derangements and decrease of correct enzyme-protein synthesis.   In this field, investigations on cerebral enzymatic activities have been limited to 30-40 days of life[8-11] or rarely performed systematically during ageing at short recording intervals[12,13], enzymatic activities being measured at irregular time intervals[14-17].

As far as the interference of the pharmacological treatment on cerebral enzymatic activities is concerned, in untreated and treated rats, age-dependent changes of some cerebral enzymatic activities were studied in the homogenate in toto and in the crude mitochondrial fraction of the whole brain from the 16th to the 28th week of age, at 4-week intervals[18].   The pharmacological response tended to be stronger after the first four weeks of treatment, while after 12 weeks it became less evident.   This fact might be related to a reduction in drug power due to prolonged treatment (e.g. because of drug induction), or to an intrinsic inefficiency of the drugs to stop the age-dependent decrement of cerebral enzymatic activities[18]. In order to verify this latter hypothesis, we evaluated the influence of age on the same subchronic (4-week) treatment, performed daily with one dose level of several drugs, in the rat. The ages studied (20, 40, 100 and 140 weeks of life) roughly correspond to adult, mature, senescent, and old rats.

The cerebral enzymatic activities studied were: lactate dehydrogenase (L-lactate: NAD^+ oxidoreductase, EC 1.1.1.27) for the glycolytic pathway; citrate synthase (citrate oxalo-acetate-lyase, EC 4.1.3.7) and malate dehydrogenase (L-malate: NAD^+ oxidoreductase, EC 1.1.1.37) for the Krebs' cycle; total NADH cytochrome c reductase (NADH-cytochrome c: oxygen oxidoreductase, EC 1.6.99.3) and cytochrome oxidase (ferrocytochrome c: oxygen oxidoreductase, EC 1.9.3.1) for the electron transport chain.   Some of the enzymatic activities were evaluated in both the homogenate in toto and the mitochondrial fraction, since many of them are variously located in the cytoplasm[19].

The substances tested were chosen on the basis of the pharmacological action which is generally attributed to them.   We therefore used: (a) papaverine and theophylline, as drugs exerting mainly a vascular action; (b) cytidine diphosphate choline, as an agent exerting mainly a metabolic action; (c) nicergoline, trimetazidine and vincamine, as drugs exerting both a vascular and metabolic action.   Obviously this subdivision is a vague and superficial one. This work is intended to overcome it, by detecting during youth, maturity and senescence the action of these substances on several enzymatic activities related to energy transduction, because no data are available in the literature on this subject.

MATERIALS AND METHODS

The study was carried out in male rats (Sprague-Dawley strain) fed a standard diet as pellets and water ad libitum and housed three, and subsequently two, per cage under optimal environmental conditions: 22°; 55-60% relative humidity; 12-hr day cycle (light from 7:00 a.m. to 7:00 p.m.) until they reached the age of 16-140 weeks. The rats were housed at first for one month under the fixed dark-light rhythm. In fact, the circadian activities of various enzymes are stable only after about 3 weeks in the adult rat at the same time of the nycthemeron. No observations were made on the possibility that circadian enzymatic fluctuations are significantly modified as the animal gets older. The initial allocation of animals to the different lots was made by random-ization. The time course of the examinations performed in the lots was established by means of permutation tables.

The drugs were administered between 9:10 and 10:40 a.m. Even though the time interval of daily treatment was the maximum on which to base comparative statistical analysis, this was at least partially excused by the number of animals utilized in the experiment. Within this time interval, the treatments were alternated so as to balance the circadian rhythm effect. Treatment was carried out daily (6 days a week), by intraperitoneal adminis-tration, using one dose of drug. The dose was strictly within the range used experimentally in order to characterize the pharamco-dynamic action of the different substances tested: (a) papaverine hydrochloride (papaverine) = 1 mg/kg as papaverine base; (b) vincamine theophyllinylpropane sulfonate (vincamine TPS) = 5 mg/kg as vincamine base; (c) sodium theophyllinylpropane sulfonate (theophylline PS) = 5 mg/kg as theophylline base; (d) cytidine diphosphate choline (CDP-choline) = 5 mg/kg; (e) trimetazidine dihydrochloride (trimetazidine) = 5 mg/kg as trimetazidine base; (f) nicergoline tartrate (nicergoline) = 1 mg/kg as nicergoline base. Control animals were given the vehicle only by the same route. Blind biochemical evaluations were performed after 4 weeks of treatment at 20, 60, 100 and 140 weeks of age.

All animals were killed between 9.30 and 10:10 a.m., 48 hours after the last injection. This interval is very important to differentiate the drug interference with basic cellular components or activities from the immediate effect of a sustained treatment with a drug. At the set time the animals were sacrificed by decapitation and their brains removed from the skull within 15 sec in a precooled box at -5°. The 0.32 M sucrose washed and weighed brains (without cerebellum) were homogenized in 0.32 M sucrose for 30 sec (precooled Potter-Braun S homogenizer). The homogenate obtained was diluted with 0.32 M sucrose (10% w/v) and an aliquot of each sample was taken for the assay of enzymatic activities. The remaining homogenate was submitted to a series of centrifug-

ations (Sorvall RC-5 Supercentrifuge) for the preparation of the
crude mitochondrial fraction[20] obtained at 14,000 g for 20 min.
On both the homogenate and the mitochondrial preparation samples,
protein content was evaluated[21] and the following enzymatic
activities were measured: malate dehydrogenase[22]; total NADE-cyto-
chrome c reductase[23]; cytochrome oxidase[24,25]. The activity of
lactate dehydrogenase was evaluated only in homogenate samples[26]
while that of citrate synthase was;measured only in the mitochondrial
preparation samples[27]. Enzymatic activities were recorded
(Beckman 25 Spectrophotometer Recorder) and calculated using the
straight portion of the reaction curves. Results were expressed as
specific activities: $\mu moles.min^{-1}$. $(mg\ protein)^{-1}$. Two statistical
tests (Anova and Dunnett's tests) were applied to these results
after checking the homogeneity of variance by the Bartlett's test.
Anova was employed to evaluate: (a) the enzymatic activities by
times interactions (i.e. to detect a possible difference in the
activity of the various enzymes as a function of age); (b) the
treatments by times interactions (i.e. to detect a possible difference in effect of the various drugs as a function of treatment time).
The Dunnett's test was used, at each individual time, to assess
differences between the cerebral enzymatic activities of controls
and those of treated rats.

RESULTS

        As shown in Figure 1, in control rats aging (from young to
adult animal, up to maturity or senescence) caused a significant
decrease in the cerebral enzymatic activities tested, senescence
values (140 weeks of age) corresponding to 1/2 - 1/4 of those found
in young adults (20 weeks of age). As for the "treatments by
times" interactions, Anova failed to exhibit any significant
difference between the values of the enzymatic activities evaluated
in the brain of control rats and of those treated for one month with
the various drugs, at the different ages. Therefore, the physiological behaviour of cerebral enzymatic activities during treatment
time prevailed over the changes which could be altogether induced by
drugs. The results reported below are thus described with regard
to significant differences pointed out by the Dunnett's test at each
tested time (20, 60, 100 or 140 weeks of age) as shown in Figures
2, 3, 4 and 5.

        The results show that each pharmacological treatment induced
different and specific changes in some of the enzymatic activities
tested in the homogenate in toto and/or in the crude mitochondrial
fraction. However, the range of the pharmacological interference
progressively narrowed with age. Thus one month treatment with
papaverine caused, at 20 weeks of age, an increase of lactate
dehydrogenase, of malate dehydrogenase and of cytochrome oxidase,
as evaluated in the homogenate in toto. At 60 and 100 weeks, the

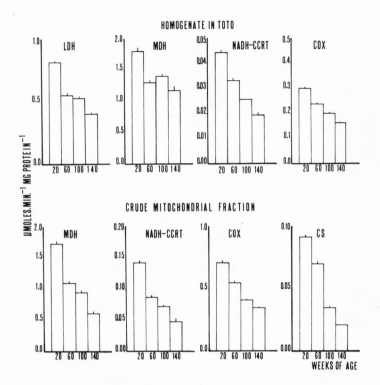

Fig. 1. Rat cerebral enzymatic activities related to the energy
         transduction evaluated at 20, 60, 100 and 140 weeks of
         age.

same treatment led only to the enhancement of lactate dehydrogenase
and cytochrome oxidase, no effect being observed at 140 weeks. One
month treatment with theophylline PS affected only lactate dehydrogenase at 20, 60 and 100 weeks of age, no effect being detected at
140 weeks.

     One month treatment with <u>trimetazidine</u> caused, at 20 weeks,
the inhibition of all the enzymatic activities tested in the
mitochondrial fraction, and the inhibition of cytochrome oxidase in
the homogenate <u>in toto</u>. Subsequently (60 and 100 weeks of age)
only cytochrome oxidase in the homogenate <u>in toto</u> and citrate
synthase were found to be inhibited. No inhibition was observed
at 140 weeks of age. After one month treatment with nicergoline,
at 20 weeks, cytochrome oxidase (homogenate <u>in toto</u>), malate
dehydrogenase (homogenate <u>in toto</u> and mitochondrial fraction) and
citrate synthase (mitochondrial fraction) were inhibited, while the
activity of NADH-cytochrome c reductase (homogenate <u>in toto</u> and
mitochondrial fraction) appeared to be magnified. At 60 and 100
weeks of age, these effects concerned only the total NADH-cytochrome

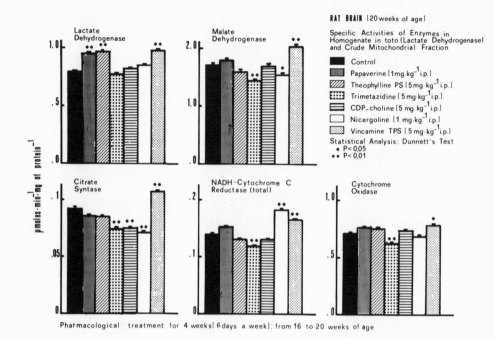

Fig. 2. Cerebral enzymatic activities related to the energy
         transduction.   Effect of the intraperitoneal pharma-
         cological treatment in rat for 4 weeks (6 days a week):
         from 16 to 20 weeks of age.
         The enzymatic activities (μmoles.min^{-1}. mg protein^{-1})
         were evaluated both in the homogenate in toto and in the
         mitochondrial fraction from the rat brain (age: 20 weeks)
         and are expressed as the mean values $\pm$ S.E.M. for each
         group of 6 animals.

c reductase and citrate synthase, while at 140 weeks they were no
longer detected.   On the other hand, one month treatment with
CDP-Choline led to the inhibition of mitochondrial citrate synthase,
this activity being still evident at 140 weeks.   Finally, one
month treatment with vincamine TPS was responsible at 20 weeks for
a magnification of the activity of lactate dehydrogenase and
cytochrome oxidase (as measured in the homogenate in toto) and of
citrate synthase, malate dehydrogenase, NADH-cytochrome c reductase,
cytochrome oxidase (as measured in the crude mitochondrial fraction).
At 60 and 100 weeks of age, one month treatment with vincamine TPS
led to a magnification of lactate dehydrogenase and cytochrome
oxidase enzymatic activities, as measured in the homogenate in toto,
as well as of citrate synthase and malate dehydrogenase in the
mitochondrial fraction.   At 140 weeks, after one month treatment
with the same drug, only lactate dehydrogenase and cytochrome

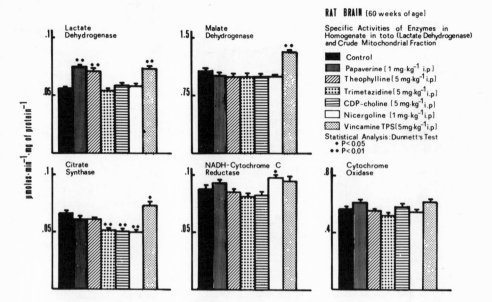

Fig. 3.   Cerebral enzymatic activities related to the energy
          transduction.   Effect of the intraperitoneal pharmaco-
          logical treatment in rat for 4 weeks (6 days a week):
          from 56 to 60 weeks of age.
          The enzymatic activities (μmoles.min^{-1}.mg protein^{-1})
          were evaluated both in the homogenate in toto and in the
          mitochondrial fraction from the rat brain (age:60 weeks)
          and are expressed as the mean values $\pm$ S.E.M. for each
          group of 6 animals.

oxidase (as measured in the homogenate in toto) were found to be
affected.

DISCUSSION

     Before discussing the present data, some problems must be
pointed out.   In the first place, these data refer to the cerebral
tissue as a whole: therefore, some important areas of the brain
might undergo different biochemical adjustments, because of the
functional and anatomical heterogeneity of the organ.   Secondly,
the methods used to determine the enzymatic activities abolish the
autoregulative interactions that maintain the cell as a functionally
integrated system.   Finally, it should be emphasized later, in
untreated rats, the different cerebral enzymatic activities
exhibited a dramatic drop as a function of age (Figure 1).

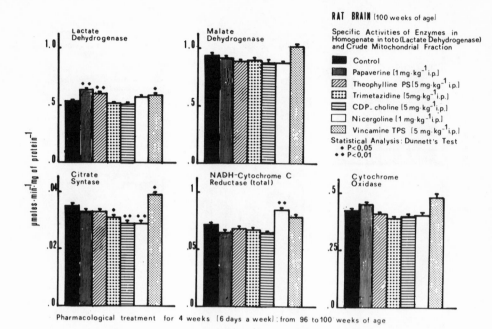

Fig. 4.   Cerebral enzymatic activities related to the energy
          transduction.   Effect of the intraperitoneal pharmaco-
          logical treatment in rat for 4 weeks (6 days a week):
          from 96 to 100 weeks of age.
          The enzymatic activities (μmoles.min^{-1} . mg protein^{-1})
          were evaluated both in the homogenate in toto and in the
          mitochondrial fraction from the rat brain (age:100 weeks)
          and are expressed as the mean values $\pm$ S.E.M. for each
          group of 6 animals.

Previous observations[13] showed this decrease (e.g. between the 14th
and the 60th week), observed at two-week intervals, to exhibit a
sinusoidal pattern.   At any rate, the overall decrease can be
ascribed both to enzyme alteration and to a reduction in the number
of enzyme molecules.   Our measurement of activity per mg of protein
in crude homogenate and in the crude mitochondrial fraction is not
suitable to give a direct answer to the problem, as a lower
activity could simply reflect the presence of fewer rather than
altered enzyme molecules.   However, the characteristics of "young"
and "old" enzymes appear to be unchanged, at least superficially,
no differences in molecular weight, charge, Km or behaviour towards
inhibitors being observed in any of the enzymes evaluated (enolase,
creatine kinase, aldolase, etc.)[28].

     Altered enzymes must arise by synthetic or post-synthetic
modification.   The case for synthetic alteration is highly

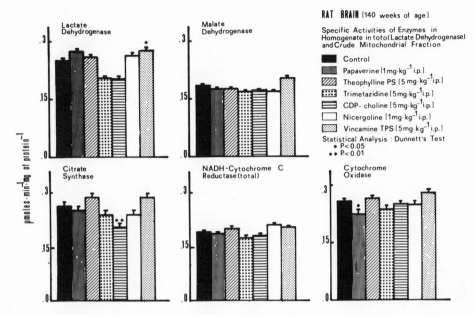

Pharmacological treatment for 4 weeks(6 days a week): from 136 to 140 weeks of age.

Fig. 5. Cerebral enzymatic activities related to the energy
transduction. Effect of the intraperitoneal pharmacological treatment in rat for 4 weeks (6 days a week):
from 136 to 140 weeks of age.
The enzymatic activities (μmoles.min^{-1}.mg protein^{-1}) were
evaluated both in the homogenate in toto and in the
mitochondrial fraction from the rat brain (age: 140 weeks)
and are expressed as the mean values $\pm$ S.E.M. from each
group of 5 animals.

questionable, since there can be no net change of charge and sequence
changes would have to be limited to replacing amino acids with
equivalent products. On the other hand, if the alteration is
related to the gene activity, each gene (representing an enzymatic
protein to be altered) would require a change in coding with this
limitation, and the error at translation level would affect all
proteins. Now, empirical data indicate that not all proteins are
affected[29-31]. The case for post-synthetic changes has more ground,
even if changes of properties through loss of end groups by proteo-
lysis[32] or modifications brought about by cellular factors[33] are
unlikely, as the concurrent changes in isoelectric focusing are not
found with altered enzymes. The change of conformation without
covalent changes is supported by several experimental observations
[28,34] and can be related to a slowing of protein turnover in old
organisms. The resulting increase in the "dwell time" for the

enzymes in cells would provide the opportunity for both a functional change and a subtle denaturation to occur without replacement of the altered enzymatic molecules[35,36]. On this basis: (a) stable enzymes would not become altered by age; (b) other enzymes would have one or more intermediate forms with partial activity (functional change or subtle denaturation; (c) other enzymes would denature directly to inactive forms (complete denaturation). With regard to a typical post-mitotic tissue, such as the cerebral one, the modifiable change in enzymatic conformation appears quite plausible for explaining the enzymatic activity changes (activation or inhibition) which are induced by the above-mentioned one month treatments in the rat.

However, it should be noted that, even when these changes in cerebral enzymatic activities do occur, their range is wider in youth (20 weeks of age) than during maturity and senescence (60, 100 and 140 weeks of age). It is therefore evident that, as the animal gets older, the pharmacological treatment has a reduced spectrum of action on enzymatic activities, without affecting those enzymatic activities which were not modified at earlier times. Each drug therefore exerts a specific and typical effect on cerebral enzymes, while ageing progressively narrows this effect, which becomes almost null during late senescence (140 weeks of age in the rat). The fact that, at given ages (i.e. 60 or 100 weeks), the pharmacological effect concerned some of the enzymatic activities affected at earlier times (i.e. 20 or 60 weeks) demonstrates that the phenomenon can not be ascribed to a lack of drug bioavailability. Indeed, if this were the case, all drug-modifiable enzymatic activities would have to be affected to the same extent. Since it is not so, it appears more reasonable to assume that the phenomenon concerns primarily enzyme structure. The hypothesis can therefore be put forward that the "dwell time" for the enzymes increases with age[28], the intermediate (partly active) forms of the respective enzymes, first exhibiting drug-modifiable changes of steric conformation appearing at different times, as the animal gets older. At subsequent times, these intermediate forms may exhibit small, drug-unmodifiable changes in steric conformation (subtle denaturation) and then reach complete denaturation. In the case of unaltered enzymes it should be assumed that they are reduced in amount, without turning into stable intermediate forms of these enzymes. The four age-dependent enzymatic stages described in Table 1 can therefore be hypothesized. In the absence of other empirical data, they can be regarded as a biological basis for physiopathological and pharmacological prospections.

In conclusion, this investigation of the interference in vivo of drugs on cerebral enzymatic activities goes deeper into the relationships between chronic pharmacological treatment and cerebral metabolism. Indeed, this study confirmed that the trend of interference on enzymatic activities can be established for each

Table 1. Hypothetical Sequences of Enzymatic State as a Function
Of Age

State I	:	young enzyme	=	full activity
State II	:	old enzyme with steric conformation change; modifiable by drugs	=	partial drug-modifiable activity
State III	:	old enzyme with steric conformation change; not modifiable by drugs (subtle denaturation)	=	partial drug un-modifiable activity
State IV	:	old enzyme with complete denaturation	=	inactivity

drug, thus providing a first approach to define the possible mode of action at the level of subcellular systems related to intermediary metabolism[13].

Furthermore, a more detailed description of pharmacological interference on tissue enzymatic systems also requires their evaluation in more homogeneous subcellular fractions. At any rate, this study showed that: (a) age progressively narrows the range of drug effects on the enzymatic activities tested; (b) this event can be tentatively related to steric conformation changes of the enzyme molecule; (c) a classification of drug action on cerebral enzymatic activities must necessarily take into account the age of the animal. The present preliminary data should be confirmed in at least two other rat strains.

ACKNOWLEDGMENTS

We thank Dr. M. L. Riva for assistance in the preparation of the manuscript, Mrs. G. Garlaschi, Mr. L. Maggi and Mr. G. Ariòli for technical assistance.

REFERENCES

1. C. E. Finch, Enzyme activities, gene function and ageing in mammals, Expl. Gerontol. 7:53 (1972).
2. R. C. Adelman, Age-dependent effects in enzyme induction - A biochemical expression of ageing, Expl.Gerontol. 6:75 (1971).
3. P. D. Wilson, Enzyme changes in ageing mammals, Gerontologia 19:79 (1973).
4. R. C. Adelman, An age-dependent modification of enzyme regulation, J.Biol.Chem. 245:1032 (1970).

5. J. L. Haining and W. W. Correll, Turnover of tryptophan
    induced tryptophan pynolase in rat liver as a function of
    age, J.Geront. 24:143 (1969).

6. H. P. Von Hahn, The regulation of protein synthesis in the
    ageing cell, Expl.Gerontol. 5:323 (1970).

7. W. Meier-Ruge, A. Enz, P. Gygax, O. Hunziker, P. Iwangoff and
    K. Reichlmeier, Experimental pathology in basic research
    of the ageing brain, in: "Aging", Gershon and Raskin eds.,
    Raven Press, New York (1975).

8. M. Hamburg and L. B. Flexner, Biochemical and physiological
    differentiation during morphogenesis, J.Neurochem. 1:279
    (1957).

9. R. E. Kuhlman and O. H. Lowry, Quantitative histochemical
    changes during the development of the rat cerebral cortex,
    J.Neurochem. 1:173 (1956).

10. K. L. Sims, J. Witztum,C. Quick and F. N. Pitts, Brain 4-
    aminobutyrate:2-oxoglutarate aminotransferase. Changes in
    the developing rat brain, J. Neurochem. 15:667 (1968).

11. G. Benzi, E. Arrigoni, P. Strada and R. F. Villa, Enzymatic
    activities in the senescent brain and interference with
    S-adenosyl-L-methionine, Expl.Gerontol. 14:183 (1979).

12. L. W. Kellogg III and I. Fridovich, Superoxid dismutase in
    the rat and mouse as a function of age and longevity,
    J. Geront. 31:405 (1976).

13. G. Benzi, Enzymatic activities related to energy transduction
    in the mature rat brain, Interdiscipl.Topics Geront.15:104
    (1979).

14. J. Hollander and C. H. Barrows jr., Enzymatic studies in
    senescent rodent brains, J. Geront. 23:174 (1968).

15. F. N. Pitts jr. and C. Quick, Brain succinate semialdehyde
    dehydrogenase, J.Neurochem. 14:561 (1967).

16. B. P. F. Adlard and J. Dobbing, Phosphofructokinase and
    fumarate-hydratase in developing rat brain, J.Neurochem.
    18:1299 (1971).

17. M. H. Epstein and C. H. Barrows jr., The effects of age on the
    activity of glutamic acid decarboxilase in various regions
    of the brains of rats, J.Geront. 24:136 (1969).

18. G. Benzi, E. Arrigoni, F. Dagani, F. Marzatico, D. Curti,
    A. Manzini and R. F. Villa, Effect of chronic treatment with
    some drugs on the enzymatic activities of the rat brain,
    Biochem.Pharmacol. 28:2703 (1979).

19. E. C. Weinbach and J. Garbus, Age and oxidative phosphori-
    lation in rat liver and brain, Nature Lond. 178:1225 (1956).

20. E. De Robertis, A. Pellegrino De Iraldi, G. Rodriguez de
    Lores Arnaiz and L. Salganicoff, Cholinergic and non-
    cholinergic nerve endings in rat brain - I, J. Neurochem.
    9:23 (1962).

21. O. H. Lowry, N. J. Rosebrough, A. L. Farr and R. J. Randall,
    Protein measurement with the folin phenol reagent,
    J.biol.Chem. 193:265 (1951).

22. S. Ochoa, Malic dehydrogenase from pig heart, in: "Methods in
     in Enzymology", S. P. Colowick and N. O. Kaplan eds.,
     Academic Press, New York (1955).

23. A. Nason and F. D. Vasington, Lipid-dependent DPNH-cytochrome
     c reductase from mammalian skeletal and heart muscle, in:
     "Methods in Enzymology", S. P. Colowick and N. O. Kaplan,
     eds., Academic Press, New York (1963).

24. L. Smith, Spectrophotometric assay of cytochrome c oxidase,
     in: "Methods of Biochemical Analysis", D. Glick, ed.,
     Wiley-Interscience, New York (1955).

25. D. C. Wharton and A. Tzagoloff, Cytochrome oxidase from beef
     heart mitochondrial, in: "Methods in Enzymology",
     R. W. Estabrook and M. E. Pullman, eds., Academic Press,
     New York (1967).

26. H. U. Bergmeyer and E. Bernt, Lactate dehydrogenase, in:
     "Methods of Enzymatic Analysis", H. U. Bergmeyer, ed.,
     Academic Press, New York (1974).

27. P.H. Sugden and E. A. Newsholme, Activities of citrate synthase,
     NAD^+-linked and $NADP^+$-linked isocitrate dehydrogenases,
     glutamate dehydrogenase, aspartate amonitransferase and
     alanine aminotransferase in nervous tissues from vertebrates
     and invertebrates, Biochem.J. 150:105 (1975).

28. M. Rothstein, The formation of altered enzymes in aging animals,
     Mech.Ageing Dev. 9:197 (1979).

29. A. Weber, C. Gregori and F. Schapira, Aldolase B in the liver
     of senescent rats, Biochem.Biophys.Acta. 444:810 (1976).

30. E. Steinhagen-Thiessen and H. Hilz, The age-dependent decrease
     in creatine kinase and aldolase activities in human
     striated muscle is not caused by an accumulation of faulty
     proteins, Mech.Ageing Dev. 5:447 (1976).

31. S. K. Gupta and M. Rothstein, Triosephosphate isomerase from
     young and old Turbatrix aceti, Arch.Biochem.Biochys.174:333
     (1976).

32. A. Kahn, O. Bertrand, D. Cottreau, P. Boivin and J.-C. Dreyfus,
     Evidence for structural differences between human glucose-
     6-phosphate dehydrogenase purified from leukocytes and
     erythrocytes, Biochem.Biophys.Res.Commun. 77:65 (1977).

33. A. Kahn, P. Boivin, H. Rubinson, D. Cottreau, J. Marie and
     J.-C. Dreyfus, Modifications of purified glucose-6-phosphate
     dehydrogenase and other enxymes by a factor of low molecular
     weight abundant in some leukemic cells, Proc.Nat.Acad.Sci.
     U.S.A. 73:77 (1976).

34. H. K. Shauma, S. K. Gypta and M. Rothistein, Age-related
     alteration of enolase in the free-living nematode,
     Turbatrix aceti, Arch.Biochem.Biophys. 174:324 (1976).

35. U. Reiss and M. Rothstein, Heat-labile isozymes of isocitrate
     lyase from aging Turbatrix aceti, Biochem.Biophys.Res.
     Commun. 61:1012 (1974).

36. M. Rothstein, Recent developments in the age-related alter-
     ation of enzymes: a review, Mech.Ageing Dev. 6:241 (1977).

BRAIN CATECHOLAMINE RECEPTOR FUNCTION DURING AGING

M. Memo, P. F. Spano, H. Kobayashi and M. Trabucchi

Dept. of Pharmacology
University of Milan
Milan, Italy

INTRODUCTION

Catecholaminergic transmission presents a reduced response to physiological and pharmacological stimuli during aging. Alterations have been described at various steps of catecholaminergic metabolism[1,2,3,4]. On the other hand, Cotzias et al[5] reported that mice fed from birth with a diet containing L-DOPA displayed a longer mean life span and an improvement of many behavioural functions during aging; moreover other authors[6] indicated a positive effect of dopaminergic agents on motor activity and intellectual performance.

These data suggest that dopamine (DA) and norepinephrine (NE) are primarily involved in biological processes altered by aging, and may indirectly induce changes on other neurotransmitter systems and hormonal functions.

It has been suggested that many molecular events, which follow the interaction between a neurotransmitter and its recognition sites, change with increasing age[7]. Because of the strong correlation between neurotransmitter receptor function and the ability to evoke a physiological event, it may be hypothesized that the altered behavioural and mental patterns observed during aging are due to an alteration in synaptic biochemical mechanism. Neurotransmitter receptor functions have been examined by various authors[3,4,8]; in most of these studies the number of neurotransmitter recognition sites were found to decrease with increasing aging. The biochemical data may be correlated with anatomical observations; in fact reduction of dendrites and neuronal loss have been described in various brain regions of aged animals and men[9,10].

15

The regulation of cerebral processes, however, is dependent not only upon the number of receptor sites, which are strictly related to number of neurones, but also upon the capacity of receptors to respond under different physiological and pharmacological conditions.

MATERIALS AND METHODS

Mature (3 - 4 months) or senescent (20 - 24 months) male Sprague-Dawley rats were used in our study. The animals were randomly caged to avoid environmental differences, housed at constant temperature and humidity, and exposed to a light cycle of 12 hrs (from 6.00 to 18.00) a day. Animals had free access to food and water. Senescent rats with pathological affections were excluded from our study.

Adenylyl cyclase activity was measured as described by Kebabian et al[11] using (8-^{14}C)-adenosine triphosphate as substrate. ^{3}H-Spiroperidol binding was measured according to Burt et al[12] with minor modifications. H-Sulpiride binding was measured following the method indicated by Spano et al[13]. Dihydroxyphenylacetic acid (DOPAC) levels were measured according to the micromethod described by Argiolas et al[14]. Cyclic AMP concentrations in brain micro-vessels formed in incubation with or without NE were measured according to Herbat et al[15].

RESULTS

In Table 1 DA-stimulated adenylyl cyclase activity in various brain areas and retina of mature and aged rats is reported. A significant decrease in cyclic AMP formation induced by DA was detected in striatum nucleus accumbens and tuberculum olfactorium of aged rats compared with mature rats. Interestingly, cyclic AMP accumulation induced by DA was about twice as great in the retina of aged rats in comparison to mature animals. In order to better explore the alteration in dopaminergic receptor function, we have performed radioreceptor binding assay using two different ligands: ^{3}H-Spiroperidol and [3]H(-)Sulpiride. The later compound has been proposed as selective antagonist at those dopaminergic receptors not linked to adenylyl cyclase. As reported in Table 2, in aged rats, ^{3}H-Spiroperidol binding sites were reduced by about 40% while the kinetics parameters of [3]H(-)-Sulpiride were unmodified. The decrease in ^{3}H-Spiroperidol binding was ascribed to a reduction in B_{max} values (from 169 $\pm$ 8 to 101 $\pm$ 9 fmol/mg prot.) rather than K_D modifications. To study the functionality of the dopaminergic system, aged rats were treated with haloperidol (1 mg/kg) and DOPAC levels were measured. As shown in Table 3, DOPAC steady state levels were reduced in aged rats in comparison to mature. Moreover, the increase of DOPAC levels after haloperidol injection was

Table 1. DA-Stimulated Adenylyl Cyclase Activity in Various Brain
          Areas of Aged (20-24 months) and Mature (3-4 months) Rats

Area	Adenylyl cyclase activity					
	mature		%	aged		%
	− DA	+ DA	changes	− DA	+ DA	changes
Striatum	236 ± 16	465 ± 20	+ 97	276 ± 14	380 ± 21	+ 38
Nucleus Accumbens	211 ± 10	410 ± 16	+ 94	230 ± 16	314 ± 14	+ 36
Tuberculum Olfactorium	132 ± 8	269 ± 14	+ 104	142 ± 12	210 ± 12	+ 47
Retina	10 ± 1	18 ± 2	+ 80	12 ± 1	33 ± 2	+ 175

DA was added at the concentration of $5 \cdot 10^{-6}$ M.

Values are the mean ± S.D. of six experiments run in triplicate and are expressed as
pmol of cyclic AMP formed/min/mg protein.

significantly smaller in aged animals in respect to those of mature
rats.

In Fig. 1, cyclic AMP formation induced by NE in brain microvessels of aged and mature rats is reported. Performing identical
purification procedures, microvessels obtained from aged rats were
lesser than those from mature. After stimulation with 10 M NE
cyclic AMP accumulation was significantly higher in mature in
respect to aged rats.

DISCUSSION

The results presented indicate relevant changes in the function
of catecholaminergic receptors in neurons and microvessels of aged
rats. In fact adenylyl cyclase activity stimulation induced by DA
is significantly lower in various brain areas, indicating a
functional decrease of the system responsible for the formation of
cyclic AMP. The first problem raised from this observation is
whether the diminished response may be ascribed to an alteration in
the DA recognition sites or to the mechanisms involved in enzymatic
cyclic AMP synthesis. The supersensitive response to DA observed
in the retina of aged rats is an indirect demonstration that the
neuronal systems responsible for the cyclic AMP formation are
operative and may even perform a transduction of stimuli in supersensitive way.

Table 2. Kinetic Parameters of ^{3}H-Spiroperidol and ^{3}H(-)Sulpiride Bindings in Striatum of Aged (20-24 months) and Mature (3-4 months) Rats

	^{3}H-Spiroperidol specific binding	
	B_{max}	K_D
mature	169 ± 8	0.12 ± 0.02
aged	101 ± 9 *	0.09 ± 0.01

	^{3}H(-)Sulpiride stereospecific binding			
	High affinity		Low Affinity	
	B_{max}	K_D	B_{max}	K_D
mature	187 ± 21	16.5 ± 1.3	914 ± 34	247 ± 12
aged	191 ± 18	15.5 ± 1.0	954 ± 31	251 ± 19

* $p < 0.01$ in respect to correspondent values of mature rats.

Values are thge means ± S.D. and were extrapolated from a representative experiment using 5 concentrations of radioligand.

However, it remains to be better clarified the possible alterations at the level of DA recognition sites. We have recently defined as D_1 and D_2 receptors those DA receptors associated or unassociated with a cyclic AMP generating system, respectively. The development of new classes of drugs as selective agonists or antagonists at dopaminergic receptors provided the most convincing

Table 3. Effect of Acute Treatment with Haloperidol on Striatal DOPAC Levels in Aged (20-24 months) and Mature (3-4 months) Rats

	DOPAC levels (ng/mg tissue)		
	Saline	Haloperidol	% increase
mature	2.4 ± 0.2	4.6 ± 0.3	+ 92%
aged	1.8 ± 0.1 *	2.5 ± 0.2 **	+ 39%

* $p < 0.01$ in respect to saline treated mature rats values.
** $p < 0.01$ in respect to haloperidol treated group of mature animals.

Animals were killed 1 h after the injection of 1 mg/kg haloperidol.
Values are the mean ± S.D. of three experiments with 5 animals for each group.

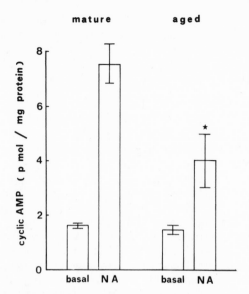

Fig. 1.   Cyclic AMP accumulation in response to NE in brain
          microvessels from mature (2-3 months) and aged
          (20-24 months) rats.

    * p<0.05 in respect to the value in the presence of NE
      of mature rats.

    Values are the mean + S.E. of six determinations run in
    triplicate.

evidence for the existence of multiple classes of DA receptors.
In particular Sulpiride, which appears to be an antidopaminergic
agent in various animal tests and clinical conditions[16], does not
antagonize the stimulation of cyclic AMP formation elicited by DA
in in vitro or in vivo experiments[17].   However,3(-) Sulpiride,
the pharmacologically active form of the drug, may interact with DA
recognition sites labelled by ^{3}H-Dopamine and ^{3}H-Spiroperidol[17].
On the basis of this observation, we suggested that ^{3}H(-)Sulpiride
may specifically bind to a group of DA-receptors in brain membrane
preparations identified as D_2 receptors which are not coupled to a
cyclic AMP system[18].

    The experiments performed with ^{3}H-Spiroperidol and ^{3}H(-)
Sulpiride indicate a decreased function during aging of the specific
group of DA-receptors named D_1.   In fact the lack of changes in
^{3}H(-)Sulpiride sterospecific binding suggests that the decreased
^{3}H-Spiroperidol binding is ascribed to a reduction in D_1 receptors.

    The kinetic parameters regarding the changes of D_1 receptors
indicate a decrease of the total number of recognition sites rather

than modifications in receptor affinity.   These results, which are
at variance with our previously reported data , suggest that the
changes in binding sites and cyclic AMP synthesis which occur in
rat dopaminergic areas during aging are most likely related to cell
loss.   The difference from our previous reports may be ascribed to
the different aging conditions among the groups of animals in the
experiments.

    The lack of changes in the D_2 receptor function is an
interesting result in view of a possible selective localization of
the cell loss during aging.   These results may be compared with
those obtained after intrastriatal injection of kainic acid.   In
these experimental conditions, striatal ^{3}H-Spiroperidol binding is
significantly decreased with a reduction in the total number of
binding sites reflecting the cell loss induced by kianate[19].
In the same lesioned animals we did not find any change in the
kinetic parameters of striatal [3]H(-)Sulpiride binding[20].   A
possible explanation for these results may be related to the
growth of glial tissue observed after kainate injection and to the
possible localization of D_2 receptors in these cells.   In this model
the loss of [3]H(-)Sulpiride sites located in the neurones is compen-
sated by the increase of those possibly present in the glia.   The
similarities of kainate lesion with aging may be discussed: however
it seems of significant importance that in old rats the behaviour
of H-Spiroperidol and [3]H(-)Sulpiride bindings may be compared to
that induced by kainate lesion and that gliosis is among the most
relevant anatomical features in the aging brain[21].   On the other
hand, neurotransmitter recognition sites have been identified in
glial cells[22,23].   The physiological role of these receptors and
their contribution to the restoration of a normal cerebral function
in pathological, physiological or experimental conditions are still
unclear.   As shown in Table 3, the increase of DA turnover induced
by haloperidol is less pronounced, but still present, in aged rats
suggesting that the regulatory loop for DA turnover is operative.
Moreover, it has been recently reported that chronic haloperidol
treatment does not induce supersensitivity of D_1 receptors in aged
rats[24] confirming the presence of alterations in the regulatory
mechanisms of DA receptor function.

    The data reported on the cyclic AMP formation in brain micro-
vessels is of particular relevance in the light of the possible
characterization of catecholaminergic recognition sites.   Various
authors[15,25] reported that cyclic AMP accumulation may be elicited
by NE in brain microvessel preparations; this fact strongly supports
the existence of β-adrenergic receptors in mammalian cerebral
vasculature.   The decreased effect of NE on brain microvessels
obtained from the cortex of aged rat is in line with the previously
reported decrease of neuronal β-adrenergic receptor function during
aging[2].   Recently Greenberg and Weiss[26] indicated that β-adrenergic
receptors of old rats does not adequately respond to changes in the

sympathetic input induced by pharmacological agents such as reserpine or desmethylimipramine.   Our results on the decreased response to NE suggest that in the aging brain the microvessels are unable to respond adequately to environmental modifications which require rapid adjustments.   For these reasons it appears that in the aged brain blood flow regulatory mechanisms are, at least partially, impaired and may respond less rapidly to brain needs.

REFERENCES

1. D. J. Reis, R. A. Ross and T. H. Joh, Changes in the activity and amount of enzymes synthesizing catecholamines and acetylcholine in brain, adrenal medulla, and sympathetic ganglia of aged rat and mouse, Brain Res. 136:465 (1977).
2. L. H. Greenberg and B. Weiss, Beta-adrenergic receptors in aged rat brain: reduced number and capacity of pineal gland to develop supersensitivity, Science 201:61 (1978).
3. S. Govoni, P. Loddo, P. F. Spano and M. Trabucchi, Dopamine receptor sensitivity in brain and retina of rats during ageing, Brain Res. 138:565 (1977).
4. S. Govoni, M. Memo, L. Saiani, P. F. Spano and M. Trabucchi, Impairment of brain neurotransmitter receptors in aged rats, Mech. Ageing and Develop. 12:39 (1980).
5. G. C. Cotzias, S. T. Miller, L. C. Tang, P. S. Papavasiliou and Y. Y. Wang, Levodopa, fertility and longevity, Science 196:549 (1977).
6. J. F. Marshall and N. Berrios, Movement disorders of aged rats: reversal by dopamine receptor stimulation, Science 206:477 (1979).
7. G. S. Roth, Hormone receptor changes during adulthood and senescence: significance for aging research, Fed.Proc. 38:1910 (1979).
8. A. Maggi, M. J. Schmidt, B. Ghetti and S. J. Enna, Effect of aging on neurotransmitter receptor binding in rat and human brain, Life Sci. 24:367 (1978).
9. M. L. Feldman, Aging changes in the morphology of cortical dendrites, Aging 3:211 (1976).
10. S. J. Buell and P. D. Coleman, Dendritic growth in the aged human brain and failure of growth in senile dementia, Science 206:854 (1979).
11. W. Kebabian, G. L. Petzold and P. Greengard, Dopamine sensitive adenylate cyclase in nucleus caudate of rat brain and its similarity to the dopamine receptors, Proc.Natl.Acad.Sci. 69:2145 (1972).
12. D. R. Burt, I. Creese and S. H. Snyder, Antischizophrenic drugs: chronic treatment elevates dopamine receptors binding in brain, Science 196:326 (1977).

13. P. F. Spano, M. Memo, S. Govoni and M. Trabucchi, Similarities and dissimilarities between dopamine and neuroleptic receptors, in:"Adv. in Biochem.Pharm.", F. Cattabeni, G. Racagni, P. F. Spano and E. Coata, eds., Raven Press, New York (1980).

14. A. Argioloas, F. Fadda, E. Stefanini and G. L. Gessa, A simple radioenzymatic method for determination of picogram aumonts of 3,4-dihydroxyphenilacetic acid (DOPAC) in rat brain, J.Neurochem. 29:599 (1977).

15. J. J. Herbat, M. E. Raickle and J. A. Ferrendelli, β-Adrenergic regulation of adenosine 3',5'-monophosphate concentration in brain microvessels, Science 204:330 (1979).

16. P. F. Spano, M. Trabucchi, G. U. Corsini and G. L. Gessa, Sulpiride and other benzamides, It.Brain Res.Found., Milan, Italy (1979).

17. P. F. Spano, S. Govoni and M. Trabucchi, Studies on the Pharmacological properties of dopamine receptors in various areas of the central nervous system,"in: Adv. Biochem. Psychopharmacol." P.J. Roberts, G. N. Woodruff, L. L. Iversen eds., Raven Press, New York (1978).

18. P. F. Spano, M. Memo, E. Stefanini, P. Fresia and M. Trabucchi, "Detection of multiple receptors for dopamine. Receptors for Neurotransmitters and Peptide Hormones", G. Pepeu, M. J. Kuhar and S. J. Enna, eds., Raven Press, New York (1980).

19. S. Govoni, V. R. Olgiati, M. Trabucchi, L. Garau, E. Stefanini and P. F. Spano, ^{3}H-Haloperidol and ^{3}H-Spiroperidol receptor binding after striatal injection of kainic acid, Neurosci.Lett. 8:207 (1978).

20. M. Memo, P. F. Spano and M. Trabucchi, Intrinsic neostriatal neurons do not contain D_2 dopaminergic receptor. Submitted for publication.

21. K. R. Brizzle, J. M. Ordy, J. Hanscke and B. Kaack, Quantitative assessment of changes in neuron and glia cell packing density and lipfuscin accumulation with age in the cerebral cortex of a nonhuman primate (Macaca mulatta), Aging 3:229 (1976).

22. A. Schousboe, Differences between astrocytes in primary cultures and glial cell-lines in uptake and metabolism of putative amino acid transmitters, in: "Cell Tissue and Organ Cultures in Neurobiology," S. Fedoroff and L. Hertz, eds., Academic Press, New York (1977).

23. F. A. Henn, B. Oderfold-Nowake and R. Roshoski, Receptor binding to astroglia cells Soc.Neurosci.Abstr. 5:590 (1979).

24. J. W. Ferkony and S. J. Enna, Alterations in striatal neurotransmitter receptor binding following chronic administration of psycho-active agents, Soc.Neurosci.Abstr. 5:647 (1979).

25. J. A. Nathanson and G. H. Glaser, Identification of β-adrenergic-sensitive adenylate cyclase in intracranial blood vessels, Nature, 278:567 (1979).

26. L. H. Greenberg and B. Weiss, Ability of aged rats to alter
        Beta Adrenergic Receptors of brain in response to repeated
        administration of reserpine and desmethylimipramine,
        J.Pharm.Exp.Ther. 211:309 (1979).

NEUROCHEMISTRY OF AGING AND SENILE DEMENTIA

D. M. Bowen and A.N. Davison

Institute of Neurology
National Hospital for Nervous Diseases
London

With normal aging there is slow decline of cognitive function (especially short-term memory). Such changes have been ascribed to an accumulated loss of nerve cells. It has been calculated that we lose 100,000 neurons per day. Since there are 20,000 million cerebral neurons, even if this is correct, we still have cells to spare and quite massive loss on the non-dominant side may have relatively minor effect. However, brain weight loss with age seems to be greater in those (e.g. men) with heavier brains, suggesting to Torack[1] that the loss of nerve cells is not a random process common to all individuals.

Programmed destruction is also evident in the predilection of certain parts of the brain for atrophy. In this respect, the frontal poles and the temporal lobes have consistently been regarded as the sites of most prominent shrinkage[2]. Thus, significant neuronal loss may occur in the hippocampus, which according to Ball[3] predisposes that memory co-ordinating area to damage but loss in other areas (e.g. in the inferior olive)[4] appears to be less. Tomlinson[5] showed that histological changes were present in nerve cells of normal elderly people, - although the number of affected neurons was few. The Scheibels[6], using the Golgi method, later demonstrated reduction in dendritic domains in various neurons from elderly people - suggesting in at least the area of the hippocampal gyri that there is a reduction in nerve terminal population. Attention was therefore directed to the possibility of reduced synaptic contacts, resulting in more subtle aging changes in neurotransmitter activity.

The major alterations reported are up to 50% loss in dopamine and noradrenaline transmitter systems[7], possibly correlating with

altered sleeping habits, depressive illness and dyskinesia.   With
increasing age there is, too, a loss of choline acetyltransferase
activity especially in the hippocampus (50% for 60-95 years).
This enzyme is responsible for the biosynthesis of acetylcholine -
the transmitter thought to be particularly involved in memory.

SENILE DEMENTIA OF THE ALZHEIMER'S TYPE

About 10% of the elderly population suffer from failing brain
function.   Although in some cases of multi-infarct dementia there
is clearly a vascular pathology in the majority of patients with
senile dementia there is no known aetiology.   In these cases of
senile dementia of the Alzheimer's type (SDAT) there is generally
atrophy of the brain and typical histopathological changes are
evident.   Such patients suffer from global slow loss of intellectual function, particularly of memory and there is no pathological or
clinical difference between the pre-senile (before 65 years of age)
and senile type of dementia.   The question is, is this simply an
exaggerated form of aging or are we dealing with a disease process?
There are doubts about the general extent of nerve cell loss,
despite atrophy frequently seen by computerized tomography.
Neuropathologists generally agree that there is an increased loss of
neurons in the hippocampus[3] (Fig. 1).   Dr. David Bowen and his
colleagues[8] in my department have attempted to assess neuronal loss
by measuring changes in biochemical markers in the whole temporal
lobe.   These findings suggest that in comparison to age matched
controls, there is about a 30% loss of nerve cells or alternatively
there is a substantial shrinkage of nerve cell cytoplasm.   However,
it is considered that functional changes are greater than that
expected from gross atrophy.   Indeed, there is no good correlation
between ventricular volume assessed by CT and the degree of dementia.
There may be a connection between blood supply, oxygen uptake and
physiological function, for with decreased oxygen tension short-term
memory is quickly affected.

BIOCHEMISTRY OF DEMENTIA

There is a reduction in total blood flow to the demented brain
from about 50 ml/100 g/min to 30 ml/100 g/min[9,10], and accompanying
this a reduced perfusion of grey matter[11].   Nevertheless, a small
increase in blood flow by use of vasodilators has not been found to
improve intellectual function.   Unlike the response of controls
inappropriate or even reduced changes in regional cerebral blood
flow are seen in senile dements.

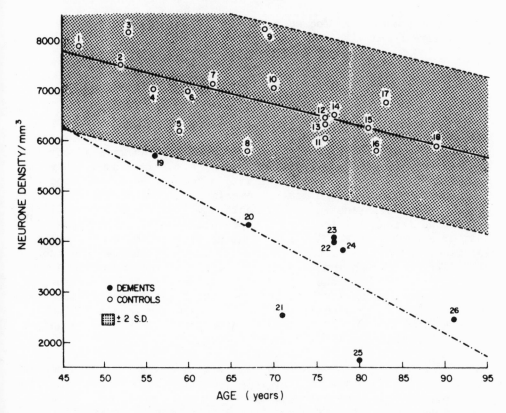

Fig. 1. Density of neurons in the hippocampal cortex at different
ages. The best linear regression is shown for Alzheimer
cases (dotted line), and controls (solid line _____)

(With permission of Dr. M.J. Ball and Acta Neuropath.).

NEUROPATHOLOGY

Perhaps a clue to the pathogenesis may come from histopathological changes which interfere with neuronal metabolism - granulovacuolar degeneration, paired helical filaments and plaques.
Paired helical filaments found in the tangles may be derived from
neurofilaments. De Boni and Crapper[12] have found a factor in SDAT
brain which induces paired helical filaments formation in tissue
culture. Experimental neurofibrillary structures probably derived
from neurofilaments are induced by aluminium. Thus, axonal flow
from neuronal perikaryon to terminal may be impeded by the presence
of abnormal axonal filaments. Although the neuritic plaque contains
some intact synapses, Buell and Coleman's[13] work on Golgi preparations shows marked reduction in synaptic contacts in dementia.

Our interest therefore focusses on the synapse and the production of neurotransmitters within the presynaptic terminal.

NEUROTRANSMITTER SYNTHESIS

Although measurement of neurotransmitter concentration in post-mortem tissue is unreliable, useful information can come from assay of the activity of enzymes synthesizing neurotransmitters. There must be careful control for age and drug intake, as well as terminal and post-mortem artefact. Many enzymes (e.g. tyrosine hydroxylase or glutamate decarboxylase) are little affected, but of all the changes so far found, the most interesting and reproducible is that of choline acetyltransferase (CAT) activity[14]. The enzyme activity is reduced throughout the brain, especially in the hippocampus[15,16]. Another enzyme involved in acetylcholine metabolism, acetylcholin-esterase, is also diminished in activity. Both enzymes are known to be synthesized in the nerve cell bodies and then accumulate in the terminal. The reduction in activity relates to the intensity of neuropathology[14,17] and to the pre-mortem intelligence rating of the patient[17]. Interestingly, the post-synaptic receptor binding site remains intact, and so there is the possibility of interaction at the cholinergic site[17,18].

Since CAT may not be a rate controlling enzyme in human nervous tissue, it is important to correlate reduction in CAT activity with the acetylcholine synthetic ability of fresh human brain. Thus crude synaptosomal preparations from biopsy samples under stimulated and resting conditions have been examined (Table 1). In all cases where intense tangles and plaque formation indicate a diagnosis of senile dementia of the Alzheimer's type, there is reduction in CAT activity which is paralleled by loss of acetylcholine synthetic capacity under both stimulated and resting conditions[19]. No significant change in glucose utilization has been found and in two biopsy samples examined, glutamate decarboxylase activity was unchanged.

This possibility has led to the use of choline, or better still lecithin as a means of stimulating acetylcholine synthesis. Thus, for example, twenty minutes after giving a titrated subcutaneous dose of physostigmine together with oral lecithin, Peters and Levin (1978)[20] found enhanced and sustained memory storage and retrieval in patients with senile dementia (ages 58 - 79). In a small proportion of biopsy specimens from cases of dementia, no Alzheimer's type histopathology is evident. In these, except in one instance, CAT activity is not reduced. Some years ago, Gottfries and his colleagues[21] in Sweden reported reduction in the concentration of the dopamine metabolite - homovanillic acid in the CSF of demented patients. There is, thus, the interesting possibility that examples of other neurotransmitter defects (e.g. in dopamine or indole

Table 1. Production of (^{14}C) - ACh and $^{14}CO_2$ in Tissue Prisms from Biopsy Samples of Control and Alzheimer Disease Patients

Biopsy	n	Acetylcholine synthesis (pmol / min / mg protein)		$^{14}CO_2$ produced (dpm / min / mg protein)		CAT activity (pmol/min/mg protein)
		5 mM K$^+$	31 mM K$^+$	5 mM K$^+$	31 mM K$^+$	
Control	9$^+$	4.0 $\pm$ 2.0	7.2 $\pm$ 1.4	143 $\pm$ 28	390 $\pm$ 97	82 $\pm$ 15
Alzheimer's Disease	5	1.3 $\pm$ 0.3*	2.9 $\pm$ 1.2*	197 $\pm$ 63	488 $\pm$ 101	29 $\pm$ 13*

Mean $\pm$ standard deviation.

+ Control values for CAT were measured in 5 cases; glucose metabolism at 5 mM K+ was determined for 8 cases.

* Significantly different from control, $p < 0.01$ (Wilcoxon rank test).

(after Sims et al)[19]

metabolism) may also give rise to dementia.

REFERENCES

1. R. M. Torack, "The Pathologic Physiology of Dementia", Springer-
   Verlag, Berlin (1978).
2. M. Critchley, The neurology of old age, Lancet 1:1119 (1931).
3. M. Ball, Neuronal loss, neurofibrillary tangles and granulovacuolar degeneration in the hippocampus with aging and
   dementia, Acta Neuropath.(Berlin) 37:111 (1977).
4. W. Meier-Ruge, O. Hunziker, P. Iwangoff, K. Reichlmeier and
   P. Sandoz, Alterations of morphological and neurochemical
   parameters of the brain due to normal aging, in: "Senile
   Dementia: A Biomedical Approach", K. Nandy, ed., Elsevier/
   North Holland, New York (1978).
5. B. E. Tomlinson, Morphological changes and dementia in old age,
   in: "Aging and Dementia", W. L. Smith and M. Kinshorne, eds.,
   Spectrum, New York (1977).
6. M. E. Scheibel and A. B. Scheibel, Structural changes in the
   aging brain, in: "Aging. Clinics, Morphological and Neurochemical Aspects in the Aging Central Nervous System",
   H. Brody, D. Harman and J. M. Ordy, eds., Raven Press,
   New York (1975).
7. B. Winblad, R. Adolfsson, C. G. Gottfries, L. Oreland and
   B. E. Roos, Brain monoamines, monoamine metabolites and
   enzymes in physiological aging and senile dementia, in:
   "Recent Developments in Mass Spectrometry in Biochemisty and
   Medicine", A Frigerio, ed., Plenum Press, New York (1978).
8. D. M. Bowen, P. White, J. A. Spillane, M. J. Goodhardt,
   G. Curson, P. Iwangoff, W. Meier-Ruge, and A. N. Davison,
   Accelerated aging or selective neuronal loss as an
   important cause of dementia, Lancet 1:11 (1979).
9. R. L. Gubb, M. E. Raichle, M. H. Gado, J. O. Eichling and
   C. P. Hughes, Cerebral blood flow, oxygen utilization and
   blood volume in dementia, Neurology 27:905 (1977).
10. D. J. Wyper, C. J. McAlpine, K. Jawad and B. Jennett, Effects
   of a carbonic anhydrase inhibitor on cerebral blood flow in
   geriatric patients. J. Neurol.Neurosurg.& Psychiat. 39:885
   (1976).
11. V. C. Hachinski, M. Iliff, E. Zilkha, G. H. du Boulay,
   V. L. McAllister, J. Marshall, R. W. Ross Russell and L.
   Symon, Cerebral blood flow in dementia, Arch.Neurol.
   32:632 (1975).
12. U. De Boni and D. R. Crapper, Paired helical filaments of the
   Alzheimer type in cultural neurons, Nature 271:566 (1978).
13. S. J. Buell, and P. D. Coleman, Dendritic growth in the aged
   human brain and failure of growth in senile dementia,
   Science 206:854 (1979).

14. D. M. Bowen, C. B. Smith, P. White and A. N. Davison, Neuro-
    transmitter-related enzymes and indices of hypoxia in
    senile dementia and other abiotrophies, <u>Brain</u> 99:459 (1976).
15. E. K. Perry, P. H. Gibson, G. Blessed, R. H. Perry and B. E.
    Tomlinson, Neurotransmitter enzyme abnormalities in senile
    dementia - Choline acetyltransferase and glutamic acid
    decarboxylase activities in necropsy brain tissue,
    <u>J.Neurol.Sci.</u> 34:(2) 247 (1978).
16. P. Davies, Neurotransmitter-related enzymes in senile dementia
    of the Alzheimer type, <u>Brain Res</u>. 171:319 (1979).
17. E. K. Perry, B. E. Tomlinson, G. Blessed, K. Bergmann,
    P. H. Gibson and R. H. Perry, Correlation of cholinergic
    abnormalities with senile plaques and mental test scores in
    senile dementia, <u>Br.Med.J.</u> 2:1457 (1978).
18. P. White, C. R. Hiley, M. J. Goodhardt, L. Carrasco, J. P. Keet,
    J.E.J. Williams and D. M. Bowen, Neocortical cholinergic
    neurones in elderly people, <u>Lancet</u> 1:668 (1977).
19. N. R. Sims, C. C. T. Smith, D. M. Bowen, R. H. A. Flack,
    A. N. Davison, J. S. Snowden and D. Neary, Glucose metabolism
    and acetylcholine synthesis in relation to neuronal activity
    in Alzheimer's disease, <u>Lancet</u> 1:333 (1980).
20. B. H. Peters, and H. S. Levin, Effects of physostigmine and
    lecithin on memory in Alzheimer disease, <u>Ann.Neurol</u>. 6:219
    6:219 (1979).
21. C. J. Gottfries, J. Gottfries and B. E. Roos, The investigation
    of homovanillic acid in the human brain and its correlation
    to senile dementia, <u>Br.J.Psychiat</u>. 115:563 (1969).

NEUROENDOCRINOLOGY AND AGING OF THE BRAIN

U. Scapagnini, P.L. Canonico, F. Drago, M. Amico-Roxas,
G. Toffano*, P. Valeri** and L. Angelucci**

Department of Pharmacology, Faculty of Medicine,
University of Catania
* FIDIA Laboratory of Research, Abano Terme
**Department of Pharmacology, Faculty of Medicine,
  University of Rome, Italy

NEUROENDOCRINE SYSTEM

The basis of the neuroendocrine system is the production of
specific neurohormones, mainly in the hypothalamus which through
the capillary portal system reach the adrenohypophyseal cells
stimulating or inhibiting the synthesis and/or the release of the
pituitary trophic hormones.   The adenohypophyseal hormones act on
the peripheral endocrine organs but the hormone production of the
target glands influences also the activity of the corresponding
pituitary cells as well as the hypothalamic peptidergic neurons
(external or long-loop feedback).   A similarly operating, so-
called short-loop or internal feedback system works between the
adenohypophysis and the hypothalamus, the anatomical basis of which
is the retrograde blood flow from the pituitary to the hypothalamus.
The most restricted hormonal feedback regulation is realized in the
neurosecretory cells which can control their own hormone producing
activity (ultra-short feedback).

From this general regulatory schema, of course, there are
exceptions, for instance the lack of peripheral negative feedback
for growth hormone (GH) and prolactin (PRL), because of the absence
of target glands for these hormones.   In addition, gonadal steroids
do not act only by negative feedback, but also by positive feedback
action directly on the brain.

The most important part of the brain controlling the adenohypophyseal function is the medio-basal hypothalamus (MBH) where

33

the highest concentration of neurohormones can be found. The MBH, by itself, is able to maintain the basal hormone secretion, however, intact afferents to the MBH are needed for physiological cyclic events, such as ovulation and diurnal hormonal rhythms[1].

The essential neural inputs to the MBH include the axons of those releasing hormone producing perikarya which are out of the MBH (like luteinizing hormone-releasing hormone, LHRH) and fibres originating from monoamine (MA) or other transmitter synthesizing neurons.

Among the neurotransmitters the MA have a key role in the hormonal regulation. The possible interaction between MA and neurosecretory cells is summarized in Fig. 1.

Norepinephrine (NE) synthesizing perikarya are located in the pons and medulla oblongata (A1, A2, A5 and A7 cell groups) as well as in the locus coeruleus (A6). The axons of the A1, A2, A5 and A7 cells forming the ventral NE bundle innervate the MBH while fibres coming from the locus coeruleus carry NE afferents to the anterior hypothalamic region[2]. The main effects of NE in the hormonal control are as follows: tonic inhibition of the hypo-thalamo-hypophyseal-adrenal axis, induction of ovulation, stimulation of cold-induced TSH secretion, increase of GH secretion both in laboratory animals and humans, and stimulation of PRL release[3].

In the neuroendocrine control mechanism the tuberoinfundibular dopaminergic (TIDA) system[4] has an extreme importance while the other dopaminergic neurons do not seem to be involved in this regulation. TIDA neurons are located within the MBH, in the arcuate and periventricular nuclei sending their axons directly to the superficial layer of the median eminence (ME). The main role of dopamine (DA) is the tonic inhibitory control of PRL secretion. Other hormonal effects of DA are generally inhibitory, except for GH which is stimulated by DA[3].

The serotonin (5-HT) cell bodies are localized in the mesen-cephalic (B7, B8, B9) and pontine raphe nuclei (B5, B6) and reach the hypothalamus, ME and other brain structures through the medial forebrain bundle[5] . The effect of 5-HT seems to be inhibitory on the hypothalamo-pituitary-gonadal axis, not clear its effects on TSH and GH secretions and rather stimulatory than inhibitory on the HHAA as well as on the PRL secretion.

5-HT does not induce usually well defined hormonal changes in the resting conditions. However, reduction of brain 5-HT level blocks the diurnal circadian hormonal rhythms and inhibits the hormone rises induced pharmacologically or by certain physiological

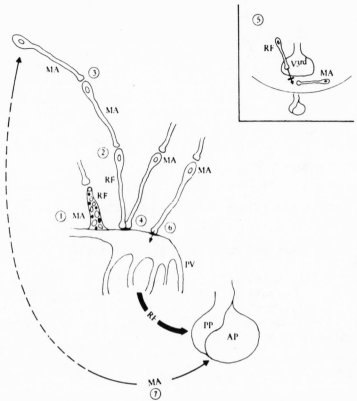

Fig. 1. Possible sites at which monoaminergic (MA) neurons could
         interact with neurosecretory cells (RF, neurosecretory
         cells; AP, anterior pituitary; PP, posterior pituitary).

states, such as pregnancy and lactation.   It seems probable that
5-HT does not act directly on the peptidergic neurons like presumably
DA and NE but 5-HT might operate on other transmitter producing
perikarya influencing their synthesis or modifying their transmitter
release through 5-HT synapses on their axons or terminals.   This
interaction of serotoninergic neurons with especially CA and
cholinergic neurons could be the basis of its modulatory effect on
the hormonal regulation.

     Beside the catecholamines other neurotransmitters, such as
acetylcholine (ACh), GABA and certain substances the nature of which
is not yet established, like endorphins, histamine, prostaglandins,
substance P and some amino acids have also hormonal effects (Fig.2).

     To complete the regulatory mechanisms we have to mention a
self-regulatory mechanism of the CA neurons.   This "one-neuron"
regulatory system operates through the presynaptic receptors controlling the transmitter release towards the postsynaptic receptors[6].

	PRL	LH-FSH	ACTH	TSH	GH
DA	(−)	(−) ?	(+) (−)?	(−) ?	(+)
NE	(+) (−)?	(+)	(−)	(+)	(+) ?
5-HT	(+)	(−)	(+)	(−) ?	(+) (−)?
GABA	(−) ?	(+) ?	(−) ?	(−) ?	(+) (−)?
ACh	(+) (−)?	(+) ?	(+)	(−) ?	——

Fig. 2. Neurotransmitters and anterior pituitary hormone secretion

Similar "autoreceptors" have been postulated also in 5-HT neurons.

Turning back to neurotransmitter-hormonal interactions, in order to keep the system equilibrated, not only do the brain transmitters influence the hormone secretion but changes in the hormonal milieu induce also modifications, usually turnover changes, in the transmitter producing neurons.

NEUROENDOCRINE FUNCTION IN THE AGING

It is possible that alterations in steady state concentration, metabolism or receptor activity of hypothalamic catecholamines, indolamines, GABA and other neurotransmitters may be in part responsible for the modification of endocrine patterns which occurs with the age.   In fact many recent studies suggest that the endocrine changes in the aging process may not be only attributed to intrinsic changes in peripheral glands or even senility of the pituitary gland, but also to changes at some higher neural center that exerts a regulatory influence on endocrine axes.

In Tables 1, 2 and 3 are summarized the age related changes in the tonus of some neurotransmitters in areas of the central nervous system (CNS) endocrinologically implicated.   Also changes in the activity of other neurotransmitters, i.e. ACh[7,8,9], endorphins[10], etc. have been shown to be involved in the failure of endocrine organs which occurs with age in animals and humans.

The aim of this paper is to review the main alteration in some neuroendocrine axes related to aging. Furthermore we have tried to relate these changes in neuroendocrine patterns to the modification of neurotransmitters activity which occurs in the aging brain (see M. Trabucchi in this book).

Table 1. Age Related Changes of DA-Ergic Tonus in
Central Areas Endocrinologically Implicated

Area of Brain	Observation	References
MBH (rat)	decreased DA	Simpkins et al., 1977
Whole hypothalamus (rat)	decreased DA	Simpkins et al., 1977
MBH (rat)	decreased DA turnover	Simpkins et al., 1977
Hypothalamus (rat)	decreased DA	Miller et al., 1976
Hypothalamus (rat)	decreased DA	Clemens et al., 1978
Pituitary (rat)	increased ^{3}H–spiroperidol binding	Govoni et al., 1979
Diencephalon (rat)	decreased TH activity	Algeri et al., 1977
Hypothalamus (human)	increased MAO activity	Robinson, 1975
hypothalamus (mice)	decreased conversion DOPA to DA	Finch, 1973

HYPOTHALAMIC-HYPOPHYSEAL-PROLACTIN AXIS (HHPRLA)

The secretion of prolactin by the anterior pituitary gland is
under the tonic inhibitory control of the hypothalamus. It is now
widely accepted that this inhibition is mediated by DA[11]. DA,
secreted from TIDA neurons into the portal blood vessels leading to
the pituitary gland directly inhibits the secretion of prolactin.
The measurement of DA in portal blood[12,13] reinforced the concept
that these neurons release DA directly into the portal circulation
which selectively perfuses the anterior pituitary with high
concentrations of this catecholamine. Furthermore, PRL release
from the anterior pituitary in vitro is depressed by DA agonists,
and the inhibition is reversed by DA antagonist in in vivo and
in vitro conditions[14,15,16]. Also other catecholamines such as NE
and epinephrine have been found to inhibit prolactin secretion but
less efficiently than DA[17].

The decrease of hypothalamic DA content (Table 1) and the
apparent decrease in metabolism might be responsible for the
increased PRL levels which occur during aging in the rat[18,19,20].
In contrast, other authors reported in studies on aged male rats
that PRL was not elevated[21,22].

In our study we compared plasma PRL levels in 3 - 5 month old,
10 - 12 month old and 22 - 26 month old female rats. The rats
were treated for 20 days with 0.5 mg/kg of bromocryptine (CB-154)
daily injected intraperitoneally. Corresponding control groups
were treated daily for 20 days with intraperitoneal injections of

Table 2.   Age Related Changes of NE-Ergic Tonus in
            Central Areas Endocrinologically Implicated

Area of Brain	Observation	References
MBH (rat)	decreased NE	Simpkins et al., 1977
Whole hypothalamus (rat)	decreased NE	Simpkins et al., 1977
Hypothalamus (rat)	decreased DBH activity	Reis et al., 1977
Diencephalon (rat)	decreased TH activity	Algeri et al., 1977
Hypothalamus (mice)	decreased conversion DOPA to DA and NE	Finch, 1973
Hypothalamus (human)	increased MAO activity	Robinson, 1975
Hypothalamus (human)	decreased NE	Bertler, 1961

physiologic solution. At the end of the treatment, the rats were
sacrificed, on the day of diestrus, by decapitation, immediately
after the removal from the cages.

As it can be seen in Fig. 3, serum PRL levels were significantly
elevated with age.   It is interesting that by 10 - 12 months serum
PRL was already elevated when compared with the levels of younger
rats.   Treatment with the DA mimetic agent CB-154 is able to
dramatically reduce the PRL values at all ages explored suggesting
that the impairment in DA-ergic function at hypothalamic level,
more than a reduced receptorial sensitivity of the lactotrophs,
might be responsible for PRL enhancement observed in aged rats.

This idea is substantiated by the finding that the ^{3}H spiro-
peridol binding at the pituitary level has been found increased in
senescent animals, in contrast with several brain areas[20].   We
propose therefore that, due to the decreased production of DA from
TIDA neurons in old animals, PRL levels are increased and simult-

Table 3.   Age Related Changes of 5HT- and Gaba-Ergic Tonus in
            Central Areas Endocrinologically Implicated

Area of Brain	Observation	References
Hypothalamus (rat)	increased 5HT turnover	Simpkins et al., 1977
Hypothalamus (rat)	decreased ^{3}H GABA binding	Govoni et al., 1979
Thalamic areas (rat)	decreased GAD	McGeer and MCGeer, 1978
Many cerebral areas (human)	decreased GAD	Perry et al., 1977

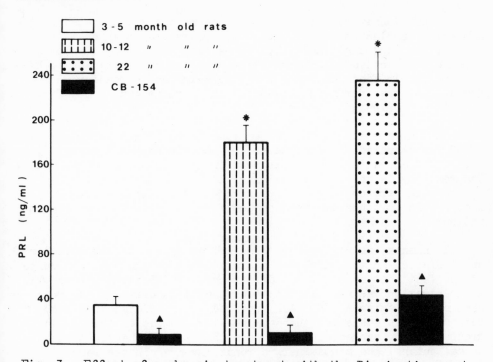

Fig. 3.   Effect of a chronic treatment with the DA-mimetic agent
          CB-154 on serum prolactin concentrations in younger and
          older rats.
          Vertical bars represent mean ± SE of 6 animals for each
          group.
          *p < 0.01 if compared to 3 - 5 month old rats
          ▲p < 0.01 if compared to the untreated animals of the
                    corresponding groups.

aneously DA receptor supersensitivity at the lactotrophs level
develops; in some instances[21,22], this supersensitivity is able to
compensate for the lack of the inhibitory neurotransmitter coming
from the hypothalamus.

The picture, however, is complicated by the fact that, besides
the classical DA-not sensitive adenylate cyclase receptors[23,24],
a population of DA-sensitive adenylate cyclase receptors has been
identified at the pituitary level[25].   These receptors appear to be
associated with PRL stimulatory activity[25] and different authors
have also reported that an increase of cyclic AMP levels in the
pituitary is followed by an increased PRL secretion[26,27].   Further
experiments are therefore needed to clarify whether or not also a
change in the ratio between DA-inhibitory and DA-stimulatory
receptors could account for the changes of PRL found in aging.

As previously reported, also 5HT neurons are implicated in the control of prolactin secretion. It appears that changes in central serotoninergic activity can be responsible for the modulation of the activated PRL secretion, through an inhibition of TIDA neurons[3]. In fact, during stress, suckling and circadian PRL variations, the increased release of the hormone can be modified by antiserotoninergic agents[3].

In old rats it has been shown that there is a greater turnover of 5HT if compared to young males[28]. As can be seen in Fig. 4, PRL circadian variations are reduced in amplitude in old male rats. The tendency is an abolition at high levels, suggesting that an increase in serotoninergic tonus in the hypothalamus might be responsible for this age-related modification of PRL phasic fluctuation.

In humans, aging, per se, does not appear to be associated with an alteration in the basal PRL secretion[29,30]. However, in spite of the fact that quantitative differences in PRL release after TRH administration was not demonstrated, the time-course of the response changed with aging. PRL peak appeared to be delayed in a high percent of old subjects; the decline also was slower than in the group of young subjects. It is possible to hypothesize that, in normal humans, the monoaminergic tonus involved in the regulation

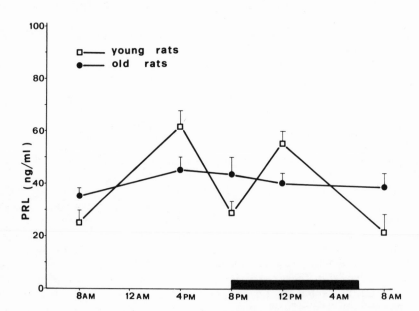

Fig. 4.   Circadian prolactin variations in old male rats. Vertical bars represent $\pm$ SE of 6 animals for each group.

of PRL secretion is sufficient to maintain low basal PRL concentrations.   Only during physiological or pharmacological activations,
can we see changes in PRL secretion, probably due to DA-ergic or
5HT-ergic alterations which occur at hypothalamic level during the
aging process.

HYPOTHALAMIC-HYPOPHYSEAL-GONADAL AXIS (HHGA)

It is now well established that the activity of the HHGA is
under a multiple central monoaminergic control.

The two CA innervating the MBH areas implicated in the release
of Gonadotropins-Stimulating-Hormone (LHRH) appear to play a
different and, in some instances, opposite role.

Neuropharmacological and neurosurgical experiments suggest the
existence of a NE-ergic synapse lying in the preoptic or anterior
hypothalamic area which makes contact with LHRH neurons and which
mediates the acute discharge of LHRH and LH occurring in the pre-
ovulatory phase or in response to removal of steroid negative
feedback[3,31]. The role of DA is more controversial. The current
view, however, is that the DA endings of the TIDA neurons establish
at the level of the lateral palisadic zone of MBH an axo-axonal
synapse with the LHRH ending; the DA released in this site could
play a tonic inhibitory role upon the constant flow of LHRH into
the portal system[32,33]. 5HT appears to modulate in an inhibitory
manner[3] or to determine the circadian fluctuation of activity of
HHGA[34].

In our study we have investigated in several experimental
situations the integrity of HHGA in old rats trying also to correlate the changes found to the modification of neurotransmitters
occurring with aging.

In Fig. 5 are reported the levels of LH at morning of diestrus
in female rats at different ages.   A general tendency to a
reduction of basal LH is present, but no statistical significance
can be demonstrated.   A finding of remarkable interest comes,
instead, from the observation that in old animals the post-castra-
tion surge of LH is remarkably blunted (Fig.6).   This finding can
be explained by the impaired NE-ergic hypothalamic tonus present in
old animals (Table 2).

The decrease of NE-ergic hypothalamic activity can also occur
from the lack of LH surge in response to progesterone in ovariectomized estrogen primed old female rats[35].

In a final experiment two years and 3 - 4 months old rats were
ovariectomized, treated for 7 days with 5 μg of estradiol in oil
and sacrificed at 9:00 a.m. and 4:00 p.m.

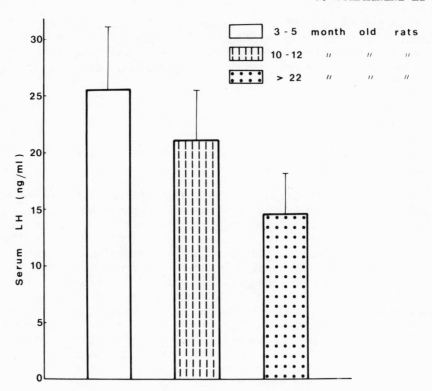

Fig. 5.   Serum LH levels at morning of diestrus in female rats at
          different ages.
          Vertical bars represent mean ± SE of 6 animals for each
          group.

          As can be seen in Fig. 7 estradiol induces a more marked
elevation of LH in young than in old rats.   Furthermore, in the
young rats, in contrast to aged animals, significant afternoon
surges of serum LH are present.   The difference found in this
experiment between old and young animals resembles that seen in
normal versus NE depleted animals[3].

          In conclusion the findings reported in the literature and those
presented by us are in agreement with a possible impairment with
age of the NE dependent mechanism regulating hypothalamic-hypophyseal-gonadal axis.

HYPOTHALAMO-HYPOPHYSIS-ADRENAL AXIS (HHAA)

          There is considerable evidence in the literature that suggests
that NE is responsible for the inhibition of corticotropin (ACTH)

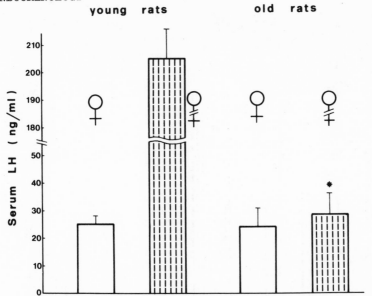

Fig. 6.   Serum LH levels in young and old female ovariectomized
            rats.
            Vertical bars represent mean $\pm$ SE of 6 animals for each
            group.
            *p < 0.01 if compared to corresponding younger group.

secretion and that NE is acting via an α- adrenergic mechanism[36,3].
Some evidence suggests that NE has an excitatory role, but these
results are mainly drawn from experiments in which CA, that barely
cross the blood-brain-barrier, were administered systemically and
might therefore exert their effects by the peripheral stimulation
of α- adrenergic receptors.   Experiments where CA has been
administered into the third ventricle have consistently shown
inhibition of ACTH secretion[36].   Collectively, the results, when
dealing with acute manipulations of brain CA levels and/or turnover,
favour the hypothesis of a central NE-ergic mechanism tonically
inhibiting the corticoliberin (CRF)-ACTH secretion[36,3].

     Unlike NE, serotoninergic neurotransmission does not seem to
be involved predominantly in tonic regulation of HHAA or in acute
response to stress, but, instead, in the circadian periodicity of
adrenal cortex secretion.   A positive correlation between 5HT
content in the limbic system and plasma B rhythm (characterised by
a peak at 8:00 p.m.) was found in the rat, suggesting that 5-HT may
play a role in the regulation of diurnal fluctuation of ACTH secretion[37,38].

     Because of changes in NE and 5-HT metabolism at hypothalamic
level (Tables 2 and 3) which occur with aging and in view of some

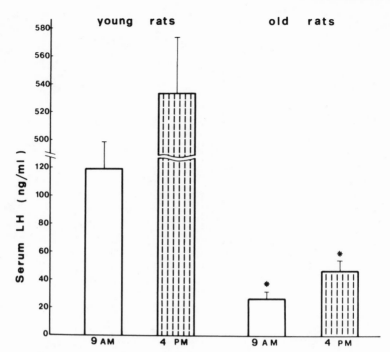

Fig. 7. Estradiol-induced response of serum LH in old and young
         rats at 9.00 a.m. and 4 p.m.
         Vertical bars represent mean $\pm$ SE of 6 animals for each
         group.
         * p < 0.01 if compared to corresponding groups.

evidence suggesting that in aged rats the adrenal circadian variations are modified[39], we decided to evaluate the plasma corticosterone (B) basic and circadian secretion and brain corticosterone
specific uptake (SCU) during the lifespan of male rats. As can be
seen in Table 4, 30 - month old rats present at morning significantly higher plasma B levels than 3-, 9-, and 18 - months old rats;
conversely, the SCU in the hippocampus appears significantly reduced
with age (Table 4).   This is in contrast with the direct correlation between plasma B levels and SCU that is found in the hippocampus and septum of young rats[40], suggesting that the aging process
can alter this kind of balance.

     We have also found that in old rats the circadian rhythms of
plasma B is abolished at intermediate-high levels, (Fig. 8) stressing the possibility that a modification of central 5-HT-ergic
pattern could account for part of the modification of HHAA related
to the aging.   This hypothesis of 5-HT involvement in the impairment of HHAA in old animals is in line with the well demonstrated
interaction between 5-HT-ergic tonus, plasma B circadian rhythms[36],
and SCU in the hippocampus[41].

Table 4.   Plasma Corticosterone (B) Levels and Specific
           Corticosterone Uptake (SCU) in the Hippocampus
           in Young and Old Male Rats

AGE	PLASMA B (µg/100 ml)	SCU (fmoles/100 mg prot. hippocamus)
3 months	6.93 ± 0.58	369.6 ± 24.57
9 months	8.93 ± 1.06	358.7 ± 27.75
18 months	7.35 ± 0.65	346.1 ± 12.72
30 months	11.19 ± 0.36 *	312.8 ± 21.93 *

* p<0.05 if compared to the 3 month group

HYPOTHALAMO-HYPOPHYSIS-THYROID AXIS (HHTA)

The role played by hypothalamic CA in the process of TRH-TSH
release has also received considerable attention.   Both in vivo
and in vitro studies have indicated that these neurotransmitters
may influence TRH-TSH secretion[42,43,44,45]. In vitro studies
provided the evidence that both DA and NE added to rodent hypothalamic fragments[43] or to isolated hypothalamic synaptosomes[46]
increase TRH release into the medium.   In vivo studies clearly showed
that only drugs able to modify central catecholaminergic tonus
produce changes in HHTA activity in basal or stimulated conditions.

The NE appears to exert a stimulatory or at least permissive
role only during acute activation of HHTA.   In fact, in rats,
inhibition of NE synthesis by α- methyltyrosine (α-MT) before stimulation of the axis is able to prevent the TSH surge elicited by a
sudden and severe cold exposure (CE)[47].

Therefore, the decrease of hypothalamic NE activity (Table 3)
which occurs during aging appeared to us to be responsible for the
impairment of thyroid function described in old animals[28].   In order
to verify this possibility, we have evaluated TSH response to cold
stress in old versus young animals in normal condition or after NE
depletion.   The rats were placed for 30 min in a cold room (4°C).
Sacrifice was performed within a minute from the end of CE, when TSH
has reached the highest plasma stimulated level[48].   Younger
(5 - month old) and older (20 - month old) male rats were treated
with saline alone or with α- MT methylester (250 mg/kg intraperitoneally) at 11:00 p.m.   Different groups were sacrificed 2, 3, 6 and
9 h after the treatment.

Our results show that TSH basal levels were similar in younger
and older male rats: the exposure of the animals, previously

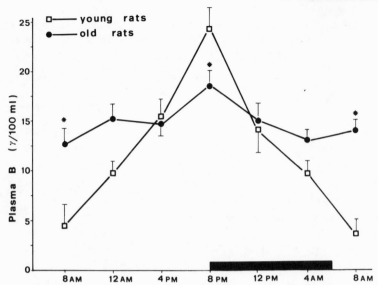

Fig. 8.   Circadian plasma corticosterone (B) variations in old
           male rats.
           Vertical bars represent mean ± SE of 6 animals for each
           group.
           *p < 0.01 if compared to younger group of animals.

acclimatized to a high environmental temperature, produced in
younger rats an acute release of TSH, as previously reported[47].
In older male rats the increase of thyrotropin secretion after CE
was statistically lower than in younger animals.   Pretreatment with
α - MT 1 h before CE was able to prevent this stimulated secretion
of TSH in all groups of animals (Fig.9).

        These data suggest that the decreased hypothalamic content of
NE which occurs with the age is able to modify the response of TSH
to stress conditions (CE), while secretion of the thyrotropin is
unaffected in basal conditions.   On the other hand, CE response is
higher than in α - MT pretreated animals, suggesting that NE-ergic
tonus is not completely abolished at hypothalamic level in aging
process, but is still able to exert, at least in part, a permissive
role on the stimulated hypothalamic-hypophyseal-thyroidal function.

        The conclusions that can be made from the present chapter can
be applied to all the neuroendocrine axes we have explored.   Aging
produced profound changes in the function of HHPRLA, HHGA, HHAA and
HHTA.   These changes can, at least in part, be ascribed to the age-
dependent modifcation in the tonus of the brain neurotransmitters
responsible for the phasic or tonic regulation of neuroendocrine
loops.

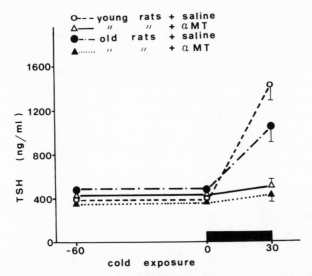

Fig. 9.   TSH response to cold stress in young and old male rats
          treated or not with α-methyltyrosine (α-MT) (250 mg/kg ip)
          60 min before cold exposure.
          Vertical bars represent mean $^{+}_{-}$ SE of 8 animals for each
          group.

REFERENCES

1. B. Halasz in: "Frontiers in Neuroendocrinology", W.F. Ganong and
      L. Martini, eds., Oxford Press, London (1969).
2. K. Fuxe, Z. Zellforsch, Mikrosk, Anat. 65:573 (1975).
3. E. E. Muller, G. Nistico and U. Scapagnini, "Neurotransmitters
      and Anterior Pituitary Function", Acad.Press, New York (1977).
4. W. J. Shoemaker, in: "Frontiers in Catecholamine Research",
      E. Usdin and S. Snyder, eds., Pergamin Press (1973).
5. K. Fuxe and G. Jonsson, in: "Serotonin - New Vistas", E. Costa
      et al, eds., Raven Press, New York (1974).
6. K. Starke, Naturwissenschaft 58:420 (1971).
7. G. Kaur and M. S. Kanungo, Can.J.Biochem. 48:203 (1970).
8. V. V. Frolkis, V. V. Bezrukov, Y. K. Duplenko, I. V. Shcheguleva,
      V. G. Shevtchk and N. S. Verkhratsky, Gerontologia 19:45
      (1973).
9. N. S. Verkhratsky, Expl.Geront. 5:49 (1970).
10. E. R. Baizman and B. M. Cox, Life Sci. 22:519 (1978)
11. R. M. MacLeod and E. H. Fontham, Endocrinology 86:863 (1970).
12. N. Ben-Jonathan, C. Oliver, H. J. Weiner, R. S. Mical and
      J. C. Porter, Endocrinology 100:452 (1977).
13. P. M. Plotsky, D. M. Gibbs and J. D. Neill, Endocrinology
      102:1887 (1978).
14. R. M. MacLeod, in: "Frontiers in Neuroendocrinology", L. Martini
      and W. F. Ganong, eds., Raven Press, New York (1976).

15. M. G. Caron, M. Beaulieu, V. Raymond, B. Cagne, J. Drouin,
    R. J. Lefkowitz and F. Labrie, J.Biol.Chem. 253:2224 (1978).
16. R. I. Weiner and W. F. Ganong, Physiol.Rev. 58:905 (1978).
17. R. M. MacLeod, I. Nagy, I. S. Login, H. Kimura, C. A. Valdenegro
    and M. O. Thorner, in: "Central and Peripheral Regulation
    of Prolactin Function", R. M. MacLeod and U. Scapagnini,
    eds., Raven Press, New York (1980).(In Press)
18. J. F. Bruni, S. Marshall, H. H. Huang, H. J. Chen and J. Meites,
    IRCS Med.Sci. 4:265 (1976).
19. P. M. Wise, A. Ratner and G. T. Peake, J.Reprod.Fer.47:363
    (1976).
20. S. Govoni, M. Memo, L. Saiani, P. F. Spano and M. Trabucchi,
    Mechanism of Ageing and Development (1979).
21. C. J. Shaar, J. S. Euker, G. D. Reigle and J. Meites,
    J. Endocrinol. 66:45 (1975).
22. B. E. Watkins, J. Meites and G. D. Riegle, Endocrinology
    97:543 (1975).
23. P. F. Spano, S. Govoni and M. Trabucchi, in: "Advances in
    Biochemical Psychopharmacology", P.J. Roberts et al, eds.,
    Raven Press, New York (1978).
24. G. M. Brown, P. Seeman and T. Lee, Endocrinology 99:1407
    (1976).
25. M. Y. Cronin, C. Y. Cheung, J. E. Beach, N. Faure, P. G.
    Goldsmith and R. I. Weiner in: "Central and Peripheral
    Regulation of Prolactin Function", R. M. MacLeod and U.
    Scapagnini, eds., Raven Press, New York (1980). (In press)
26. A. Lemay and F. Labrie, FEBS Lett. 20:7 (1972).
27. Y. C. Clement-Cormier, J. J. Heindel and G. A. Robinson,
    Life Sci. 21:1357 (1977).
28. J. W. Simpkins, G. P. Mueller, H. H. Huang and J. Meites,
    Endocrinology 100:1672 (1977).
29. T. Yamaji, K. Shimamoto, M. Ishibashi, K. Kosaka and H. Orimo,
    Acta.Endocrinol. 83:711 (1976).
30. L. Murri, T. Barreca, A. Gallamini and R. Massetani,
    Chronobiologia IV:135 (1977).
31. U. Scapagnini, B. Marchetti, A. Prato and I. Gerendai,
    Acta Eur.Fertilitatis 8:4 (1977).
32. K. Fuxe, Triangle 17:1 (1978).
33. U. Scapagnini, G. Clementi, A. Prato, P. L. Canonico, B.
    Marchetti and I. Gerendai, in: "Long-term Effects of
    Psychotropic Drugs", G. Racagni et al, eds., Raven Press,
    New York (1980).
34. M. Hery, E. Laplante and C. Kordon, Endocrinology 99:496 (1976).
35. J. A. Clemens, R. W. Fuller and N. V. Owen, in: "Adv.Exp.Med.
    Biol.", Finch et al, eds., Plenum Press, London (1978).
36. P. Preziosi and U. Scapagnini, in: "The Endocrine Function of
    the Human Adrenal Cortex", V.H.T. James et al, eds.,
    Acad.Press, London (1978).
37. U. Scapagnini, G. P. Moberg, G. R. Van Loon, J. De Groot and
    W. F. Ganong, Neuroendocrinology 10:155 (1971).

38. M. L. Simon and R. George, Neuroendocrinology 17:125 (1975).
39. G. Y. Nicolau and S. Milcu, Chronobiologia IV:136 (1977).
40. L. Angelucci and P. Valeri, Ann.Ist.Sup.Sanita XIV: 40 (1978).
41. U. Scapagnini, L. Angelucci, I. Gerendai, P. Valeri,
    P. L. Canonico, M. Palmery, F. Patacchioli and B. Tito,
    in: "Interaction within the Brain-Pituitary-Adrenocortical
    System", M. T. Jones et al, eds., Acad.Press, London (1979).
42. M. Kotani, T. Onaya and T. Yamada, Endocrinology 92:288 (1973).
43. Y. Grimm and S. Reichlin, Endocrinology 92:626 (1973).
44. J. Tuomisto, T. Ranta, A. Saarinen, P. Mannisto and J.
    Leppaluoto, Lancet 2:510 (1973).
45. L. Annunziato, G. F. Di Renzo, G. Lombardi, P. Preziosi and
    U. Scapagnini, Br.J.Pharmacol. 52:442 (1974).
46. G. W. Bennet and J. A. Edwardson, Nature 257:323 (1975).
47. L. Annunziato, G. F. Di Renzo, G. Lombardi, F. Scopacasa,
    G. Schettini, P. Preziosi and U. Scapagnini, Endocrinology
    100:738 (1977).
48. J. Leppaluoto, T. Ranta, H. Lybeck and R. Varis, Acta Physiol.
    Scand. 90:640 (1974).

CLINICAL PSYCHIATRY OF THE ELDERLY

T. Arie

Professor of Health Care of the Elderly
Sherwood Hospital
Nottingham

The subject allotted to me is broad and much of it (including the crucial areas of dementia and depression) is dealt with by other speakers.   I shall pick a few topics, using these as pegs on which to hang some general consideration in the clinical psychiatry of old age.   First, I will make some comments aimed particularly at aspects which distinguish old age mental disorders from equivalent disorders in younger adults.   Then I shall say a little about the rapid recent growth of interest in old age psychiatry in Britain, the reasons for it, and the associated developments in services.   My special interest is in psychiatric services for the elderly, and that dimension colours my thinking on clinical matters.

OLD AGE MENTAL DISORDERS

It has often been believed, and sometimes still is, that all but the organic degenerative disorders become less prevalent in old age.   In fact, at least 25% of all old people suffer from frank mental disorders[1].   If to these one adds "minor" but significant neuroses and character disorders, the figure may be as high as 40%[2]. Formidable problems are posed by the dementias, which are central to this field, and by acute confusional states; but the bulk of mental disorder in old age is functional.   The commonest mental illness in old age is depression, and affective components also shape many of the neurotic states of the elderly.

There is a complex relationship between physical and mental disease in old age.   Old people tend to somatise their disorders. Old age is a time of increasing awareness of bodily function which

51

in younger life is normally taken for granted; and such awareness
easily becomes preoccupation.   Consequently, physical symptomatology looms large even in functional old age disorders and, indeed,
is often the presenting feature.   Actual physical pathology is,
of course, common in old people; and much functional mental disorder
in old age is causally linked with physical disease.   There is
probably more danger of allowing physical features to mask the
psychological, than vice versa.

   Old age is a phase of life that challenges the resources of the
personality as perhaps nothing else[3].   Along with physical
limitations and impairments of perception, the changes of old age
threaten not only security and independence, but self-image and
self-esteem.   And death itself becomes a prospect, rather than a
remote contingency.

   Death of others is commonplace in old age and there are other
bereavements too - of components of the view of oneself which one
has constructed throughout a lifetime and on which one has come to
depend for self-esteem and security.   The manner in which the
individual reacts to loss of what Murray Parkes has called "world
models"[4] varies between individuals; some adjust reasonably well,
others may react with anxiety, others with depression, others with
paranoia.   Paranoia, so pervasive of the mental disorders of old
age, may even have some adaptive quality, for the world in which the
elderly live is not on the whole a friendly world, and to be
suspicious in such a setting may not be inappropriate[5].   Paranoia,
for some, protects the self-image by projecting as faults or
hostility of others, its dissolution or impairment.

   Deafness and paranoid states are also associated[6].   Around a
third of all people over 65 have some hearing difficulty[7].   Among
these hard-of-hearing old people, paranoid disorders are much more
common and the evidence is that it is conductive deafness rather
than nerve deafness that is chiefly associated with paranoid states[8].

   The evidence, often implicitly accepted by the public and by
politicians, of causal links between poverty, low social class or
social isolation with mental disorders in old age is, in fact,
very slender.   But such associations may be important; that noxious
components of these states may be masked by a too simple
characterisation of these privations is suggested by George Brown's
work with young women[9].   He and his colleagues have shown that it
is not so much stress and loss itself that produces breakdown, but
rather the operation of such pressures on individuals with
particular vulnerability - e.g. lack of an intimate relationship,
or early loss of a parent.   In the privations and bereavements of
old age, similar vulnerability factors must surely operate but they
have not yet been studied in this way.   It may well be that it is
change in one's circumstances, when it appears to have lasting

consequences, rather than merely the objective quality of those
circumstances, that is the critical threat; and that the impact of
these factors depends on a constellation of features in the life
history, the personality, and the current life situation which
together make the individual vulnerable.   It is encouraging that
at least one of our younger colleagues is embarking on a study that
will test some of these hypotheses[10].

I want now, extremely briefly, to look at some of the ways
that depression, the prime functional disorder of old age, differs
in older people from its counterpart in younger life.   Depression
is a good paridigm of many of the issues that arise in clinical
psychiatry of the elderly, and particularly in the provision of
services.   I may even here repeat some of the things that
Professor Williamson has already told us - but there is no harm in
hearing a physician and a psychiatrist saying the same things!
And I will be returning in a moment to the theme of collaboration
between psychiatrists and geriatricians.

Old age depressions have, as we have heard, excellent
prognoses for the individual episode, but recurrence is common[11].
Therefore, many and perhaps most depressive patients require
continuing surveillance.   Suicide is common in the depressions of
the elderly, and so-called "attempted suicide" becomes less common[12].
Even more common than suicide is the danger of dehydration and
metabolic derangement due to depressive slowing and refusal of
fluids; sometimes its onset is so dramatic that one wonders whether
there may not even be more direct neurophysiological pathways.   No
facility for the treatment of depressive states in the elderly
which cannot confidently diagnose and promptly arrange treatment
for such crises has the right to admit depressed old people.

Depression is not the only old age functional disorder which
may present as "pseudodementia".   Meticulous history taking and
observation is crucial and may be lifesaving in confused old
people; nuggets of specific symptomatology, identified commonly
by the nursing staff, in a general confusional background may
suggest a diagnosis and justify a trial of treatment.   The coexistence of several different processes of psychological disorder
in the elderly is, of course, common and commonest of all is the
coexistence of cognitive degeneration and functional disorder.
Felix Post has shown that in old people schizophrenic and affective
disorder also not infrequently coexist[13].   "Schizo-affective
disorder" is a diagnosis which one has learnt whenever possible to
eschew in younger patients, but there seems to be a significant
group - 29 patients in three years of Post's clinic - in whom the
two conditions clearly coexist.

Responses to physical methods of treatment in old age mental disorders are often excellent but we know well that drugs, of which we will hear much in this meeting, carry special risks in the elderly. For instance, recent evidence about the nephrotoxicity of lithium (itself an excellent drug in the elderly) may be particularly relevant to the elderly[14]; and there is now evidence that in old age electroplexy should almost always be given unilaterally to the non-dominant hemisphere and that this is associated with a quicker recovery time and less memory impairment[15].

Such special consideration in the clinical psychiatry of old age further underpin the case for a subspecialty of old age psychiatry such as is developing in our country. A specialist in old age psychiatry finds himself not only working closely, but thinking closely, with his colleagues in geriatric medicine; in so far as old age psychiatry is a builder of bridges between psychiatry and medicine - bridges which some feel have often decayed in recent generations - there is a further bonus which comes from the development of our subspecialty. It is at this development that I wish now briefly to look.

OLD AGE PSYCHIATRY IN BRITAIN[16]

The last ten years have seen a surge of interest, and rapid development, in this field. A small "coffee-house group" of psychiatrists specially interested in the elderly had begun to meet in the early 1970s; there were some half-a-dozen of us. Now, at the end of the decade, we have a thriving Specialist Section in the Royal College of Psychiatrists devoted to Old Age Psychiatry, which numbers its members in hundreds, which through the Royal College has established a dialogue with the Government on policy in this field, and which has made close links with its sister discipline of medical geriatrics. A survey currently being conducted in our department in Nottingham of all those consultant psychiatrists whose special interest is in the elderly has identified well over a hundred - amounting to some 10% of all consultant psychiatrists in our National Health Service. Meetings, papers, books and policy documents abound, and psychiatric and other trainees are increasingly seeking experience in this field - though our recruitment problems are far from solved. There is an atmosphere of interest, enthusiasm and movement; and in keeping with national policies, services dealing specifically with the whole range of old age psychiatry are being established in many localities, normally in close association with the local service in geriatric medicine. Sometimes local services contain a facility run jointly by geriatricians and psychiatrists in which patients with mixed disorders, or with disorders that need the joint care of both teams, can be admitted and offered collaborative assessment and care.

In Nottingham we have gone a stage further and have established a University Department of Health Care of the Elderly which brings together the medicine and the psychiatry of old age. We aim to give patients easy access to the staff and facilities of both specialties and to offer a unified teaching programme. We have now, in addition to my own post, two full-time academic posts in old age psychiatry - and these may well be the first such academic posts in this country. The department thus has a medical wing and a psychiatric wing, each with its discrete facilities, but working together as one service, readily supporting each other, geographically situated side by side in the same general hospital campus, and functioning together in almost all respects as a single team.

This team is also responsible for a student teaching programme. All Nottingham medical students spend a month full-time in our department, and during this time are concerned entirely with the care of the elderly; they receive a planned course of teaching, as well as a clinical "apprenticeship" to all aspects of the department's work.

One can identify a variety of factors which have contributed to these developments in our country, and of them, obviously the most important is the first:

1. The enormous upsurge in the numbers of the elderly, and particularly of the very old, which has been the experience of industrial societies during this century - and not only of industrial societies. Everywhere the pressure of heavily disabled old people on services, and particularly on families, is making itself felt; and the same features of prosperity which have contributed to the increase in the number and proportion of the elderly have led to increased expectations of help from state and other services on the part of those who look after them.

2. Our understanding of the psychiatry of old age was transformed by the work of pioneers such as Felix Post[17] and Martin Roth[18]. There is now widespread recognition of the heterogeneity of old age mental disorders, and we have a much better understanding of causation and natural history; and we know that the prognosis of most functional disorder in the elderly is good, and despite the salience of organic degenerative conditions, functional disorders are the commonest mental illnesses of the elderly.

2. Old age psychiatry raises most of the issues of modern psychiatry, indeed, one might say of modern medical care, in sharper and more challenging form[19]: and by the same token it has benefited massively from modern treatments, physical, social and psychological. Yet the dementias continue to baffle us, and are now the main generators of need for long term care.

4. Concurrently, there has been a growing awareness, particularly in countries where comprehensive provision for health is chiefly the responsibility of the state, of the need to direct resources preferentially towards those groups of the population who have previously not had an equitable share.   The elderly are such a group par excellence.   Defeatism and pessimism, and until recent decades a real lack of effective treatments, has caused old people suffering from these disorders to be under-recognised and even where recognised, to be inadequately or inappropriately treated, or to be denied access to properly equipped specialist services.   Old age mental disorder has mainly been seen as a social nuisance, calling for alleviation of trouble for others, and posing issues merely of "disposal", preferably at minimum cost in effort, resources and quality of care.   As more people recognise such problems in their own parents, and as the public becomes more vocal, something better is increasingly demanded.

5. The association between physical and mental disorder, common enough in younger people, is now seen to be of the essence of the psychiatry of the elderly[20,21].   Hence the community of interest between physicians, psychiatrists and other medical specialties, the growth of joint facilities, and the origin of such ventures as our unified department in Nottingham.

6. The development of British geriatric medicine, and its wide contribution to education and debate, have further stimulated interest in and recognition of the potential contribution of psychiatry and of other specialties to this field.

7. Amid the pressure on inevitably scarce resources, an important role of the geriatric psychiatrist is to act as an "advocate" for the elderly - to ensure that amid local psychiatric facilities a voice is constantly heard for the needs of old people. At the same time, such a psychiatrist and his team acts as an identifiable focus to which other local services with an interest in the elderly can relate.

So, plenty of reasons combine to give these developments their importance.   The cost of decent services is formidable. Despite efforts almost everywhere to bring services out of hospitals, and to develop community support networks, day hospitals and the apparatus of "community care", health services of all countries which recognise the burden are staggering beneath its weight; and they are finding that effective community services are not necessarily always cheaper than institutional care[22,23].   In addition to securing a share of resources for this work, the geriatric psychiatrist has a special responsibility to husband those resources effectively, to experiment and to innovate with a view to improving their effectiveness; and to help to generate enthusiasm in the staff, and to ensure that this field of work gets its share of the best people in all the

health disciplines.   In our country, so far, there is every reason
to be pleased with the extent to which these objectives are being
achieved.

REFERENCES

1. D. W. K. Kay, P. Beamish and M. Roth, Old age mental disorders
        in Newcastle upon Tyne, Br.J.Psychiat. 110:146 (1964).
2. D. W. K. Kay, Depression and neuroses of later life, in:
        "Recent Advances in Clinical Psychiatry" 2., K. Granville
        Grossman, ed., Churchill Livingstone, London (1976).
3. F. Post, Learning from old age, Proc.Roy.Soc.Med. 63:359,
        (1969).
4. C. M. Parkes,"Bereavement", Tavistock, London (1972).
5. E. W. Busse and E. Pfeiffer, Functional psychiatric disorders
        of old age, in: "Behaviour and Adaptation in Late Life",
        Busse and Pfeiffer, eds., Little, Brown & Co.,Boston (1977).
6. A. F. Cooper, Deafness and psychiatric illness, Br.J.Psychiat.
        129:216 (1976).
7. L. Fisch, The ageing auditory system, in: "Textbook of Geriatric
        Medicine and Gerontology",J. Brocklehurst, ed., Churchill
        Livingstone 2nd Edition, London (1978).
8. A. F. Cooper, A. R. Curry, D. W. K. Kay, R. F. Garside and
        M. Roth, Hearing loss in paranoid and affective psychoses of
        the elderly, Lancet, 2:851 (1974).
9. G. W. Brown and T. Harris, "Social Origins of Depression",
        Tavistock Publications, London (1978).
10. E. Murphy, Personal Communication.
11. F. Post, The management and nature of depressive illness in late
        life. A follow through study, Br.J.Psychiat. 121:393 (1972).
12. K. Shulman, Suicide and parasuicide in old age: A review,
        Age & Ageing 7:201 (1978).
13. F. Post, Schizo-affective symptomatology in late life,
        Br.J.Psychiat. 118:437 (1971).
14. J. Hestbech and M. Aurell, Lithium-induced uraemia, Lancet
        1:39 (1979).
15. R. M. Fraser and I. Glass, Recovery from E.C.T. in elderly
        patients, Br.J.Psychiat. 133:528 (1978).
16. T. Arie and A. D. Isaacs, The development of Psychiatric
        services for the elderly in Britain, in: "Studies in
        Geriatric Psychiatry", A. D. Isaacs and F. Post, eds.,
        J. Wiley & Sons, London (1978).
17. F. Post, "The Clinical Psychiatry of Late Life", Pergamon Press,
        Oxford (1965).
18. M. Roth, Mental diseases of the aged, in: "Clinical Psychiatry"
        by Mayer, Gross, Slater and Roth. E. Slater and M. Roth
        eds., 3rd Edition, Bailliere, Tyndal & Cassell, London (1969).

19. D. J. Jolley and T. Arie, Organisation of psychogeriatric
    services, Br.J.Psychiat. 132:1 (1978).
20. K. Bergmann and E. J. Eastham, Psychogeriatric ascertainment
    and assessment for treatment in an acute ward setting,
    Age & Ageing 3:174 (1974).
21. T. Dunn and T. Arie, Mental disturbances in the ill old person,
    Br.Med.J. 4:413 (1973).
22. L. J. Opit, Domiciliary care for the elderly sick: Economy
    or neglect? Br.Med.J. 1:30 (1977).
23. D. Hawks, Community care: An analysis of assumptions,
    Br.J.Psychiat. 127:276 (1975).

DEMENTIA: EPIDEMIOLOGICAL ASPECTS

K. Bergmann

Consultant Psychiatrist
Maudsley Hospital
London

INTRODUCTION

"La demence senile est la suite de progrès de l'age"
(Esquirol)[1].

If this observation be true can we assume that all old people eventually dement or is there a separate disease entity age related but essentially separate from normal ageing?   Henry Maudsley[2] already grasped the nature of this problem in his discussion concerning the testamentary capacities of the aged;

> "The natural decline of mental faculties which in greater
> or less degree commonly accompanies the bodily decline of
> old age, should be distinguished from the greater loss of
> mental power which is known as senile dementia, notwith-
> standing that in the least degree of the former and the
> worst stages of the latter there are all degrees of
> transition".

As in Britain today 95% of all the population over 65 years of age live at home.   We need to know what proportion of the elderly living at home do resemble institutional demented patients and whether there are a large number of the elderly in "A degree of transition".

The tragic downhill course associated with the diagnosis of senile dementia in hospital samples is well established[3,4], but frequently this effect has been ascribed to 'labelling'[5].   One might expect that the largely undiagnosed dementias living at home[6] would escape these consequences and therefore, show a relatively benign course and outcome.

59

Another important reason for adopting more of a community orientated epidemiological approach to the study of dementia is that the effect of various types of selection for admission to different institutions may very powerfully influence the clinical picture, prognosis, and possibly also the neuropathological findings correlating with the clinical diagnosis of dementia. Thus statements about the behavioural characteristics of dementia, the course and prognosis would benefit from observations based on epidemiological studies.

THE PREVALENCE OF DEMENTIA OF AT LEAST MODERATE SEVERITY

Many surveys carrying out prevalence studies on community resident elderly have been reported and the degree of agreement between various authors is on the whole good. The percentage prevalences in studies where institutional populations are also included are not markedly higher.

For example, community resident elderly subjects studied in Japan and England gave prevalences of 7.1 and 6.2% respectively for moderate or severe dementia[7,8]. Total population studies include those of Essen Möller[9] and Nielsen[10] yielded prevalence of 5.9% and 5% respectively though case selection for the former study and inclusion of younger subjects from the age of 60 years for the latter may account for these differences.

The age related prevalences in the community for dementia is shown (see Table 1). It can be seen that the main age groups at risk are the over 75 year olds and especially the over 80s when the prevalence is 22%. Nevertheless even in this age group it has to be pointed out that there is little evidence that dementia is the normal condition for any group of the aged, and we would be in great danger if we assumed that the types of dysmnestic syndrome and higher cortical dysfunctions by which dementia are characterized were 'normal' for even the very old.

What proportion of these diagnosed as having dementia suffer from the two commonest conditions: senile or arteriosclerotic psychosis? The study by Kay et al[8] suggest that 61.5% of all cases of dementia are of the degenerative senile type, other studies[5,10,11] confirm the more frequent occurrence of senile dementia.

Males under 75, however, predominantly suffer from arteriopathic dementia.

In Britain today where 95% of all elderly people (over 64 years of age) live at home, as many as 86.5% of those elderly with dementia also live at home[9].

Table 1.   Prevalence of Chronic Brain Syndromes by Age and Sex.
           Newcastle upon Tyne, 1960 and 1964 Samples

Age	Both sexes			Males		Females	
	N	CBS	% ± S.E.	N	%	N	%
65-69	253	6	2.3±0.9	110	3.6	143	1.4
70-74	243	7	2.8±1.1	91	3.3	152	2.6
75-79	144	8	5.5±1.9	51	5.9	93	5.4
80 +	118	26	22.0±3.8	39	20.5	79	22.8
Total	758	47	6.2±0.9	291	6.2	467	6.2

Age related prevalence of community sample from Kay et al
(see reference no. 8)

Such figures are dependent on the social provisions, available from
institutional places and family acceptance of elderly relatives
with dementia.   In Denmark Nielsen[11] reported only 66.7% of his
demented sample as living at home, illustrating the effect of such
differences.

MILD DEMENTIA AND OTHER BORDERLINE ORGANIC STATES

Mild dementias are rarely seen in hospital samples, but are
reported in community samples as "early", "mild", or "borderline"
states.   Allocation to this group may be by judgements as diverse
as an indefinite clinical picture, the brevity of the history, the
benign nature of the course or, in the absence of clinically
diagnostic features, a poor quality of life and an apparent inability to cope socially.

The prevalence of such conditions varies widely from 52.7% in
Japan[7], surely implying a different set of criteria, 15.4% in
Denmark[10] to 5.7% in England[9].

An analysis of these mild and borderline states by Bergmann
et al[12] suggests that they can be divided into several groups:
suspected senile dementia, cerebral arteriopathic disease without
dementia, and miscellaneous conditions such as myxoedema or subacute
delirious states e.g. those accompanying carcinomatosis.

THE INCIDENCE OF DEMENTIA

An important question which also needs an epidemiological
approach is the problem of incidence of dementia.   There are two
main approaches to the study of incidence: first consultations in

hospital case register studies, and measuring the incidence of new cases in longitudinal community surveys, where case detection is based on personal interview.

Case Register Studies

Case register studies give a remarkable degree of agreement for average annual incidence per 1000 of the population; 2.0 (Adelstein et al[13]), 2.25 (Wing et al[14]) and 2.4 (Helgasson[15]).

Two samples based on population samples personally interviewed and followed up[12,16] yielded annual incidence figures per 1000, uncorrected for death, of 15.0 and 15.6 respectively.

Longitudinal Studies

Longitudinal studies of dementia are also of value in order to determine the importance of the predictive power of the diagnosis of dementia on the likelihood of dying over a given period, and the need for institutional care among community resident subjects, with dementia.

Kay et al[17] examined the original sample seen in 1960 and found that the single assessment of "poor memory" correlated more highly with death 4 years later ($r = 0.35$ $p < .001$) than did age ($r = 0.32$), physicians' rating of disabilities ($r = 0.25$) and poor living conditions ($r = 0.15$). Comparing the mortality of various major diagnostic groups; normal, functional and organic: Kay and Bergmann[18] employed actuarial expectations of survival (see Figs. 1, 2, and 3) to demonstrate that organic subjects exceeded the expected mortality to a greater degree than did the population as a whole, or normal and functional subjects.

A further follow-up study of the Newcastle material[8] combining 2 samples consisting of nearly 800 subjects showed the increased need of the demented subjects for institutional care even when they were matched with normal subjects by age and sex. The demented subjects spent most time in geriatric hospitals, homes for the elderly and psychiatric hospitals and more than 10 times as many weeks per person at risk than did normal subjects (see Table 2).

The longitudinal study of "early", "mild" or "borderline" states is a matter of practical and theoretical interest. Kral[19] has suggested that a distinction can be made between the benign senescent memory defect, characteristic of normal ageing and malignant amnestic syndromes which progress to dementia. A study following up suspected dementias or arteriopathic cerebral disorders[12], tried to examine these questions (see Table 3).

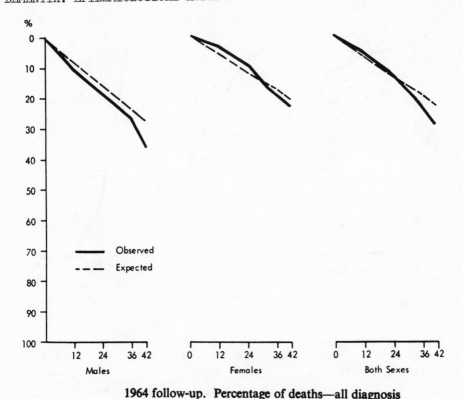

1964 follow-up. Percentage of deaths—all diagnosis

Fig. 1.   Expected and observed mortalities for all subjects
          within a community sample. Kay and Bergmann (see
          reference no. 18)

Of the suspected dementia group a significantly higher than chance
number developed a definite dementing syndrome (30%), but of the
remainder about one third were found to have been misdiagnosed as
dementia because of their low I.Q., subcultural environment and
consequently poor performance on the memory and information test
which was employed[20].   Another third of this sample remained
'borderline' at the three year follow-up point.

The greatest number of respondents who developed dementia
however came from the normal group among whom three years earlier,
no predictive signs of dementia could be found.

This suggests the need for a more careful and detailed
psychiatric evaluation of people in the early stages of dementia
combined with a longitudinal follow-up study examining clinical and
prognostic features.

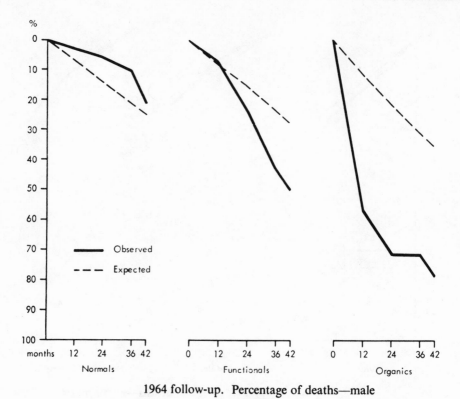

1964 follow-up. Percentage of deaths—male

Fig. 2. Expected and observed mortality for male subjects
by diagnosis Kay and Bergmann (see reference no. 18)

The main problem of epidemiological surveys involving clinical
psychiatric assessment is that a psychiatrist would have to
interview 94 respondents without dementia to obtain six subjects
with evidence of organic cerebral impairment, such a method of case
finding to obtain a demented sample of any size would be
uneconomical.

Screening Questionnaires

The need for an effective screening questionnaire for use by
psychiatrically untrained community workers is apparent, and a
pilot study suggested feasible methods of constructing such a
questionnaire[21], and such a questionnaire is currently being
developed. This has provided an opportunity to ascertain the
psychopathological profile of elderly patients with dementia in the
community and compare it with hospital patients suffering from
dementia, but relatively well preserved. (See Fig. 4).

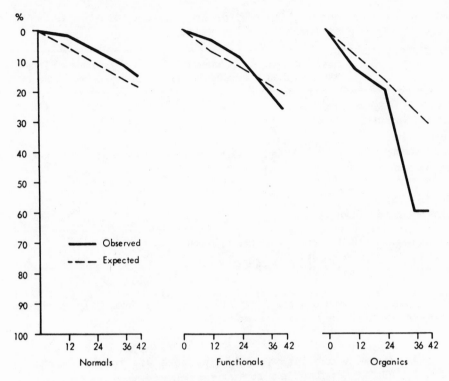

1964 follow-up. Percentage of deaths—females

Fig. 3.   Expected and observed mortality for female subjects
by diagnosis Kay and Bergmann (see reference no. 18)

Table 2.   1960 and 1964 Samples: Matched Groups; Numbers Admitted
and Duration of Stay in Various Types of Care During
Period of Follow-up

	N	Normal controls (N=73) Total Duration and Mean in Weeks		Total Stay N	N	CBS subjects (N=47) Total Duration and Mean in Weeks		Total Stay N
Acute	15 ⎫	95 ⎫			8 ⎫	43 ⎫		
Geriatric	2 ⎬16*	7 ⎬102	(6.4)	1.4	19 ⎬26*	298 ⎬390	(15.0)	8.3
Mental	0 ⎭	0 ⎭			5 ⎭	49 ⎭		
Local Authority Homes	0 ⎫ 3	0 ⎫			4 ⎫ 8*	288 ⎫596	(74.5)	12.7
Private Homes	3 ⎭17*	59 ⎬59	(19.7)	0.8	5 ⎭	205 ⎭		
Total		161	(9.5)	2.2	28*	986	(35.2)	21.0

*Some patients admitted to more than one type of care.

Normal and demented subjects matched for age and sex: weeks of
bed occupancy in various types of institutional care over 2-4 year
period. Kay et al (see reference no. 8).

Table 3.   Development of Chronic Brain Syndrome in all Diagnostic
           Groups at Follow-up (2-4 years)

	% Chronic brain syndrome dead	% Chronic brain syndrome alive	Total % chronic brain syndrome	Total no.
Normal	1.3	1.5	2.8 †	471
Functional syndrome	2.4	2.4	4.8	164
Suspected chronic brain syndrome	15.8	15.8	31.6 **	19
Cerebrovascular insufficiency	4.5	9.1	13.6 *	22
Total	2.1	2.4	4.5	676
Subjects not traced at follow-up				17

† $p < 0.01$ $x^2 = 10.31$ df 1   significantly lower.
** $p < 0.01$ $x^2 = 33.9$ df 1 $\Big\}$ significantly higher.
* $p < 0.05$ $x^2 = 4.53$ df 1

Development of dementia in a random sample of community residents.
Bergmann et al (see reference no. 12).

The questionnaire employed for this purpose was the Geriatric Mental
State Schedule (Copeland et al[22]).

     It is evident that the degree of memory impairment is not the
only significant differentiating feature between community and
hospital organics, but also disorientation, lack of insight, and a
fragmentary collection of psychotic symptoms.   Also of note is the
significant degree of affective disturbance found in dementing
patients and community residents.   However, surprisingly lack of
skills and perceptuo-motor and language dysfunctions subsumed under
the "Cortical dysfunction" factor is not significantly different
between the hospital and community groups.   A more detailed study
comparing various aspects of impaired function and psycho-pathological profiles might well provide new insights into the important
clinical features of different types of dementia.

CONCLUSION

     What does epidemiology have to contribute in the future in
studying dementia?   Two main points come readily to mind.

     The first is an extension of point already developed in this
contribution: namely it seems no more likely that dementia, even of
the 'senile' type will remain in the future, a single syndrome, than
did Bright's disease of the kidney.

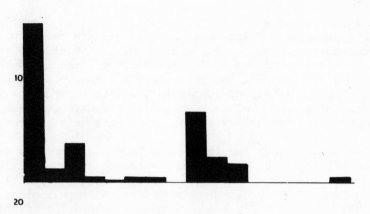

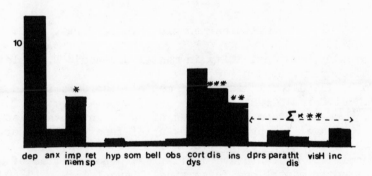

Fig. 4. Symptom profile comparing hospital patients and community resident subjects with mild or moderate dementia. (Bergmann et al, Bayer-Symposium VII. Brain Function in Old Age. Springer-Verlag, Berlin, 1979).

*Key to abbreviations from left to right in Depression Anxiety, Impaired memory, Retarded speech, Hypochondriasis, Somatic preoccupation, Belligerence, Obsessionality, Cortical dysfunction, Disorientation, Insight, Depersonalisation, Paranoid traits, Thought disorder, Visual Hallucinations, Incontinence.

It is likely that the demented patients we study with their small shrunken brains will not yield a great deal of new clinical and pathological information.   Clearly defined clinical studies on early cases, including longitudinal studies of a variety of courses may correlate with differing biochemical and pathological findings in those subjects coming to investigation or autopsy.

The second point which is of importance is that currently available prevalence and incidence figures already permit us, for any known population, to calculate how many of the elderly in any given community make up the hidden "iceberg" of organic psychiatric disorder, and by surveying the percentage of those residents with or without similar levels of impairment, within various institutions for longterm care of the elderly, new priorities in assessment, placement and care can be developed even within existing resources.

It is unlikely that the increasing burden of dementia can be understood without an epidemiological approach and where epidemiological considerations are not taken into account, 'experts' will be allowed to "solve" the problem by transferring it to another section of institutional care or into the community, where its effect is not being measured.

REFERENCES

1. E. Esquirol, Des maladies mentales. Vol.2.Paris, J. B. Ballière
    (1838).
2. H. Maudsley,"Responsibility in mental disease", D. Appleton
    and Company, London (1874).
3. M. Roth, The natural history of mental disorders in old age,
    J.Ment.Sci. 102:281 (1955).
4. D. W. K. Kay, Outcome and causes of death in mental disorders
    of old age: A longterm follow-up of functional and organic
    psychoses, Acta Psychiat.Scand. 38:249 (1962).
5. D. Macmillan, Preventative geriatrics: Opportunities of a
    community mental health service, Lancet 1439   (1960).
6. J. Williamson, I. H. Stokoe, S. Gray, M. Fisher, A. Smith,
    A. McGhee and E. Stephenson, Old people at home: their
    unreported needs.   Lancet 1:1117 (1964).
7. Z. Kaneko, Epidemiological studies on mental disorders of the
    aged in Japan, in:"Proceedings of 8th International Congress
    of Gerontology, 1. Abstracts of Symposia and Lectures",
    International Association of Gerontology, Washington D.C.
    (1969).
8. D. W. K. Kay, K. Bergmann, E. M. Foster, A. H. McKechnie and
    M. Roth, Mental illness and hospital usage in the elderly:
    a random sample followed up, Compr.Psychiat. 2:26 (1970).
9. D. W. K. Kay, P. Beamish and M. Roth, Old age and mental
    disorder in Newcastle upon Tyne, I. A study of prevalence.
    Br.J.Psychiat.110:146 (1964).

10. J. Nielsen, Geronto-psychiatric period prevalences investigation in a geographically delimited population, Acta Psychiat.
Scand. 38:307 (1962).

11. H. O. Akesson, A population study of senile and arteriosclerotic psychoses, Human Heridity 19:546 (1969).

12. K. Bergmann, D. W. K. Kay, E. M. Foster, A. A. McKechnie and
M. Roth, A follow-up study of randomly selected community
residents to assess the effects of chronic brain syndrome
and cerebrovascular disease, Psychiat. Pt. II.
Excertpa Medica Int. Congress Series No. 274, Amsterdam:
Excerpta Medica. (1971).

13. A. M. Adelstein, D. Y. Downham, Z. Stein and M. W. Susser,
The epidemiology of mental illness in an English city,
Social Psychiat. 3:47 (1968).

14. J. K. Wing, A. Hailey, E. R. Bransby and T. Fryers, The
statistical context: comparisons with national and local
statistics, in: Evaluating a community psychiatric service,
J. K. Wing and A. M. Hailey, eds., The Camberwell Register,
1964-1971, New York for Nuffield Provincial Hospital Trust
by the Oxford University Press (1972).

15. L. Helgason, Psychiatric services and mental illness in
Iceland, Acta Psychiat.Scand.Suppl. 268:54 (1977).

16. O. Hagnall, Disease expectancy and incidence of mental illness
among the aged, Acta Psychiat.Scand.Suppl. 219:83 (1970)

17. D.K.W. Kay, K. Bergmann, E. Foster and R. F. Garside, A four
year follow up of a random sample of old people originally
seen in their own homes. A physical, social and psychiatric
enquiry. Proceedings of the 4th World Congress of Psychiatry
Psychiatry, Excerpta Medica Int.Congress Senes No. 150,
Amsterdam: Excerpta Medical (1966).

18. D. W. K. Kay and K. Bergmann, Physical disability and mental
health in old age. A follow-up of a random sample of
elderly people seen at home, J.Psychosomatic Res. 10:3(1966).
(1966).

19. V. A. Kral, Senescent forgetfulness: benign and malignant,
Canadian Med.Assoc.Journ.86:257 (1962).

20. M. Roth and B. Hopkins, Psychological test performances in
patients over sixty. I. Senile psychosis and the affective
disorders of old age, J.Ment.Sci. 99:439 (1953).

21. K. Bergmann, L. B. Gaber and E. M. Foster, The development of
an instrument for early ascertainment of psychiatric
disorder in elderly community residents: a pilot study,
in:Janssen Symposien: Gerontopsychiatrie, 4, R. Degwitz
and P. Schulte, eds., Janssen, Dusseldorf (1975).

22. B. J. Gurland, J. L. Fleiss, K. Goldberg, L. Sharpe (U.S.team)
with J. R. M. Copeland, M. J. Kelleher, J. M. Kellett
(U.K. team): A semi-structured clinical interview for the
assessment of diagnosis and mental state in the elderly.
The Geriatric Mental State Schedule II. A factor analysis.
Psychol.Med. 6:451 (1976).

DEMENTIA: CLINICAL ASPECTS

L. Bracco and L. Amaducci

I° Clinica Neurologica dell'Università degli Studi
Florence
Italy

INTRODUCTION

Dementia is described as a progressive deterioration of mental functions, without clouding or disturbances of perception, resulting from diffuse or disseminated disease of the cerebral hemispheres during the adult life.   The inclusiveness of this definition is emphasized in order to stress some typical aspects of the syndrome:

1. The degree of altered function and the severity of the disorder which ranges from a barely discernible deviation from normal to a virtual cerebral death.

2. The organic and not focal damage of cerebral hemispheres which underlies the clinical features.

3. The description of change in mental functions as a "loss"of previously acquired capacities, so that in order to obtain an accurate assessment of functional deterioration it is necessary to judge the patient's symptoms in comparison with premorbid activities.

This global brain dysfunction is a common finding in healthy old people.   Nevertheless, the deterioration in personality is so pathological an aspect, that it is not difficult to distinguish it from a physiological aging process.

CLINICAL ASPECTS: A LONGITUDINAL STUDY OF MENTATION AND BEHAVIOUR

The components of mentation and behaviour that are to be examined in a demented patient are: sensorial activity i.e. the processes of sensation and perception which, generally, are not affected; the cognitive functions (capacity for memorizing; the ability to think, reason and form logical conclusions), the affective component (temperament, mood and emotion), the conative functions (initiative, impulse and drive) and insight which, more or less, are all severely affected.

In order to give a better explanation of the interesting evolution of these aspects, the clinical description will adopt a criterion which must be a predominantly chronological one.

Early in the process of dementia a variety of barely perceivable disorders may present such as alteration in drive, mood, enthusiasm, capacity to give and receive affection and creativity; thus one early and frequent complaint is that the patient is not "himself".   Already in this stage the examiner may note a vagueness in the presenting story, an imprecision in detail or a lack of order in the mental processes that do not fit well with the patient's background and previous performance, bringing to mind the possibility of an early dementia[1].

Hollander[2] suggests several behaviour observations which should alert the examiner to the possibility of dementia; the patient who expresses disinterest in a topic, allowing that his sons or daughters keep up with that sort of thing; the patient who expresses too much satisfaction from trivial accomplishments; the patient who struggles too hard to turn out performances which earlier would have required little effort; the patient who says he will come back to that question later, but neglects to do so.

Over and over one may read in the notes that prior to admission money or cherished possessions were "lost" and relatives are accused of taking them; that they turned on gas jets, but forgot to light them; that the patients wandered away from home and could not find their way back...[3].

In the incipient phases of cerebral degeneration - according to Well's description which we have selected from among others because of its effectiveness and copiousness of details -

"the individual experiences diminished energy and enthusiasm. He has less interest and concern for vocational, family and social activities.   Lability of affect is common often with considerable increase in the overall anxiety level, particularly as the individual becomes aware of his failing powers. He has less interest in goals and achievements, diminished creativity, less incentive to stick to a task, trouble in concentrating, and difficulty in screening out disturbing environmental stimuli.   Failures, frustrations, changes, postponements, and troublesome decisions produce

more annoyance and internal upheaval than usual, and it is
harder to recover equilibrium after such disturbances. The
individual's characteristic defence mechanisms are utilized
more frequently and more blatantly, often with less than
normal effectiveness. The individual becomes increasingly
absorbed with himself and his own problems and less
concerned with feelings and reactions of others".

The patient often complains conspicuously of <u>memory loss</u>
which, on the contrary, cannot be demonstrated by available
testing techniques; this idea is confirmed by Kahn's observation
according to which complaints of memory loss correlate well with
depression and poorly with evidence of brain damage[4].

The amnestic syndrome[5] is characterised by shortened retention
time, an inability of the subjects to recall events in the recent
past, and an inability to recall not only relatively unimportant
facts of an experience, but the experience itself. The loss of
recent memories is accompanied by disorientation, at first in time
and place and later also as to person and by retrogressive loss of
remote memories. The subjects frequently produce, at least in the
beginning, confabulations that usually are in keeping with their
habitual premorbid activities. When the disease progresses, the
so-called "partial memories" also are lost; eventually, all areas
of sensory and kinesthetic function become involved, with one
notable exception: the memory for pain sensation.

As damage proceeds - Well's description[1] goes on -

"the individual has trouble making plans, dealing with new
  situations, and initiating activity. He avoids choices
  and decisions. Delayed recall and unreliability in
  calculations may be troublesome, as are slowed speech and
  understanding. Judgement suffers. Frustration
  tolerance is usually even further reduced. The person's
  characteristic defence reactions are utilized excessively
  in dealing with his environment, but they are less well
  regulated and less effective.

With worsening of the condition, drive and feelings diminish.
  The drive for achievement vanishes, and the patient may even
  lose interest in other's opinion of him. Appropriate dress
  and personal cleanliness may be ignored. There is usually
  a diminution of anxiety with progressive flattening of
  affect. Personal warmth and concern for others often
  disappear.

Now the effects of functions characteristic of syndrome
  become readily apparent: defective memory, particularly for
  recent events is blatant; time and space orientation are

faulty and the patient is easily lost. Learning ability is
markedly impaired.   He is unable to function in complex
situations, has trouble understanding and following
directions, and often loses the train of thought.

Some patients are restless and overactive; others, lethargic
and lacking in energy.   Failure cannot now be ignored.
With diminished drives and appetites, frustrations are not so
likely to arise.   The individual's characteristic premorbid
defence mechanisms are now obvious, and reliance upon them
may be damaging rather than protective".

At this stage the patients may displace objects and then
accuse others of theft.   They frequently will roam at night, wander
away from home, or, in the hospital, enter other patients' room;
they may become incontinent or satisfy their toilet needs in
inappropriate places[6].

As the disorder increases there may be occasional bursts of
excitement, but the characteristic course is one of increasing
apathy and indolence; the humane substance of personality is lost;
danger, loss, pain, even impending death may rouse only the feeblest
and briefest of responses; frustrations may be hardly noted or not
perceived at all, and defence mechanisms are dissolved[1]. The
patients show severe dementia, failure to comprehend and communicate
with their environment, they are grossly disorientated as to place
and time, and recent and remote memory are defective.   Changes in
level of consciousness are seldom conspicuous, however, until
advanced stages of degeneration are reached, but the perceptions can
be blurred and distorted.

Survival provides opportunity for the appearance of paralysis,
mutism, incontinence, stupor and coma.

Death usually occurs from 5 to 10 years after the diagnosis of
the disease and the most common cause is a respiratory infection.
Important changes may occur in the clinical presentation of dementia
because of the differing frequencies, intensities and order of
appearance of symptoms.   Nevertheless, the most authors agree upon
some points: memory impairment is always present and, with impaired
orientation it is seen to be clearly the most consistent and
frequent symptom over time[7].   Affective disorders, such as
anxiety, irritability and depression are typical signs of mild
dementia, while psychotic features appear in severe dementia[8].

CLINICAL ASPECTS: THE NEUROLOGICAL EXAMINATION

The dichotomization of symptoms into psychiatric and neurological is artificial, of course, but has value in terms of such

actual clinical practice factors as referral criteria, assessment
procedures and choice of treatment modalities.

As far as strictly neurological symptomatology is concerned,
the most frequent feature is an aphasic-apraxic-agnosic syndrome;
disorientation in space may result in the patient's becoming lost
in familiar surroundings, even in his own home. Disorders of
symbol utilization, such as dyslexia and dysgraphia are common
though not usually isolated phenomena and may be coupled with
repetitive utterances (logoclonia) or repetitive actions[6].   The
vast majority of patients has at least one abnormal finding on
neurological examination.

Frequently such individuals <u>walk</u> in a clumsy, graceless
fashion; rocking movements of the trunk and neck, stereotypes of
the hand and feet are common findings and there is some tendency
toward an overall flexion.   The inconstant rigidity of opposition
is in contrast to the plastic rigidity of parkinsonism, which is
consistent and usually best demonstrated in the neck and shoulder[9].
In patients with dementia and diffuse cerebral dysfunctions may
also appear <u>primitive reflexes</u> (snout and sucking reflex, glabella,
palmomental, corneomandibular, rooting, nucocephalic and grasp
reflexes); these responses might be particularly valuable to judge
demented patients.

Frequent are also phenomena of motor impersistence and
perseveration.

In early dementia it is possible to find decreased muscle
strength, extensor plantar responses, hyperactive deep tendon
reflexes; later hemiparesis, extrapyramidal rigidity, tremors,
spasticity and seizures appear.

CLINICAL ASPECTS: DIFFERENTIAL DIAGNOSIS

A) <u>Between Dementias and other Psychiatric Disorders</u>

Whenever the physician encounters a patient whose history,
symptomatology, or clinical behaviour reveals a disorder of the
highest functions, the first differential diagnostic question must
be: "Is the disorder functional or organic?"   Early in its course,
in fact, the clinical picture of dementia may specifically suggest
<u>neurosis</u>; later <u>psychosis</u>[1].

The patient with mild to moderate brain disease often displays
symptoms that suggest an <u>anxiety neurosis</u> or a <u>depressive neurosis</u>[1].
This time there may be a picture which comprises a combination of
anxiety or depression associated with apathy, forgetfulness, poor

intellectual performance, and lower scores on the shortened form of
the WAIS than in sympton-free subjects; the patients are especially
women handicapped by low initial intellectual endowment, responding
with emotional disturbance to the many social and familiar
vicissitudes, and the impaired physical health that old age brings
in its wake[10]. Frequently, however, neurotic features are present
in old people; more common are the hypochondriac states accompanied
by restlessness and mild depression. Nevertheless it is possible
to distinguish between neurosis and dementia: the neurotic patient
almost never has impairment of orientation. Though he often
complains of poor memory, its intactness can usually be demonstrated
satisfactorily by persistent questioning; similarly, the patient
often complains of impaired intellectual functions which can rarely
be substantiated.

Wells[1] emphasizes the different aspects of judgement and
affective disorders in both neurosis and dementia:

> "In neurosis one senses usually that poor judgement is
> secondary to inner conflicts and emotions (particularly
> feelings such as anger and fear) that so preoccupy the
> individual and distort his perception that he is incapable
> of making logical dispassionate judgement. In dementia
> poor judgement usually results from the individual's
> inability to attend to details and to assimilate the multiple
> factors that are usually weighed in decision-making. The
> nature of affective change is also often a distinguishing
> feature. In neurosis, a pervasive feeling, such as anxiety
> or depression or anger, is likely to dominate the clinical
> picture, whereas in dementia the lability of affect is more
> striking. Thus, the demented patient appears, in turn,
> and often with great rapidity, fearful, then sad, then
> angry, and in each situation the affectual response may be
> appropriate though excessive for the situation. This
> lability may not, however, be prominent early, when a
> pervasive feeling of sadness or fearfulness is more common".

With more severe dysfunction, dementia may be confused with
psychosis, especially with depression and schizophrenia. The
differentiation of dementia from depressive psychosis, particularly
in the aged patients, is often challenging. In fact, approximately
15 to 20% of elderly depressive people in various series have been
found to exhibit transient cognitive impairment and, more often, a
combination of defective orientation, patchy impairment of memory
for recent events, and vague or inappropriate responses to questions
Retardation may be extreme so as further to reinforce the impression
that one deals with a dementing disorder[11].

Even though cooperation is hard to obtain and responses come with ponderous slowness, the examiner is usually able to demonstrate preservation of orientation, memory and other intellectual functions in depressive illness, though their demonstration may demand unusual tenacity of the examiner.   Certain biological features, such as anorexia, weight loss, constipation, early waking are more typical of depression than of dementia . McHugh[12] stresses the more significant points to make a diagnosis of <u>dementia or depression</u>: it becomes important to recognize the primary psychological manifestations of depression such as mood change and change in self-attitude especially self-blame, worthlessness, or hopelessness, along with a few secondary changes, including early morning wakening and vegetative signs.   Moreover, a crucial point is the rate at which the syndrome appears: the onset is, in fact, rapid in depression whilst in dementia it is rather insidious and slowly progressive.   Secondly, if there is a history of previous attacks of mania or depression in the patients, one should try treating them empirically as psychotics, particularly if their problems are acute. There is a group of depressed patients, in fact, who shows a cognitive impairment whose progress parallels the progress of depressive syndrome; in these patients the cognitive disabilities respond to treatment of the affective disorder[13].   The third important finding for the differential diagnosis is that the patients with depression never show a jargon aphasia, while many patients with Alzheimer's disease do[12].

As far as <u>mixed organic-functional pictures</u> are concerned, in approximately 20% of patients with arteriosclerotic dementia and less often with senile dementia, a depressive syndrome is present in the early stages, often causing these patients to appear much more demented than they really are[11].   Behaviours denoting a "catastrophic reaction" or indicating an anxious-depressive orientation of mood were found to be statistically more frequent among left brain damaged patients.   On the contrary, symptoms denoting an opposite emotional reaction such as minimization, indifference reactions and tendency to joke were more frequent among patients suffering from a lesion of the minor hemisphere[14]. According to these results it is possible that organic cerebral disease can show features typical of functional disorders.

For some reasons that are not yet understood, the conjunction of severe depression with organic disturbance in the form of clouding of consciousness and memory defect is associated with considerable suicidal risk; in the case of multi-infarct dementia (M.I.D.), there may be violent, self-destructive attempts by slashing the throat or hanging[11].

Although disorders of thought are prominent in dementia, it can usually be differentiated clinically from a primary disorder

of thought, i.e. schizophrenia.   In the latter, in fact, it is
possible to demonstrate preservation of orientation, memory, and
intellectual functions; the patient's thought is affected by sudden
and inexplicable twists, tolerance for conflicts is striking and
conclusions are reached on logically unacceptable bases.   In
dementia the thoughts expressed take more of a meandering course,
one can follow the progression of thoughts but they appear to lead
nowhere[1].   In schizophrenia one can also observe a flatness or
inappropriateness of affect in contrast with the labile, perhaps
poorly regulated, but nevertheless appropriate, affective responses
of dementia[1].

B) Between Dementias and other Organic Disorders

If in response to the question "Is the disorder functional or
organic?" the disorder is identified as organic, differentiation as
to whether the clinical presentation results from diffuse or focal
dysfunction of the cerebral hemispheres must first be made.

A mosaic of defects, formed from bits of dysfunctions in
multiple mentative capacities, is typical for the patient suffering
from diffuse brain disease.   In other patients, however, defects
are limited to certain specific memory, language, recognition, or
performance functions, with relative preservation of the spectrum
of other mentative capacities; such dense dysfunction in one area
with good function in other is typically found in the patient with
focal and circumscribed brain disease[1].

Initially the clinician must ask himself if the patient's
observed mentative incapacity might be attributed to an amnestic,
agnosic, apraxic, aphasic, or other specific defect alone[1].

If the disorder be identified as organic and due to diffuse
brain dysfunction, then the clinical differentiation lies between
delirium and dementia.   Delirium can usually be distinguished from
the chronic dementing diseases on the basis of both history and
examination.   The majority of patients who have delirious and
clouded states suffer from a specific physical illness, most
commonly cardiac failure, acute or chronic respiratory diseases,
other infectious illness, pernicious anaemia, intoxication with
drugs or other substances, electrolyte disturbance, or vitamin and
mineral deficiencies[11].

Theoretically, progressive cerebral disease could first declare
itself by means of an acute delirious or confusional state due, for
example, to a small cerebral infarction.   At times, effectively
the onset of dementia is related by family members to an acute
emotional or physical stress, such as an infection or operation[6].
However, unless there is evidence of a focal neurological lesion

or a specific psychological deficit, <u>an abrupt onset is a rare
phenomenon</u>. Exacerbation of the degree of confusion may certainly
occur in the course of a dementing process owing to the advent of
some new aetiological agent such as an acute infection that evolves
"silently" without fever and little in the way of physical signs;
the underlying dementia, however, is clearly identifiable in most
such cases[11].

There are various aspects which can be useful to distinguish
dementia from delirium. Men predominate in a ratio of 2:1 among
clouded states, whereas women predominate among the dementias.
The duration differs: a few weeks or even 2 to 3 months for a
clouded state, at least 1 to 2 years for dementias[11]. The onset
is usually abrupt and precise in delirium, whereas a stuttering
or gradual course of uncertain onset is typical of dementia[1].
Alterations in consciousness are commonly present in delirium while
they are uncommon until the terminal phases of dementia.
Depression of consciousness in delirium varies from an inattentive,
dazed, dreamy state to stupor, so that normal orientation may return
in lucid intervals that may last minutes or sometimes hours; such
marked contrasts are not observed in dementia[11].

Mood also differs: demented patients are often emotionally flat
or indifferent, whereas a more positive emotional disturbance in the
form of intermittent fear, perplexity, or bewilderment may be
clearly evident in clouded patients. The disorientation of clouded
patients shows a more lively, active and constructive intelligence
that tries to make some coherent sense out of the world around.
The delusions of clouded patient are persecutory, but there is often
a measure of order and cohesion, in contrast to the vague random and
contradictory delusional utterances of the demented. General
personality functioning is well preserved in confused patients,
whereas personal habits have deteriorated with intermittent or
lasting incontinence in the demented; impairment of memory is
confined in clouded subjects to recent events, whereas in the
generality of demented patients remote memory is also partially or
markedly impaired[11].

Physiological alterations are common and prominent in delirium:
restlessness, tremor, slurred speech, insomnia, anorexia, nausea,
vomiting, constipation, diarrhoea, sweating, pallor, flushing,
tachycardia, fever; while these may be seen individually in
dementia, they are seldom prominent and seldom dominate the
clinical picture as they commonly do in delirium[11].

In conclusion, it is possible to diagnose a dementia sydrome
by the following criteria[15]:

1. History of deterioration in intellectual capacity;
2. Evidence from the Mental State Examination of global cognitive

disturbances.
3. No disturbance in the state of consciousness.

C) Among Different Types of Dementia

Once one has reached the conclusion that the aspects of
progressive mental deterioration are to be correlated to a diffuse
organic cerebral disease of dementia type, the next problem to be
solved is to establish which type of dementia is concerned.
An accepted classification of dementias is shown in Table 1[6].

It can be easily inferred, even from a quick perusal of this
classification, that most cases may be suspected on the ground of a
careful history and an accurate clinical investigation; in a
further number of cases that diagnostical doubt can be dispelled
or confirmed as a result of suitable instrumental and laboratory
investigations. However, the correct definition of some cases may
present a few difficulties and the picture often remains unsolved:
this occurs especially in the differential diagnosis between
degenerative and vascular dementias and, within the former, between
senile and presenile dementias.

According to Constantinidis[16], it is possible to distinguish on
the ground of clinical aspects and their evolution, three types of
primitive degenerative dementia:

a) Simple senile dementia. The clinical onset is usually progressive
    during the senile period of life. Disorders of memory appear
    first, with disorientation in time and space, fixation and recall
    amnesia (first for recent events, later for earlier ones),
    confabulation, false recognitions. Language impoverishment is
    observed at this stage, and the patient cannot find the right
    word; constructual apraxia appears, but no other instrumental
    signs.   Apart from a slight hypertonia in the limbs and a
    slight or inconstant grasping reflex, the neurological examin-
    ation is normal.

b) Alzheimer's senile dementia. is characterized by the more or less
    progressive intrusion into the clinical picture described above
    of a syndrome of increasing aphasia-apraxia-agnosia, following
    the four stages previously mentioned, accompanied by prefrontal
    symptoms such as oral reflexes (visual and tactile with cardinal
    points), marked resistant hypertonia, stereotypes, oculomotor
    disturbances, and in advanced cases Balint's-like syndrome and
    verbal iterations (echolalia, palilalia, logoclonia). Epileptic
    seizures and myoclonic twitchings may occur.

c) Alzheimer's presenile dementia. is distinguished from the former
    by its onset during the presenile period, i.e. before 60-65-70

Table 1. Diseases causing Dementia

1. Diffuse Parenchymatous Diseases of the Central Nervous System
   So-called presenile dementias
      Alzheimer's Disease
      Pick's Disease
      Kraepelin Disease
      Parkinsonism-dementia complex of Guam
      Huntington's chorea
   Senile Dementia
   Other Degenerative Diseases
      Hallevorden-Spatz Disease
      Spinocerebellar Degenerations
      Progressive myoclonic epilepsy
      Progressive supranuclear palsy
      Parkinson's disease

2. Metabolic disorders
   Myxoedema
   Disorders of the parathyroid glands
   Wilson's Disease
   Liver disease
   Hypoglycaemia
   Remote effects of carcinoma
   Cushing's syndrome
   Hypopituitarism
   Uraemia
   Dialysis dementia
   Metachromatic leukodystrophy

3. Vascular disorders
   Arteriosclerosis
   Inflammatory disease of blood vessels
      Disseminated lupus erythematosus
      Thromboangiitis obliterans
   Aortic arch syndrome
   Binswanger's disease
   Arteriovenous malformations

4. Hypoxia and Anoxia

5. Normal Pressure Hydrocephalus

6. Deficiency diseases
   Wernicke-Korsakoff syndrome
   Pellagra
   Marchiafava-Bignami disease
   Vitamin B_{12} and folate deficiency

7. Toxins and Drugs
        Metals
        Organic compounds
        Carbon monoxide
        Drugs

8. Brain Tumours

9. Trauma
        Open and closed head injuries
        Punch-drunk syndrome
        Subdural haematoma
        Heat stroke

10. Infections
        Brain abscess
        Bacterial meningitis
        Fungal meningitis
        Encephalitis
        Subacute sclerosing panencephalitis
        Progressive multifocal leukoencephalopathy
        Creutzfeldt-Jakob disease
        Kuru
        Behcet syndrome
        Lues

11. Other diseases
        Multiple sclerosis
        Muscular dystrophy
        Whipple's disease
        Concentration-camp syndrome
        Kufs' disease
        Familial calcification of basal ganglia

From G.R. Haase (1977)

years of age according to various authors.   Its clinical picture
is comparable to that of the senile form described above.   A pure
amnesic stage is always observed in the beginning; the aphasia-
apraxia-agnosia syndrome seems, however, to be quicker in onset,
more serious and more regular in its chronological evolution
than in senile cases.

All these clinical pictures are characterized also by the
presence of an amnestic syndrome; therefore, it is necessary to
distinguish them from "benign senescent forgetfulness" described
by Kral[5].   This type of memory loss can be found in aged people
who are not suffering from Alzheimer's disease or other types of
dementing process.   The memory dysfunction is characterized by
an inability of the subject to recall relatively unimportant data
and parts of an experience (e.g. a name, a place, or a date),
whereas the experience of which the "forgotten" data form a part
can be recalled.   However, the same data not recalled on one
occasion may be retrieved at another time: the "forgotten" data
seem to belong to the remote rather than the recent past.   Also
the subjects are aware of their shortcomings and try to compensate
by circumlocution and they may apologise.   The author speculates
that the ageing process, when it affects the brain only mildly,
and particularly when it relatively spares the hippocampalfornix-
mammillary system, may be accompanied by the benign type of memory
dysfunction, while, when these structures are affected more
severely, Alzheimer's disease may result.   The "benign type" of
senescent forgetfulness could therefore be an expression of
physiological cerebral ageing, whereas the amnestic syndrome of
Alzheimer's dementia is an expression of pathological cerebral
ageing[5].

When discussing the differential diagnosis between senile and
multi-infarct dementia (M.I.D.), attention must first be drawn to
the marked prevalence of the latter among men.   The M.I.D., then,
is less frequent than degenerative forms: in 140 cases of dementia,
in fact, 48 - 52% were presenile forms, 20 - 22% were arteriosclerotic dementia and 12 - 13% were mixed forms[16].

The presence of clear neurological signs or suddenly appearing
psychological deficits such as aphasia, agnosia, or apraxia retains
their importance as major criteria in diagnosis.   In fact, isolated
focal psychological deficits are rarely seen in pure form in the
setting of very well preserved intelligence in elderly people; when
they are found, the aetiology is almost invariably cerebrovascular.
In senile and presenile dementia such deficits are usually obscured
in some measure by general cognitive impairment involving memory
and intellect[11].   According to Roth[11], the course of the illness
is intermittent and fluctuating, punctuated by episodes of clouding.
Emotional lability, generally in the direction of depression is

more common and pronounced than in senile dementia, and occurs independently of the syndrome of pseudobulbar palsy. The personality remains better preserved; personal habits and the outward facade may be intact until a relatively advanced stage. Convulsion and the presence of mild parkinsonism or pseudobulbar palsy suggest that the associated intellectual deterioration is of the "multi-infarct" kind.

The Hachinski's Ischemic Score[17]is a useful differential diagnostic criterion (Table 2).

As far as differential diagnosis between senile and presenile dementia is concerned, it appears probable that the clinical difference arises from focal intensification of one and the same pathological process in patients with classic Alzheimer features[11]. The clinical picture of senile dementia is similar to that of Alzheimer's disease, although, by definition, the onset is later in life and the progression is slower. Some authors[18], therefore, regard both clinical aspects as a single process and refer to them as "Alzheimer's disease".

Up to now, no notice has been taken of the alleged differences between Alzheimer's and Pick's disease.

Patients with Alzheimer's disease show rigidity of extrapyramidal type,with cogwheel phenomenon accompanied by a syndrome of "direct forward staring"; moreover, the disturbance of gait observed in patients with Alzheimer's dementia (slow, unsteady, clumsy gait) was not seen in any of the patients with Pick's disease[19].

According to Adams[20] memory, orientation, and attention tend to be well preserved in patients whose disease is limited to temporal lobes, in contrast to Alzheimer's disease. Language disorders are present in two-thirds of all cases of Pick's disease: at first the patient forgets and misuses words, and does not understand much of what he hears and reads. His speech becomes a "medley of disconnected words and phrases" and eventually is reduced to an incomprehensible jargon. Finally he is altogether mute, seemingly without impulse to speak or ability to form words. Verbal perseveration, palilalia and echolalia have been described. Symptoms indicating Alzheimer's disease are slow progress of the disease with early manifestations of spatial disorientation and amnesia for remote events, apraxia, agnosia, aphasia, logorrhea and logoclonia, while Pick's disease is indicated by a slow progress with early loss of insight, signs of disinhibition, irritability, confabulation, logorrhea, echolalia, mutism, amimia and signs of a Kluver-Bucy syndrome[21].

Table 2. Ischemic Score

1) Abrupt onset	2
2) Stepwise deterioration	1
3) Fluctuation	2
4) Nocturnal confusion	1
5) Relative preservation personality	1
6) Depression	1
7) Somatic complaints	1
8) Emotional lability	1
9) Hypertension	1
10) History stroke	2
11) Other signs of arteriosclerosis	1
12) Focal symptoms	2
13) Focal signs	2

from Hachinski et al (1975)

REFERENCES

1. C. E. Wells, Dementia: definition and description, in:
   "Dementia", C. E. Wells, ed., Edition 2, Contemporary
   Neurology Series, F. A. Davis Company, Philadelphia (1977).
2. M. H. Hollender, personal communication to C. E. Wells in:
   "Dementia", C. E. Wells, ed., Edition 2, Contemporary
   Neurology Series, F. A. Davis Company, Philadelphia (1977).
3. M. A. Neumann and R. Cohn, Epidemiological approach to
   question of identity of Alzheimer's and senile brain
   disease: a proposal in: "Alzheimer's Disease: Senile Dementia
   and Related Disorders", R. Katzman, R. D. Terry and K.L. Bick
   eds., Raven Press, New York (1978).
4. R. L. Kahn, S. H. Zarit, N. M. Hilbert and G. Niederche,
   Memory complaint and impairment in the aged. The effect of
   depression and altered brain function, Arch.Gen.Psych.
   32:1569 (1975).
5. V. A. Kral, Benign senescent forgetfulness, in: "Alzheimer's
   Disease: Senile Dementia and Related Disorders", R. Katzman,
   R. D. Terry and K. L. Bick, eds., Raven Press, New York
   (1978).
6. G. R. Haase, Disease presented as dementia, in: "Dementia",
   C. E. Wells, ed., edition 2, Contemporary Neurology Series,
   F. A. Davis Company, Philadelphia (1977).

7. E. H. Liston, Clinical findings in presenile dementia, Jour.Nerv.Ment.Dis. 167:337 (1979).

8. C. E. Wells, Chronic brain disease: An overview, Amer.J.Psychiat. 135:1 (1978).

9. G. W. Paulson, The neurological examination in dementia, in: "Dementia", C. E. Wells, ed., Edition 2, Contemporary Neurology Series, F. A. Davis Company, Philadelphia (1977).

10. C. Nunn, K. Bergmann, L .G. Britton, E. M. Foster, E. H. Hall, and D. W. K. Kay, Intelligence and neurosis in old age, Br.J.Psychiat.124:446 (1974).

11. M. Roth, Diagnosis of senile and related forms of dementia, in: "Alzheimer's Disease: Senile Dementia and Related Disorders", R. Katzman, R. D. Terry and K. L. Bick, eds., Raven Press, New York (1978).

12. P. R. McHugh, Discussion about dementia and depression in: "Alzheimer's Disease: Senile Dementia and Related Disorders" R. Katzman, R. D. Terry and K. L. Bick, eds., Raven Press, New York (1978).

13. M. Roth, Discussion about dementia and depression, in: "Alzheimer's Disease: Senile Dementia and Related Disorders" R. Katzman, R. D. Terry and K. L. Bick, eds., Raven Press, New York (1978).

14. G. Gainotti, Emotional behaviour and hemispheric side of the lesion, Cortex 8:41 (1972).

15. M. F. Folstein and P. R. McHugh, Dementia syndrome of depression,in: "Alzheimer's Disease: Senile Dementia and Related Disorders", R. Katzman, R. D. Terry and K. L. Bick, eds., Raven Press, New York (1978).

16. J. Constantinidis, Is Alzheimer's disease a major form of senile dementia? Clinical, anatomical and genetic data, in: "Alzheimer's Disease: Senile Dementia and Related Disorders", R. Katzman, R. D. Terry and K. L. Bick, eds., Raven Press, New York (1978).

17. V. C. Hachinski, L. Iliff, G. H. Du Boulay, V. L. McAllister, J. Marshall, R. W. Ross Russell, L. Symon, Cerebral blood flow in dementia, Arch.Neurol. 32:623 (1975).

18. S. Grufferman, Alzheimer's disease and senile dementia: One disease or two?, in: "Alzheimer's Disease: Senile Dementia and Related Disorders", R. Katzman, R. D. Terry and K. L. Bick eds., Raven Press, New York (1978).

19. T. Sjogren, H. Sjogren and A. G. H. Lindgren, Morbus Alzheimer and Morbus Pick. A genetic, clinical and pathoanatomical study, Acta Psychiatr.Neurol.Scand. (Suppl) 82:1 (1952).

20. R. D. Adams and M. Victor, Degenerative disorders of the nervous system, in: "Principles of Neurology", McGraw Hill, ed., (1977).

21. G. Johannesson, B. Hagberg, L. Gustafson and D.H. Ingvar: EEG and cognitive impairment in presenile dementia, Acta Neurol.Scand. 59:225 (1979).

'THERAPEUTIC' POSSIBILITIES IN 'DEMENTIA'

M. R. P. Hall

Professor of Geriatric Medicine
Southampton General Hospital
Southampton

This particular topic is one in which confusion reigns supreme!
Firstly there is often doubt concerning the therapeutic objectives
because too little is known about how a drug acts and therefore what
it is meant to do.   Secondly the therapeutic objective is not
clearly defined.   Thirdly the cause of 'dementia' is not known and
consequently a heterogenous group of conditions, giving rise to a
symptom complex, is treated.   In other words it is not known
exactly what condition is being treated.   Isaacs[1] has drawn attention to some of the problems which it is necessary to overcome in
order to evaluate drugs in Alzheimer's disease.   These involve very
careful patient assessment not only to confirm the diagnosis, but
also to record undesirable behaviour and measure cognitive performance.   It is clear that to achieve accurate results the patient
sample must be as pure as possible i.e. all must suffer from
Alzheimer's disease and patients with other causes of 'dementia'
be excluded.   In addition that function which it is hoped to
improve, must be clearly measurable, be able to be standardised and
be produced by the disease.

Most of the work done on therapeutic agents so far does not
achieve the standards which Isaacs suggests.   The rating scales
used to measure effects frequently measure other disabilities than
those produced by the disease.   Studies have been of short duration
and long term effects need to be guessed at.   Samples of patients
have been heterogenous and almost no attempt has been made to
diagnose the cause of the 'dementia', so that the significance of
results in relation to diagnosis has been impossible to assess.

It is well recognised that 'Dementia' is a term applied to a
syndrome which comprises symptoms of intellectual deterioration and

87

thereby performance and behaviour.   These symptoms may be exacer-
bated by a wide variety of conditions.   Therapeutic possibilities
are therefore many.   Firstly they may relate to treatment of any
conditions which may worsen the 'dementia'.   This can be almost
any pathology one likes to think of and treatment of these will not
be considered here.   Secondly they may relate to the modification
of symptoms which relate directly to the dementia and may improve
performance or lessen undesirable behaviour.   Examples are the
alleviation of secondary depressive symptoms or the control of
wandering.   Thirdly they may relate to attempts to relieve the
underlying brain defect which may be biochemical or perhaps
structural.   Finally they may aim at preventing the onset or
further deterioration in those cases with minimal symptoms.

        The objective of this paper is to consider the third and fourth
of these therapeutic possibilities.   Since the target of therapy
is the brain itself, drugs used as treatment should attempt to
correct brain function.   This may be achieved, theoretically, in
one of three ways, either by ensuring that brain cells receive the
appropriate nutrients, or that their internal metabolism is normal
or that unwanted metabolites are removed.   It is therefore
appropriate to look briefly at brain function and metabolism with
this in mind.

BRAIN FUNCTION

        It is now well recognised that some brain cells may be lost
with ageing so that some atrophy will occur.   However, until very
great ages are reached, the brain possesses sufficient reserves for
this loss to be insignificant, and it seems unlikely that any
therapeutic intervention will either be necessary, effective or
possible in this respect.

        Some neuronal changes do occur and these include granulovacu-
olar degeneration, neurofibrillary degeneration and lipofuscin
accumulation.   This latter change is more prominent in some sites
of the brain, e.g. the inferior olive.   Its significance is
uncertain, but it may interfere with cellular metabolism and a
possible therapeutic opportunity may exist if it can be shown that
removal improves some measure of brain performance.   Lipofuscin
has been shown to occur in animal cells and consequently an animal
model exists which can be and has been studied[2].

        Perhaps in association with the neuronal changes there is a
slight reduction in brain metabolic rate $(CMRO_2)$ with normal ageing
but there is little difference in cerebral oxygen consumption though
there is some evidence of a reduced cerebral glucose consumption.
Cerebral blood flow (CBF) may also decline slightly with age but it
is possible that this may result from diminished function rather

than cause this, for Ingvar[3] has shown in the normal brain blood
flow is controlled by neuronal activity.   There is little evidence
however that the reverse is true and consequently the therapeutic
advantages of improving cerebral blood flow must remain in doubt.

The neurochemical changes which occur in the normal brain have
been well reviewed by Bowen and Davison[4].   They found few changes
apart from RNA/DNA ratio which tended to decrease as did glutamate
decarboxylase (GAD), acetylcholinesterase (AChe) and adenosine
2'3'-cyclic nucleotide-3'-phosphohydrolase (CNP).   It would also
seem that neurotransmitter levels may also alter and their
receptor binding sites diminish.   Choline acetyltransferase (CAT)
also diminishes so that if both acetylcholine (ACh) synthesis and
its receptor sites are reduced, the resulting deficit in the
cholinergic system could account for recent memory loss.   A therapeutic possibility exists here in that improvement in ACh synthesis
might improve recent memory.

Similarly the controlling enzymes for the synthesis of other
neurotransmitters Noradrenaline (NA) and Serotonon (5HT) decrease
with age and there is an increase in Monoamine oxidase (MAO).
γ-Aminobutyric acid (GABA), the main inhibitory neurotransmitter
may however increase.   The action of the neuropeptides is only just
beginning to be studied but it would seem that these are vasoactive
substance-P and Vasopressin being dilators and Angiotensin II a
constrictor - Angiotensin may act as a neurotransmitter in cholinergic cerebral neurones while vasopressin acts in the mid-brain
limbic system and may well aid the consolidation of memory.   On a
priori grounds there are many possible avenues whereby pharmacological manipulation may alter biochemical brain function and
metabolism.   Reduction in MAO and GABA might bring improvement in
function but on the other hand imbalance of neurotransmitters may
induce unwanted side effect as occurs in Parkinson's disease when
treated with l-dopa.   Indeed as Kendall[5] points out the model of
Parkinson's disease may be a useful one to study of Alzheimer's
dementia turns out to be a deficiency disorder.

BRAIN METABOLISM IN 'DEMENTIA'

Since 'dementia' is a heterogenous condition brain metabolism
may be affected in many ways.   Figure 1 depicts one chain of events
which could affect neuronal metabolism and could therefore b
affected in many ways.

An inverse relation exists between 'dementia' and $CMRO_2$ and
CBF is reduced in vascular 'dementia'.   However when dementia is
due to neuronal disease CBF may be modified because $CMRO_2$ is
reduced rather than the converse being true.   CBF also correlates
well with glucose utilisation and possibly drugs which improve CBF

M. R. P. HALL

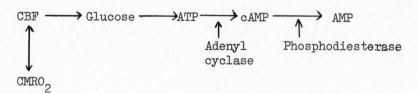

Fig.1.   Factors in brain metabolism which theoretically could
         be influenced by drug therapy.

will increase glucose concentration and thereby neuronal function.

    Neurotransmitter malfunction or deficiency has been shown to
exist in two 'abiotrophies' viz. Parkinson's disease and
Huntingdon's chorea.   Both conditions are associated with
'dementia'.   In Parkinson's disease improvement in both physical
and mental performance results from dopamine replacement.   It
would be nice if replacement therapy of some form could correct the
lesion in senile dementia.   So far searches have revealed some
dificiencies and at present are focused on choline acetyltransferase
(CAT) deficiency[6].   This work has been confirmed by others and it
would certainly seem that CAT deficiency occurs in senile dementia
due to and Alzheimer type of pathology and is linked with neurofibrillary tangle formation.   However other studies[8] have shown
reduced activity of glutamic acid decarboxylase (GAD) and l-dopa
decarboxylase (DOPAD) substances which are reduced in Parkinson's
disease and a possible explanation for this is that widespread
fallout of dopaminergic and GABA-ergic neurons occurs.   If this is
so then the pathogenesis of Parkinson's disease and senile dementia
may be similar.   Alternatively and perhaps more likely this is a
global brain disease with widespread effects and manifestations.
If this is so therapeutic possibilities are remote.

THE DEFINITION AND CLASSIFICATION OF DEMENTIA

    One of the major problems in treating 'dementia' is to define
exactly what pathology one is treating.   'Dementia' is a clinical
syndrome in which behaviour and global intellectual performance are
affected.   For the purpose of this paper it is considered as
chronic brain failure and its causes are classified in Table 1.

    Other causes of 'dementia' such as Jacob-Creutzfeld disease
and alcoholism although they may occur in old age have not been
included since they are only rarely met with in Geriatric practice.

Table 1.   The Common Causes of 'Dementia' in Old Age

Benign senile mental impairment (Dysmnesic Syndrome)

Vascular Dementia (Multiinfarct Dementia
                             Arteriosclerotic Dementia)

Alzheimer Type Dementia (Senile Dementia)

Mixed Vascular and Senile Dementia

Normal pressure hydrocephalus

Hypothyroidism (Myxoedema Madness)

Parkinson's Disease (Extrapyramidal dementia)

THERAPY OF 'DEMENTIA'

The treatment of the last three conditions mentioned in Table 1 can be easily dealt with.   Dementia due to normal pressure hydrocephalus needs neurosurgical management.   Hypothyroidism requires l-thyroxine replacement therapy but in spite of this rarely responds[9] while dementia due to Parkinson's disease may respond well to antiParkinsonian therapy.

Intellectual impairment in Parkinson's disease is common varying between 25 - 80%[10].   The effect of l-dopa on the dementia of Parkinson's disease is variable but may be very beneficial[11].   The effect of l-dopa in 'dementia' per se has been noted in only a few patients[12] and from this it would seem that further trials are indicated.   However the 'dementia' in Parkinson's disease may not be associated with just dopamine deficiency so that it may be more appropriate to improve dopamine receptor function with dopaminergic agonists such as ergoline or bromocriptine.   Alternatively improved intracellular metabolism of dopamine might be equally effective and an MAO-B metabolism of dopamine might be equally effective and an MAO-B inhibitor such as l-deprenyl might also be worthy of trial.

This leaves us with the first four causes of mental deterioration in old age to treat and this is where the confusion begins.   No study makes any attempt to differentiate between the conditions which are lumped together and treated as a group.   It is not surprising therefore that the results are variable, often unconvincing and that a large number of therapeutic 'agents' exist.   These have been very well classified by Hyams[13] under the headings of vasoactive drugs, 'metabolic improvers' and cerebral 'activators'.

Drugs With a Possible Therapeutic Value

Hyams[13] has written an excellent review of the drugs which have
been used to treat 'dementia' and it would be superfluous to attempt
to repeat this.   Consequently this review will be restricted to
those substances on which most work has been done and on others
which are purely speculative.   More recent articles[14,15] with regard
to drugs which may affect behaviour and performance have covered
almost every substance capable of influencing cerebral performance
from alcohol to Royal (bee) Jelly.   Lehman[15] has classified these
into two groups: (i) Unspecific agents which included the
psychostimulants, anxiolytics, antidepressants, antipsychotics and
euphoriates; and (ii) specific agents which include regulators of
cerebral metabolism and circulation as well as antioxidants and
"free radical" scavengers.   Included amongst these is a new
psychotropic class for which the term nootropic geriatric drugs has
been proposed.   The first of these was piracetam and others which
might fall into this category are the dihydrogenated ergot alkaloids,
the procaine drugs (gerovital and KH_3), meclofenoxate, vincamine
and naftidrofuryl.

Nootropic Drugs

A summary of some of these drugs, their effect and the mechanisms
by which they may work is shown in Table 2.

Piracetam. Perhaps the most interesting study was a double blind
study[16] on the effects of this substance upon perceptual and psycho-
motor performance at varied heart rates in patients treated with
artificial pacemakers.   Performance was retarded at slow rates and
this retardation could be prevented by ingestion of piracetam.   This
observation raises the question of the role of cardiac arrhythmias
in the causation of 'dementia'.   Another 'cause' to be excluded
and maybe the underlying reason for the 'dementia' associated with
myxoedema.

Hydergine. This probably works by improving neurone function by
modifying energy turnover and improving the Na^+/K^+-ATPase[17] mechanism
through blocking catecholamine transmission.   It is probably the
most extensively studied substance and trials have shown it to be
superior to placebo if not by very much.   It may also have a
vasodilator effect but probably needs to be given for a long time
to produce this.

Procaine. This has been used longer than any other preparation
being introduced by Aslan in 1956.   It acts as a monoamine oxidase
inhibitor and certainly has an antidepressant effect.   This may
also improve cognitive function and consequently improve memory as
a result of improved ability to concentrate.

Table 2. Some Nootropic Geriatric Drugs

Drug	Effect	Mechanism
Piracetam	Cognitive performance, memory	GABA
Ergot Alkaloids (Hydergine)	Cognitive performance, memory, mood	Neuronal metabolism CAmp
Procaine	Mood, cognitive performance, memory	MAO Inhibitor
Meclofenoxate (Centrophenoxine)	Memory	Lipofuscin dispersal
Naftidrofuryl	Cognitive performance	Glucose metabolism CBF increased

Meclofenoxate (centrophenoxine). This substance has been shown to increase drive and improve mood under conditions of anoxia and may in this way act as an antioxidant. It has also been shown to improve memory[18] and function[19]. Its action has already been described.

Naftidrofuryl. This has been shown to increase $CMRO_2$ and glucose metabolism. It also increases cerebral blood flow. Controlled clinical trials have shown it to be effective[20].

Other Substances Affecting Performance

Some of these are summarised in Table 3 and probably represent the only other drugs worth considering. The various vasodilators have been on trial for a long time. Significant improvement has been shown in many trials and yet their use is not universal, possibly because of side effects. Their efficacy may be more related to a mild antidepressant effect than vasodilation.

Thiothixene is an improved treatment for schizophrenia and one study[21] has shown it to be effective in the management of elderly patients with 'dementia'. Adverse reactions are however likely but it may be worthy of further trial. The use of l-dopa

Table 3.   Some other drugs/substances which may improve
function/behaviour in 'dementia'

Drug	Effect	Mechanism
Vasodilators (Cyclandelate, Isoxsuprine)	Cognitive performance mood	Increase CBF Increase Glucose
Thiothixene	Cognitive performance mood	Increase dopamine metabolism Block dopamine receptors
1-Dopa	Cognitive performance mood	Improve dopaminergic transmission
Vasopressin	Cognitive performance, memory	Mid brain limbic system
Choline, Lecithin Deanol (Deaner)	Cognitive performance, memory	Improve cholinergic transmission
Oestrogens	Cognitive performance, memory	?Feed back through hypothalamic-pituitary-adrenal/neuroendocrine system

has already been discussed and the use of vasopressin analogues and choline or choline analogues are as yet only in a preliminary phase though some early encouraging results have been quoted[22].

The final substance listed is oestrogens and these have been reported as improving function in elderly women[23].

CONCLUSIONS

It is fair to state that though it has been shown that many substances will improve performance in 'dementia' no substance has proved itself truly efficacious or to the extent that it is universally prescribed as a preventative to deterioration.   There has however been an enormous expansion in our knowledge of brain function particularly with regard to the chemical changes in recent years.   These have opened up new therapeutic possibilities.   To find the key to this particular puzzle we will need to be more accurate in our diagnosis, selective in our sampling and define our

objectives more clearly.   Studies will need to be of much longer
duration and substances tested and compared under controlled
conditions for periods of years.   Some of these are already underway
but there is I feel no easy answer.

REFERENCES

1. B. Isaacs, The evaluation of drugs in Alzheimer's disease,
    Age & Ageing 8:1 (1979).
2. P. Glees and M. Hasan,"Lipofuscin in neuronal aging and diseases",
    Georg Thieme, Stuttgard (1976).
3. D. H. Ingvar, Functional landscapes of the dominant hemisphere,
    Brain Research 107:181 (1976).
4. D. M. Bowen and A. N. Davison, Biochemical changes in the normal
    ageing brain and in dementia,in: "Recent Advances in Geriatric
    Medicine", B. Isaacs, ed., Churchill Livingstone, Edinburgh,
    London and New York (1978).
5. M. J. Kendall, Will drugs help patients with Alzheimers Disease?
    Age & Ageing 8:86 (1979).
6. I. White, C. R. Hilly, M. J. Goodhunt, L. H. Carrasco, J. P. Keet,
    I. E. I. William and D. M. Bowen, Neocortical cholinergic
    neurones in elderly people, Lancet 1:668 (1977).
7. Anon. Cholinergic involvement in senile dementia, Lancet 1:408
    (1977).
8. D. M. Bowen, Pamela White, R. H. A. Flack, Carolyn Smith and
    A. N. Davison, Brain decarboxylase activities as indices of
    pathological change in senile dementia, Lancet 1:1247 (1974).
9. E. H. Jellinek, Fits, faints, coma and dementia in myxoedema,
    Lancet 2:1010 (1962).
10. D. A. Crachman and S. Stahl, Extrapyramidal dementia and levo-
    dopa, Lancet 1:809 (1975).
11. G. Broe and F. I. Caird, Levodopa for Parkinsonism in elderly
    and demented patients, Med.J.Aust. 1:630 (1973).
12. C. Lewis, B. R. Ballinger, A. S. Presley, Trial of levodopa in
    senile dementia, Br.Med.J. 1:550 (1978).
13. D. E. Hyams, Cerebral function and drug therapy: the use of
    cerebral vasodilators, vasorelaxants, haemokinators and
    activators, in: "Textbook of Geriatric Medicine and Geront-
    ology",J.C. Brocklehurst, ed., Churchill Livingstone,
    Edinburgh, London and New York (1978).
14. J. M. Ordy, Geriatric Psychopharmacology. Drug modification of
    memory and emotionality in relation to aging in human and
    non-human primate brain. p.435 Bayer Symposium VII, Brain
    Function in Old Age, F. Hoffmeister, C. Muller eds.,
    Springer-Verlag, Berlin, Heidelberg, New York (1979).
15. H. E. Lehmann, Psychopharmacotherapy in Psychogeriatric Disorders
    p.456 Bayer Symposium VII, Brain Function in Old Age.
    F. Hoffmeister, C. Muller, eds. Springer-Verlag, Berlin,
    Heidelberg, New York (1979).

16. K. Lagergren and S. Levander, A double blind study on the
    effects of Piracetam upon perceptual and psychomotor
    performance at varied heart rates in patients treated with
    artificial pacemakers, Psychopharmacologia 39:97 (1974).

17. W. Meier-Ruge and P. Iwangoff, Biochemical effects of ergot
    alkaloids with special reference to the brain,
    Post.grad.Med.J. 52 (suppl.1): 47 (1976).

18. D. Marcer and S. Hopkins, The differential effects of
    meclofenoxate on memory loss in the elderly, Age & Ageing
    6:123 (1977).

19. J. L. Gedye, A. N. Exton-Smith and J. Wedgwood, A method of
    measuring mental performance in the elderly and its use
    in a pilot clinical trial of meclofenoxate in organic
    dementia (preliminary communication), Age & Ageing 1:74
    (1972).

20. T. D. Judge and A.Urquhart, Naftidrofuryl - a double blind
    crossover study in the elderly, Curr.Med.Res.Opin.1:162
    (1972).

21. D. P. Birkett, W. Hirschfield, G.M. Simpson, Thiothixene in
    the treatment of diseases of the senium, Curr.Therap.res.
    14:775 (1972).

22. W. D. Boyd, J. Graham-White, G. Blackwood, I. Glen and
    J. McQueen, Clinical effects of choline in Alzheimer senile
    dementia, Lancet 2:711 (1977).

23. H. I. Kantor, C. M. Michael, H. Shore and H. Wayne Ludvigson,
    Administration of estrogens to older women;a psychometric
    evaluation, Am.J.Obst.& Gynae. 101:658 (1968).

EVALUATION OF CEREBRAL BLOOD FLOW FOR THE DIFFERENTIAL

DIAGNOSIS OF DEMENTIA

C. Fieschi, S. Bernardi, G. L. Lenzi

III[a] Clinica Neurologica dell'Universita di Roma
Rome
Italy

The problem of dementia, which was up until a few years ago very vague and unclear, has undergone some noteworthy clarifications recently.   This improvement has occurred thanks to the introduction of clinical protocols and to the use of special clinical methods which have permitted its pathogenic classification.

We must keep in mind that for dementia we intend a deterioration of the intellectual activity (attention, memory, knowledge) caused by an organic disease of the cerebral hemispheres[1].

It is now confirmed that 60% of cases are caused by primary degeneration; only 20% by cerebral vascular disease, more precisely by multiple small infarcts and the remaining 20% of the cases have a variety of causes combined vascular and degenerative, metabolic and/or pharmacological[2].

It has been estimated that about 20% of the population, over 65 years of age suffers from dementia[3].

The evaluation of the cerebral blood flow has proved to be of particular importance in the study of dementia.   From the very first studies done by Kety[3] in 1956 the decrease in the cerebral blood flow and the oxygen consumption with aging was noted.   This decline seems to be much higher in older patients when the risk factors such as a slight atherosclerosis, a cardiopulmonary pathology and in particular hypertension are present.   The evaluation of the blood flow and mean oxygen consumption for a whole hemisphere is not relevant data; particularly not for a differential diagnosis of dementia.   It appears possible that the variations of the functional activity may be evident in well defined cerebral regions and therefore a global measurement would not bring them to light.   In contrast these variations become relevant when methods which measure both the blood flow and the local metabolism are employed.

The clinical protocol of Harrison's group[5] appears to be useful for a differential diagnosis between degenerative and multi-infarct dementia.   Table 1 gives a score for various signs and symptoms of the patient.   If the score is over 6, he is probably affected by multi-infarct dementia.

The first measurements of the regional Cerebral Blood Flow (rCBF) were taken using radioactive gases, first with Krypton-85 and then with Xenon-133[6].   A gas bubble, dissolved in physiological solution, is rapidly injected into the internal carotid artery and the radio-activity then measured with detectors placed on the outside of the skull.   Later on other non-traumatic methods were developed, in particular the use of Xenon-133 through inhalation or intravenous injection[7,8].

Hachinski studied the mean CBF in patients with multi-infarcts and degenerative dementia; both groups showing a decrease in the blood flow with respect to a control group: the most significant decrease was evident in the group of patients with multi-infarct dementia[3].

Gustafson and Risberg[9] in experimental studies using Xenon inhalation clearly showed a decrease in the blood flow in the post-central regions in cases with Alzheimer's disease; a reduction of the blood flow mainly in the frontal region in patients with Pick's disease and a non homogenous blood flow reduction between the two hemispheres in cases with multi-infarct dementia.   In research done by Lavy[10], using the same methodology, a significant decrease was shown in cases of presenile or senile dementia, as compared to subjects in the same age group.

The use of non invasive methods in the measurement of CBF is to be encouraged for their non traumatic aspect.   However, Xenon-133 injected intravenously or by inhalation does not allow a clear measurement of the deep areas or very small infarct areas.   Given the low energy level of Xenon-133, we basically measure the radio-activity in the peripheral areas of the brain, that is the cortical area.   In addition there are areas with no blood flow due to a previous softening or other pathology of the parenchyma and the isotope does not arrive there and consequently cannot be detected. Computerized Tomography has been able to overcome this problem,by obtaining tridimensional pictures of the object under examination. The method employed is the use of a series of sections of a given thickness characterized by a high resolution power, even in the

Table 1. Differential Diagnosis between Degenerative and Multi-
          infarct Dementia

MULTI-INFARCT DEMENTIA (VASCULAR)
= 10 - 20% of cases
diagnosis : score $\geqslant$ 6

Abrupt onset
Fluctuating cause
History of strokes   } x 2
Focal neurological symptoms
Focal neurological signs

Stepwise deterioration
Nocturnal confusion
Relative preservation of personality
Depression   } x 1
Somatic complaints
Emotional incontinence
History of hypertension
Evidence of associated atherosclerosis

(Harrison et al., J.Neurol.Sci.,40,97, 1979)

deep areas of the hemispheres.   The object of computerized
tomography using X-ray is essentially morphological; the Computerized Emission Tomography can supply quantitative information regarding the rCBF or metabolic parameters.   It also has the big advantage of being a noninvasive method because the tracer is either
injected intravenously or inhaled.

    Two different methods have been used in order to obtain
pictures with the Emission Tomography: 1) ECAT properly called,
detects the emission of single photons from radioactive isotopes
introduced into the organism; 2) the second is a special adaption
of ECAT, called Positron Emission Tomography (P.E.T.),which detects
the emission of positrons; the latter has a well defined field of
vision and greater resolution power, even for very deep tissue and
at any level of activity.   So for, in man N^{13}, O^{15}, F^{18} have been
used to measure the blood flow, oxygen and glucose metabolism.
Unfortunately these techniques require very costly equipment because
positron emitting isotopes have a brief half-life and must be
produced by a cyclotron[11].

    Patients with acute stroke are now being studied employing these
methods and the first results are showing a luxury flow in the
regions of the lesion and decreased metabolic activity of the paran-
chyma involved.   Preliminary studies of degenerative dementia are
showing a high extraction of O_2 in the parenchyma in some patients

and an absolutely normal extraction in others[12].

It is still not possible to draw any definite conclusions from such preliminary data.   The importance of this type of study is obvious because we have seen how different pathogenesis leads to the same kind of disturbance in demented patients.   A differential diagnosis is therefore of extreme importance in order to choose suitable therapy for cerebral vascular disease or a metabolic deficiency.

REFERENCES

1. C. E. Wells (Contemporary Neurology Series) F. Plum and F. H. McDowell, eds., Blackwell Scientific Publications, 2nd Edit. (1977).
2. V. Hachinski, N. A. Lassen, J. Marshall, Multi-infarct dementia, Lancet 2:207 (1974).
3. V. C. Hachinski, L. D. Iliff, E. Zilkha, G. H. Du Boulay, V. L. McAllister, J. Marshall, R. W. Ross Russell, L. Symon, Cerebral blood flow in dementia, Arch.Neur. 33:632 (1977).
4. S. S. Kety, Human cerebral blood flow and oxygen consumption as related to aging, Ass.Res.Nerv.Proc. 35:31 (1956).
5. M. J. Harrison, D. J. Thomas, G. H. Du Boulay, J. Marshall, Multi-infarct dementia, J. Neurol.Sci. 40:97 (1979).
6. N. A. Lassen, K. Mooedt-Rasmussen, S. C. Sorenson, E. Skinhoj, S. Cronquist, B. Bodforss, E. Eng, D. H. Ingvar, Regional cerebral blood flow in man determined by Krypton-85. Neurol. 13:719 (1963).
7. W. D. Obrist, H. K. Thompson, C. H. King, H. S. Wang, Determination of cerebral blood flow by inhalation of Xenon-133, Circ.Res.20:124(1967).
8. D. J. Thomas, E. Zilkha, S. Redmond, G. H. Du Boulay, J. Marshall, R. W. Ross Russell, L. Symon, An intravenous 133-Xenon clearance technique for measuring cerebral blood flow, J. Neurol.Sci. 40:53 (1979).
9. L. Gustafson and J. Risberg, Regional cerebral blood flow measurement by Xenon-133 inhalation technique in different-ial diagnosis of dementia.IX International Symposium on "Cerebral Blood Flow and Metabolism" Suppl. 72. Acta Neurol. Scandinavica (1979).
10. S. Lavy, Reduction of rCBF in dementia: correlation with age-matched normal controls and computerised tomography.   IX International Symposium on "Cerebral Blood Flow and Metabolism" Suppl. 72, Acta Neurol.Scandinavica (1979).
11. G. L. Lenzi and Coll. Positron tomography of the brain. Brain and Behaviour Symposium, University of Copenhagen (1979).
12. R. S. I. Prackowiak and G. L. Lenzi (unpublished results).

REVERSIBLE CEREBRAL ISCHEMIC ATTACKS IN AGED PATIENTS

C. Loeb

Department of Neurology
University of Genova
Italy

The statement made by Marshall[1] that in the field of cerebrovascular disease the recognition of the significance of transient ischemic attacks constitutes one of the most important clinical advances of the last two decades is shared by many authors.

In fact, the importance of transient attacks as warnings of impending stroke has long been acknowledged[2] , in that they undoubtedly mark the transition from the preclinical to the clinical stage of cerebral atherosclerosis[3] and offer, in addition, the unique opportunity to plan possible efficient therapeutic strategies, such as the use of anticoagulants or vascular surgery, to ward off the redoubtable evolution into the completed stroke.

An attempt to evaluate critically the clinical features and course of the transient attacks with particular reference to the elderly seems indeed appropriate, even considering the great number of papers devoted to this topic in the last twenty years.

HISTORICAL REVIEW

A glimpse at the historical evolution of the concept of transient cerebrovascular episodes turns out to be rather disappointing. Even the interesting review by Schiller[4] of the concept of stroke before and after Virchow did not touch upon this problem.

From ancient Greek and Roman students to the founders of the modern anatomicoclinical investigation (17th - 18th centuries), all who dealt with cerebrovascular disorders seemed most concerned with severe hemorrhagic apoplexies or, at the most, with the harbingers of stroke[5,6,7].

Transient episodes were either indirectly mentioned (42nd aphorism of Hippocrates), or briefly recalled in the clinical history (see Zani case of Morgagni's letter IV[6]), almost always being overlooked in favour of pathological studies of the completed stroke.

Only during the second half of the 19th century did transient cerebrovascular episodes come somewhat to the fore, their possible causes being grouped as follows: (a) massive extracranial hemorrhages[8,9,10]; (b) cardiac alterations[11,9]; (c) cerebral arterial spasms; this hypothesis, put forward by Raynaud[12] to account for the transient loss of vision, was subsequently accepted by many authors[13,14,15,16], as the most convincing pathophysiological explanation of cerebrovascular transient episodes; in fact, Wilson and Bruce[17] devoted a whole chapter of their treatise to cerebral spasm; (d) neurolues (apoplectiform episodes of general palsy[18]; syphilitic hemiplegia)[19].

The vasospasm theory held sway, although among controversies [20,21,22], up to the 1950s, being unanimously rejected when Corday and Rothenberg[23] introduced the successful label of cerebrovascular insufficiency, following a similar cardiological definition by Rein[24] and Buchner[25].

During the 2nd Princeton Conference, along with other terms such as "intermittent vascular insufficiency, recurrent ischemic attacks, recurrent focal cerebral ischemic attacks, transient cerebral ischemia", the definition of "transient ischemic attacks" was employed for the first time[26].

Such definition was popularized gradually: in 1958 the "Ad hoc Committee" established by the NIH indicated the transient episodes as "transient cerebral ischemia"; in the 1961 3rd Princeton Conference, the term "focal intermittent insufficiency or ischemic attack" was still used, with only Miller Fisher[26] employing the TIA acronym (absent from the analytical index); it was only in the 4th Princeton Conference of 1965, however, that the definition of "transient ischemic attacks" gained unanimous acceptance. With respect to the duration of the episode, as long ago as 1928 Lhermitte[27] stated that rarely it lasted more than 24 h; on the other hand Alajouanine et al[28], dealing with "transient cerebral ischemia", wrote that "in a few cases the attacks may be more prolonged, persisting for several days".

Quite recently, attention has been focused on a number of cases exhibiting transient episodes which cleared completely later than 24 h from the onset, and were labelled as "strokes with full recovery" (6th Salzburg Conference, 1972). In the same year similar cases, although devoid of a complete clinical evaluation and follow-up, were mentioned in Vinken and de Bryn's Handbook of Neurology as "reversible ischemic neurological deficits". A few years later the same terminology was employed by an official classification of the "Ad hoc Committee on Cerebrovascular Diseases", headed by Millikan[2].

CLINICAL DEFINITION

The transient ischemic attack was classically defined as a cerebral dysfunction of ischemic nature lasting not longer than 24 h, with a tendency to recur. The following were therefore the features essential to the definition: ischemic pathogenesis, complete reversibility, and 24 h duration. To conform to the definition, the completeness of the regression should, first of all, be verified by the neurological examination, which should be negative; in other words, free from even slight abnormalities - inequality of tendon reflexes, slight pronator drifts, etc.[29].

In fact, episodes which do not recede completely, often labelled as "minor strokes", can be brought about by small infarctions or even hemorrhages[30,31,32], and even large softenings, and cannot therefore be considered as proper TIAs[29].

The 24 h limit was chosen arbitrarily; in fact, quite a few authors think that the usual duration of these attacks is about 1 h[33] or even some minutes[34]. The 24 h duration was chosen particularly with a view to possible therapeutic strategies, such as anticoagulants or vascular surgery.

It seems clear that ischemic episodes which resolve later than 24 h from the onset - usually within 3 weeks - require further elucidation. Such clinical pictures are often indicated as "strokes with full recovery[35,36] or "reversible ischemic neurological deficits"[2] and more recently "protracted transient ischemic attack" (PTIA)[37].

Therefore, the classification proposed by the NINCAS ad hoc Committee on Cerebrovascular Diseases (1975) could be modified, with respect to the description of the current state of patients in reference to the whole temporal profile of cerebrovascular diseases, as follows:

(A) Asymptomatic. (B) Focal cerebral dysfunction: (1) Reversible ischemic attacks (RIA): (a) transient ischemic attack (TIA); (b) protracted transient ischemic attack (PTIA). (2) Actively changing neurological deficit: (a) improving; (b) worsening (also known as "stroke in evolution" or "progressing stroke"). (3) Prolonged neurological deficit: (a) completed stroke. (C) General cerebral dysfunction. (a) transient; (b) prolonged.

Because of the evolution of the disease it is likely that patients will be in different categories at different times. In particular, TIA and PTIA can appear in the same patient on different occasions.

CLINICAL FEATURES

In agreement with the report of the Joint Committee for Stroke Facilities[38] many symptoms, such as dysphagia, dysarthria and diplopia, amnesia, confusion, loss of consciousness and vertigo were not considered when they appeared alone, since they can be explained by causes other than reversible cerebral ischemia.

In our case material, out of 1430 (from 1962 to 1976) cases with cerebrovascular focal episodes, 131 (i.e. 9.1%) had transient episodes labelled as reversible ischemic attacks (RIA), therefore comprising both classical TIA (80 cases, i.e., 61.1%) and PTIA (51 cases, i.e. 38.9%).

The frequency of remitting cerebrovascular episodes (not more clearly definable due to the obvious lack of objective data) was particularly high in the history of cases with cerebrovascular disease verified at necropsy; in particular, in 50 out of 126 cases with white softenings, in 8 out of 33 cases with red softenings and even in 14 cases with hemorrhage.

In the majority of cases (86) the first attack occurred in the age range 46 - 65 years. Eighty-seven males and 44 females sustained RIAs; in particular, 59 cases (TIA) and 28 cases (PTIA) were males.   The carotid territory was involved more than three times as often as the vertebrobasilar one (88 vs 25), 8 cases having both carotid and vertebro-basilar territories involved in multiple, subsequent attacks.

The TIAs lasted less than 1 hour in 60 (75%) and more than 1 hour in 20 cases (25%).   PTIAS could last from 4 to 60 days, clearing within about 3 weeks (19.2 days on the average) in 45 out of 51, that is, in 88.2 per cent of cases.   A typical example is provided by case 42, a man aged 58 years with six former TIAs in whom a right hemiparesis lasted 33 days, its complete resolution occurring after surgery for a left internal carotid stenosis[35].

In the carotid territory, the onset of PTIAs was more gradual than that of TIAs, motor disturbances affecting usually one side of the body, one limb, one half of the face; aphasia was often the prevailing symptom in cases with TIAs and sensory disturbances were rather rare, while visual symptoms (unilateral amaurosis, hemianopia) were rare in cases with PTIAs.

In the vertebrobasilar territory the onset of PTIAs was more often abrupt, the usual clinical picture being represented by motor disturbances (of the four or of the lower limbs, prevailing in cases with TIA of one-half of the body or of the face) ataxia, visual (bilateral amaurosis, hemianopia) and sensory disturbances.

Involvement of consciousness, associated with other focal neurological signs, occurred only in 4 out of 8 cases, in which both the carotid and the vertebrobasilar territories were involved.

The EEG showed focal bilateral or diffuse alterations in 40 cases (25 TIAs and 15 PTIAs), the tracings reverting to normal only in 16 cases (6 TIAs and 10 PTIAs). In 11 out of 50 cases with carotid RIAs the carotid compression test brought about unilateral (5 cases) or bilateral abnormalities (6 cases). TC brain scans were positive, showing focal uptakes, in 16 cases (9 TIAs, 8 PTIAs), with subsequent normalisation in 5 - 8 days in all cases, except one with a PTIA, in which an initial occipital uptake could not be verified subsequently.

Angiography showed abnormalities in 42 out of 95 cases (44.2%) in which it was performed, such as middle cerebral artery occlusion, carotid or vertebral stenosis and, above all, absence of the terminal branches of the sylvian artery[39].

The results of computed tomography (CT) studies in cases with TIAs and PTIAs deserve particular mention. A total of 160 cases with TIAs examined with CT[40,41,42,43,44], and 89 cases with PTIAs [43,44,45], has been described so far.

Most authors maintain that in TIAs the CT examination is unrevealing; Oltenau-Nerbe et al[45], Perrone et al[43], and Ladurner et al[44], found low attenuation areas in 32%, 34% and 18% respectively in cases with TIAs; while 87%, 60%, 76% of cases with PTIAs were found positive by Oltenau-Nerbe et al[45], Perrone et al[43], and Ladurner et al[44] respectively.

In our experience, albeit small, of cases with RIAs (30 cases with TIAs and 26 with PTIAs) computed tomography was always unrevealing. It should be recalled, however, that when cases with incomplete regression, namely exhibiting even slight neurological signs, are wrongly included under the TIA or PTIA label, then the likelihood of CT disclosing small infarcts or even hemorrhages is

fairly high[29].   Paradoxically the only exception in our experience
was represented by three cases with small hemorrhages in which a
complete clinical regression could be verified 1 - 2 months after
the stroke.

   In our opinion cases initially wrongly included in the TIA
and/or PTIA group which later exhibit low attenuation CT lesions
should be considered as cases with incomplete recovery and grouped
under the label of minor or little strokes.   In other words, the
verification of a complete recovery should be based also on the
normalization of the findings yielded by the ancillary tests
(EEG, brain scan, CT).   We elected to adopt such a strict
criterion mainly in view on the substantial crudeness of the
neurological examination.

NATURAL HISTORY AND FOLLOW-UP

   A 5.3 years follow-up of a group of cases with RIA (76 cases
with TIAs and 45 cases with PTIAs) permits the following comments:-

   a) Multiple attacks are significantly more frequent in TIAs
than in PTIAs.   In particular, in only 13 out of 45 patients
(28.9%) with PTIAs were there multiple attacks, as compared with 59
out of 76 cases with TIAs (77.6%).

   b) TIAs and PTIAs tend to recur significantly in the carotid
territory (both TIA and PTIA attacks occurred in 8 cases with
multiple attacks).

   c) In 58 out of 59 cases with multiple TIAs and in all cases
with multiple PTIAs the subsequent attacks recurred in the same
district as the first one.

   d) Multiple TIAs recurred within 1 month of the first one in
36 cases and after 1 month in 15 cases while multiple PTIAs
recurred in 6 out of 13 cases at intervals of some years.

   e) Completed strokes occurred in 16 out of 121 cases (13.4%)
with RIAs (in particular, 7 out of 76 cases with TIAs and 9 out of
45 cases with PTIAs).   Ten of these cases had multiple attacks.
The completed stroke occurred in 9 cases within 1 - 12 months, and
in the remaining 7 within 2 - 6 years.

   f) A total of 95 cases (78.5%) survived through the follow-up,
death occurring in 26 cases (21.5%).   The outcome of the 95
surviving cases was: no further neurological troubles in 58 (61%),
while in the remaining 37 cases (39%) further episodes occurred.
In particular, 20 cases with TIAs and 8 cases with PTIAs had
subsequent attacks.

g) Out of the 26 deaths, 7 were brought about by a completed stroke, 9 by heart infarction, while the remaining 10 deaths were due to miscellaneous diseases (tumours, bronchopneumonia, intestinal obstruction).   In other words 61.5% of cases with RIAs died from vascular diseases, namely completed stroke (26.9%) and heart infarction (34.6%).

RIAs IN THE ELDERLY

Data regarding annual incidence rates are taken from community and hospital surveys [46].

Strokes primarily afflict the elderly; the increase of incidence with age is striking[46], only one fifth occurring in individuals under 65 years (Framingham study[3]).   Recently, however, a major decline in the incidence of strokes occurred in the population of Rochester, Minnesota, in the period 1970-74 as compared with the period 1945-1949[47].

In particular, a more pronounced reduction in rates occurred in the elderly who showed a reduction of 60% during the study period if persons aged 80 years or older were considered.

Studies on the incidence of transient attacks are scanty. The only community study available[48] gives an incidence of 31/100,000 per year for all ages.   A sharp increase in incidence rates with age was shown, namely, more than 2/1000 subjects from 65 to 74 years and about 3/1000 subjects above 75 years of age were affected each year (study period from 1955 to 1964).

The prevalence of TIAs in the elderly population (over 65 years of age) is relatively high, about 1 out of 16[49].

Data are available from other surveys[50,51], but it would be premature to generalize from these studies, as Kurztke[46] rightly pointed out.

Out of our 121 patients sustaining RIAs, 93, i.e. 76.9%, were under 65 and 28 (23.1%) over 65 years of age.   In other words, although the role of age is well known, the analytical study of these two age groups (below and above 65 years respectively) showed no significant differences regarding: occurrence of multiple attacks, involvement of carotid or vertebrobasilar territories, number of deaths and survival, further occurrence of completed stroke and/or RIAs.

Apparently, age seems therefore to have no relevance as a risk factor, but it would be premature to draw firm conclusions from the evaluation of our small group of cases.

ASSOCIATED DISEASES

In 76 (63.7%) out of 121 cases hypertension (SBP > 160; DBP > 90 and/or signs of cardiac involvement (ventricular hypertrophy, mycocardial ischemia, atrial fibrillation, heart failure, recent infarction) were present.

In our case material the difference in incidence of subsequent completed strokes (68.6%) and/or RIAs between hypertensive and/or cardiac patients with RIAs and patients without hypertension and no cardiac abnormalities (31.4%) reach the level of significance (p < 0.01).

According to Lyon[52] and Stallones[53] the completed stroke, especially in cases with previous RIAs, should be considered as a mere component of a larger complex of vascular disease of the heart and brain. In fact, the risk of all types of stroke is correlated with hypertension and some forms of heart disease[54]. We cannot offer any explanation for the discrepancy between our data and those of the literature.

Dyslipidemia, diabetes, thrombophylic states, peripheral vascular disease, obesity, cigarette smoking were factors in a small group of cases (23)[55].

DIFFERENTIAL DIAGNOSIS

Apart from the classical differential diagnoses such as focal epilepsy, migraine, hypoglycemic spells, hypertensive crises, (differentiation which PTIAs share with TIAs[56,32]), episodes quite similar to PTIAs can occur in cerebral tumours[37]. In fact, out of 168 cases of histologically verified brain tumours 17 (10.1%) had shown transient episodes of focal cerebral dysfunction. Such episodes were isolated in 10 cases and multiple in 7, totalling a number of 33 attacks in 17 patients. Of relevance to the present report is the fact that as many as 15 attacks out of 33 took 2 - 15 days to recede completely.

The clinical features of these cases consisted of aphasia and/ or hemiparesis, central facial paresis, loss of consciousness, hemiparesis and/or amaurosis. The tumours were represented by 6 metastases, 6 glioblastoma, 2 meningiomas and 3 astrocytomas. It is therefore obvious that at least a CT study is mandatory in all cases with transient cerebral dysfunctions of any duration.

CONCLUSIONS

To the arbitrary definition of TIA, mainly adopted for practical
therapeutic reasons, another one has been added, equally artificial,
that of PTIA, comprising episodes receding over an average period
of three weeks.

It is quite apparent that TIAs and PTIAs share a common pathogenesis, clinical picture and treatment in spite of some differences
which could be conveniently clarified by the study of a larger case
material[37].

Even if the occurrence of completed strokes in cases with TIAs
is generally estimated between 25 - 40%[38], in our cases with RIAs
followed for about 6 years such an incidence reaches a bare 11.6%.

It seems therefore from our experience that the prediction of
the risk of subsequent completed strokes, solely on the basis of the
previous occurrence of TIAs or PTIAs should not be overemphasized.
It is still clear, however, that TIAs or PTIAs, along with hypertension and cardiac disease, constitute prominent risk factors for
subsequent completed strokes.

No significant data on the natural history could be put forward
as regards to age groups under and over 65 years.

The epidemiology of TIAs has been studied only by a few Authors
[38,48] and that of PTIAs is completely lacking. To this, and other
equally relevant ends, investigations of larger groups of cases are
needed.

REFERENCES

1. J. Marshall, The natural history of cerebrovascular diseases,
    in: "Modern Concepts of Cerebrovascular Disease", J. S. Meyer,
    ed., Spectrum Publ., New York (1975).
2. C. H. Millikan et al., Ad hoc Committee on Cerebrovascular
    diseases. A classification and outline of cerebrovascular
    diseases. Part II., Stroke 6:565 (1975).
3. W. B. Kannel and P. A. Wolf, Risk factors in atherothrombotic
    cerebrovascular disease, in: "Modern Concepts of Cerebro-
    vascular Disease", J. S. Meyer, ed., Spectrum Publ.,
    New York (1975).
4. F. Schiller, Concepts of stroke before and after, Virchow Med.
    Hist. 14:115 (1970).
5. Caelius Aurelianus, "De acutis morbis", lib. III, Ludguni,
    Rovillium (1565).
6. G. B. Morgagni, "Delle sedi e cause delle malattie", con note
    di F. Chaussier e N. P. Adelon, Sansone Coen, Firenze (1839).

7. H. Boherave, "Opera omnia medica venetiis", Occhi (1771).
8. M. Durand-Fardell, "Traité du ramollissement due cerveau",
    Baillière, Paris (1843).
9. G. Andral, "Corso di Patologia Interna", Collected by
    A. Latour, Italian version of the fifth French edition,
    Oliva, Milano (1853).
10. L. Bouveret, Aphasie, hémiplégie, apoplexie suite d'hémorragie
    gastrique. Autopsie, Revue Méd. 19:81 (1899).
11. C. Rokitansky, "A manual of pathological anatomy", 1824 1844,
    Sydenham Society, London (1854).
12. A. Raynaud, "Thèse", Paris (1862).
13. M. Weiss (1882), Quoted by S.A.K. Wilson and A. N. Bruce,
    Neurology Vol. 3, Butterworths, London (1955).
14. W. C. Bland (1889), Quoted by S.A.K. Wilson and A. N. Bruce,
    Neurology Vol. 3, Butterworths, London (1955).
15. G. L. Peabody, Relations between arterial disease and visceral
    changes, Trans.Ass.Am.Physns. 6:154 (1891).
16. W. Osler (1896), Quoted by S.A.K. Wilson and A.N. Bruce,
    Neurology Vol. 3, Butterworths, London (1955).
17. S. A. K. Wilson and A. N. Bruce, "Neurology" vol.3,
    Butterworths, London (1955).
18. C. Neisser, "Die paralytischen Anfälle", Stuttgart (1894).
19. H. Jackson (1888), in Taylor Holmes and Walshe, Selected
    writings of J. H. Jackson, vol.2, Staples Press, London
    (1958).
20. H. Oppenheim, "Trattato delle malattie nervose", vol.2,
    SEI, Milano (1905).
21. T. H. Alajouanine and R. Thurel, La pathologie de la circulat-
    ion cérébrale, Revue Neurol. 65:1276 (1936).
22. C. Fazio and C. Loeb, Apoplessia transitoria e apoplessia
    senza focolaio, Riv.Neurol. 18:142 (1948).
23. E. Corday and S. F. Rothenberg, The clinical aspects of cere-
    bral vascular insufficiency, Ann.intern.Med. 47:626 (1957).
24. H. Rein, (1931) Quoted by K. J. Zülch, Cerebral circulation
    and stroke, Introduction, Springer, Berlin (1971).
25. F. Büchner, (1939), Quoted by K. J. Zülch, Cerebral circulation
    and stroke, Introduction,Springer, Berlin (1971).
26. C. Miller Fisher, Intermittent cerebral ischemia, in:
    "Cerebral Vascular Diseases", Wright and Millikan, eds.,
    Grune and Stratton, New York (1958).
27. J. Lhermitte, Les idées nouvelles sur la genese de l'hémiplegie
    transitoire et du ramollissement cérébral, Encéphale 23:27
    (1928).
28. T. H. Alajouanine, F. Lhermitte and J. C. Gautier, Transient
    cerebral ischemia in atherosclerosis, Neurology (Minneap.)
    10:906 (1960).
29. C. Loeb, Clinical evaluation of patients with transient
    ischemic attacks, in: "Advances in Neurology" vol. 25,
    M. Goldstein et al, eds., Raven Press, New York (1979).

30. J. H. Drift, N. K. D. van der and Kok, Transient ischaemic
    attacks, in: "Assessment in cerebrovascular insufficiency",
    Stöcker, Kuhn, Hall, Becker and Van der Veen, eds., Thieme,
    Stuttgart (1971).
31. P. O. Yates, The pathogenesis of transient attacks, in:"Stroke",
    Gillingham, Mawdsley and Williams, eds., Churchill-Livingstone,
    Edinburgh-London (1976).
32. R. W. R. Russell, "Cerebral arterial disease", Churchill-
    Livingstone, Edinburgh-London (1976).
33. E. C. Hutchinson and E. J. Acheson, "Strokes. Natural history,
    pathology and surgical treatment", Saunders, London (1975).
34. M. S. Pessin, G. W. Duncan, G. P. Mohr and D. C. Poskanzer,
    Clinical and angiography features of carotid transient
    ischaemic attacks, New Engl.J.Med. 296:358 (1977).
35. C. Loeb and A. Priano, Strokes with full recovery. A reappraisal
    in: "Cerebral vascular disease", Meyer, Lechner, Reivich
    and Eichorn, eds., Thieme, Stuttgart (1973).
36. C. Loeb and A. Priano, Accidents cérébrovasculaires. Evolution
    régressive complète tardive, Revue Neurol. 131:873 (1975).
37. C. Loeb, Protracted transient ischemic attacks, Eur.Neurol.
    (In Press, 1979).
38. A. Heymann et al, Report of the Joint Committee for Stroke
    Facilities. XI. Transient focal cerebral ischemia: epidem-
    iological and clinical aspects, Stroke 5:276 (1974).
39. A. Ring, Middle cerebral artery. Anatomical and radiographic
    study, Acta Radiol. 57:289 (1962).
40. W. Kinkel and L. Jacobs, Computerized axial transverse tomo-
    graphy in cerebrovascular disease, Neurology (Minneap.)
    26:929 (1976).
41. P. Constant, A. M. Renou, J. M. Caille and J. Vernhiet,
    C.A.T. Studies of Cerebral Ischaemia, in: "The First
    European Seminar on Computerised Axial Tomography in
    Clinical Practice", G. H. du Bouley and I. F. Moseley, eds.,
    Springer, Berlin (1977).
42. U. Büll, J. Tongendorff, R. Rothe and K. Fischer, Results of
    serial scintigraphy with[99m]Tc-pertechnetate in comparison
    with angiography and computerized tomography in cerebro-
    vascular diseases, in: "Cranial computerized tomography",
    Lanksch and Kazner, eds., Springer, Berlin (1976).
43. P. Perrone, L. Candelise, G. Scotti, D. de Grandi and G.
    Scialfa, CT evaluation in patients with transient ischemic
    attacks. Correlation between clinical and angiographic
    findings. Eur.Neurol. 18:217 (1979).
44. G. Ladurner, W. D. Sager, L. D. Iliff and H. Lechner, A
    correlation of clinical findings and CT in ischemic cerebro-
    vascular disease, Eur.Neurol. 18:281 (1979).
45. V. Oltenau-Nerbe, P. Schmiedek, E. Kazner, W. Lanksch and
    F. Marguith, Comparison of regional blood flow and computer-
    ized tomography in patients with cerebrovascular disease and
    brain tumors, in: "Cranial computerized tomography",

Lanksch and Kazner, eds., Springer, Berlin (1976).

46. J. F. Kurtzke, Epidemiology of cerebrovascular disease, in:
     "Cerebrovascular Survey Report", Whiting Press, Rochester,
     Minn. (1976).

47. W. M. Garraway, J. P. Whisnant, A. J. Furlan, L. H. Phillips,
     L. T. Kurland and W. M. O'Fallon, The declining incidence
     of stroke, New Engl.J.Med. 300:449 (1979).

48. J. P. Whisnant, Epidemiology of stroke. Emphasis on transient
     cerebral ischemic attacks and hypertension, Stroke 5:68
     (1974).

49. A. M. Ostfeld, R. B. Shekelle, H. Klawans and H. M. Tufo,
     Epidemiology of stroke in an elderly welfare population,
     Am.J.Public Health 64:450 (1974).

50. H. L. Karp, A. Hayman, G. Heyden et al., Transient cerebral
     ischaemia. Prevalence and prognosis in a biracial rural
     community, J.Amer.Med.Assoc. 225:125 (1973).

51. G. D. Friedman, W. S. Wilson, J. M. Mosier, Transient ischemic
     attacks in a community, J.Amer.Med.Assoc. 210:1428 (1969)

52. C. Lyons, Progress report of joint study of extracranial
     occlusion, in: "Cerebral vascular diseases", Millikan,
     Siekert and Whisnant, eds., Grune and Stratton, New York,
     (1965).

53. R. A. Stallones, Epidemiology of stroke in relation to the
     cardiovascular disease complex,in: "Cerebrovascular disorders
     disorders and stroke", M. Goldstein, L. Bolis, C. Fieschi,
     S. Gorini and C. H. Millikan, eds., Raven Press, New York
     (1979).

54. S. Lavy, Medical Risk Factors in Stroke, in: "Cerebrovascular
     disorders and stroke", M. Goldstein, L. Bolis, C. Fieschi,
     S. Gorini and C. H. Millikan, eds., Raven Press, New York
     (1979).

55. J. F. Toole, R. Janeway, K. Choi, R. Cordell, C. Davis,
     F. Johnston and H. S. Miller, Transient ischemic attacks
     due to atherosclerosis. A prospective study of 160
     patients. Arch.Neurol. 32:5 (1975).

56. C. H. Millikan, The transient ischemic attacks, in: "Cerebro-
     vascular disorders and stroke", M. Goldstein, L. Bolis,
     C. Fieschi, S. Gorini and C. H. Millikan, eds., Raven Press,
     New York (1970).

RISK FACTORS FOR STROKE IN THE ELDERLY

J. Grimley Evans, D. Prudham, I. Wandless

Department of Medicine (Geriatrics)
University of Newcastle upon Tyne

The cost of cerebrovascular stroke to sufferers and their
families and to the Health & Social Services is immense. Although
there is some scope for improving the care of patients with
established disease the only prospect for a radical attack on the
condition lies in primary prevention.   As with the prevention of
any disease the research approach is to define precursor conditions
or high risk groups in samples of the population and then to carry
out appropriately designed controlled trials of intervention.
All intervention will result in some costs and disadvantages to the
community and to individuals and it is therefore crucial that a
trial should be carried out in the way that preventive programmes
would be implemented in order to provide appropriate estimates of
the balance of costs and benefits.

80% of strokes occur among people over 60 but most of our
knowledge about risk factors for the disease has been derived from
younger age groups.   We have reviewed elsewhere some of the main
findings on the epidemiology of stroke[1] but the most quoted source
of data on risk factors for stroke in later life lies with the
Framingham Study[2].   There are good reasons why we should not
assume that findings among the young will necessarily apply among
the old or that findings made in one country will necessarily apply
in another.   The epidemiology of coronary heart disease has taught
us that in a multi-factorial disorder, that is to say one that has
no single necessary or sufficient cause, our ability to identify
a risk factor and its apparent importance will depend on the
background of other factors against which it has to be detected.
Furthermore, it is likely that the relationship between a factor and
the risk of disease will be a sigmoid dose-response curve rather
than a linear function so that even a highly important risk factor

113

might fail to be detected if it is examined over a range that falls
below the threshold or above the ceiling of its relationship with
disease.

Problems with extrapolating findings from younger ages or other
countries also arise when we consider preventative programmes among
the elderly.   Increasing risk with age of adverse reactions to
drugs is well recognised and this may significantly affect the
balance of advantage and disadvantage of preventive measure at
different ages.   This balance will also depend on the organisation
of primary and community care.   It is very doubtful, for example,
whether general practice as presently organised in the United Kingdo
would be a safe and suitable environment for surveillance of blood
pressure and its treatment among the elderly.   There will also be
problems in the logistics and quality control of measurement of
risk factor status of individuals in the community that may be more
easily coped with in some systems of medical care than in others.

THE NEWCASTLE STUDY

The Sample and Methods

Against this background the Newcastle Age Research Group
established in 1975 a prospective study of stroke among persons aged
65 and over in a community of North-East England.   The design was
restricted to the use of personnel that might be available by
redeployment and redistribution in primary care teams in contemporar
Britain and to using methods that could be universally available,
widely acceptable and feasible within the economic and logistic
limitations of the National Health Service.   The area chosen for
study was defined geographically and contained an unusually wide
spectrum of social milieu including a small market town, a rural
fringe and part of a major conurbation.   All general practitioners
serving the area co-operated in the study.   The primary sampling
frame for identifying persons aged 65 and over resident in the
defined area were the files of the Family Practitioner Committee bu
these were supplemented from general practitioners' own lists and
from enquiries among participants.

Following initial approval from the general practitioner all
persons aged 65 and over were visited at home by specially trained
health visitors or community nurses and invited to take part in the
study.   The study schedule included an interview, measurement of
height and weight, measurement of blood pressure (using a random-zer
sphygmomanometer) and limb-lead electrocardiogram.   These methods
had been chosen to avoid venepuncture and undressing both of which
significantly reduce response rates in community surveys of the
elderly.   We had previously shown[3] that reduction of electrocardiograms to limb-leads only resulted in approximately 25% loss of

information but it was felt that the improved response rates and
the reduced interview time provided by omission of the chest leads
would compensate for this loss.   The research field workers were
trained in the use of the schedules by simultaneously coding
interviews and similar exercises were undertaken to reduce the
degree of observer-variation in the measurement of blood pressure.
Re-training exercises in order to standardise blood pressure
measurements technique were also carried out at intervals during
the time that data were being collected.

Computerised data files were established for all participants
and for non-participants in the study and several follow-up
procedures were instituted to detect and identify episodes of acute
cerebrovascular disease among the study population.   The main
source of data were reply-paid envelopes inserted in the general
practitioner record folder with the request for general
practitioners to notify the survey team of any possible stroke
events.   On the anniversary of each person's entry to the study
the record cards were inspected for possible cerebrovascular events
which had not been already reported.   The death notices of local
newspapers were inspected daily and copies of death certificates
and relevant hospital notes sought.   At annual intervals copies of
death certificates relating to residents of the study area aged 65
and over were made available to us by Community Physicians.   At
approximately two year intervals respondents were re-visited for an
abbreviated interview and possible episodes of cerebrovascular
disease enquired for.   In a few cases some persons could not be
traced on the anniversary of their initial enrolment and in these
cases the National Health Service Central Registry at Southport was
consulted and information on general practice registration or a
copy of the appropriate death certificate were obtained.
Respondents to the study who were alive at the time of identification
of a possible stroke were visited by a medical member of the survey
team and on the basis of the clinical history and physical assessment
with inspection of any relevant hospital notes a decision on whether
a stroke had taken place and its probable type was made by the
Medical Director.   At the time of making this decision the survey
entry data were not consulted so that the diagnosis should not be
biased by knowledge of pre-existing factors.   The criteria for the
diagnosis of stroke required the acute or sub-acute onset of focal
neurological disorder involving the cerebrum or brain stem lasting
more than 24 hours and of presumed vascular origin.

RESULTS

3,036 people were identified as resident in the designated area
and aged 65 and over on 1st January, 1975.   Of these 243 died
before being contacted by the survey team.   The response rate
among persons available to the study was 89.3% with no significant

variation with age or sex (Table 1).   Table 2 lists the numbers of
persons suffering stroke during a mean follow-up period of
approximately two years.   The overall incidence rate was approximately 16 per 1,000 per year with a male to female ratio of 1.3.
These findings are comparable with other studies.   In Table 3 we
present the data from the relationship between stroke and aspects
of the medical history.   The results are summarised by the
standardised risk rate comparing the risk among those respondents
giving a particular history to the rate among those without such
a history adjusted for age, sex and duration of follow-up.   This
table shows a minor increase in risk among respondents with
diabetes, a history of heart attack and previous high blood pressure
but only the last achieved statistical significance.   In Table 4
the data are presented in relation to the answers to a standarised
questionnaire on vascular symptomatology[4] and this shows a highly
significant increase in risk of stroke among patients with intermittent claudication.

Table 1.   Response Rates

AGE	MALES		FEMALES		PERSONS	
	No.	Response %	No.	Response %	No.	Response %
65-69	413	93.2	502	88.8	915	90.8
70-74	357	85.2	526	88.4	883	87.1
75-79	212	88.2	341	91.2	553	90.1
80-84	108	93.5	194	87.6	302	89.7
85-	51	88.2	89	88.8	140	88.6
ALL	1141	89.6	1652	89.0	2793	89.3

Table 2.   Persons Developing Stroke

AGE	MALE	FEMALE	PERSONS
65-69	20	15	35
70-74	10	27	37
75-79	12	15	27
80-84	7	11	18
85-	1	1	2
ALL	50	69	119

Table 3.   Medical History and Stroke

	No. of respondents	Standardised stroke rate*	Sig.
Diabetes mellitus	+ 61 - 2287	5.9 4.1	n.s.
Record of heart attack (G.P.)	+ 267 - 2410	5.9 3.7	n.s.
"Heart trouble"	+ 323 - 1832	5.7 3.8	n.s.
Previous high B.P.	+ 478 - 1659	5.9 3.6	p < .05

* per thousand in follow-up period

In Table 5 similarly data are presented for a number of symptoms which are thought possibly to be related to incipient cerebrovascular disease but only one of these, the transient dimness or blurring of vision, is significantly associated with risk of subsequent stroke.   In Table 6 the relationship between stroke and use of drugs is explored.   The increase in risk among patients taking digoxin, diuretics and hypotensives is related to a previous history of high blood pressure and to the presence of atrial fibrillation.   The increased risk with those patients taking cerebral activators is of particular interest.   We do not suppose that it means that these drugs induce stroke but rather that some of the symptoms for which the drugs were prescribed may be precursors of stroke.

Table 4.   Vascular Symptoms and Stroke

	No. of respondents	Standardised stroke rate	
Angina	+ 442 - 1911	5.1 3.8	n.s.
Possible myocardial infarction	+ 209 - 2144	5.5 4.0	n.s.
Claudication	+ 123 - 2230	10.0 3.8	p < .001

Table 5. Symptoms and Stroke

		No. of respondents	Standardised stroke rate	Sig.
Dizzy spells	+	824	4.3	n.s.
	−	1336	3.5	
Diplopia	+	336	3.8	n.s.
	−	1868	3.8	
Sight dim or out of focus	+	545	5.3	p <.05
	−	1659	3.3	
Faints or blackouts	+	580	2.9	n.s.
	−	1628	4.2	
Weakness or numbness	+	535	4.5	n.s.
	−	1668	3.7	
Falls	+	609	4.3	n.s.
	−	1599	3.6	

In Table 7 we show the relationship between stroke and various items on the electrocardiograms. This table indicates that those features of the E.C.G. usually interpreted in Western populations as indicati▼ of coronary heart disease are significantly associated with stroke risk as is atrial fibrillation. The presence of left axis deviation or of ectopic beats, however, do not predict stroke.

In Table 8 we compare persons developing stroke with age - and sex-matched standards derived from the study respondents as a whole. There is no association between stroke and present height weight or Quetelet's Index (used as an index of adiposity)[5] but a suggestion that persons developing stroke have been fatter than average in the past.

Table 9 shows the means and standard deviations of blood pressures observed in the study. In Table 10 we share the relationship between stroke incidence rates and tertiles of blood pressure defined in terms of sex and five year age group. The table provides no evidence of an association between blood pressure as measured and stroke. In Table 11 the analysis is repeated but excluding all patients taking any medication which might modify blood pressure namely diuretics, beta-blockers and hypotensives. There is some slight suggestion of a gradient of stroke risk against systolic pressure in this table but none is found for diastolic pressure.

Table 6.   Use of Drugs and Stroke

		No. of respondents	Standardised stroke rate	Sig.
Digoxin	+	149	8.4	<.01
	−	2205	3.9	
Diuretic	+	434	6.6	<.01
	−	1921	3.6	
Hypotensive	+	231	7.7	<.01
	−	2124	3.7	
Activator	+	13	18.8	<.01
	−	2341	4.0	
Tranquilliser	+	201	3.5	n.s.
	−	2153	4.2	
Antidepressive	+	121	5.8	n.s.
	−	2234	4.0	
Sleeping tablets	+	301	3.3	n.s.
	−	2036	4.3	
Vitamin	+	88	3.4	n.s.
	−	2256	4.1	
Laxatives	+	296	3.4	n.s.
		2050	4.2	
Other	+	988	4.8	n.s.
	−	1350	3.6	

Table 7.   Observed and Expected Number of Stroke among Respondents
           with particular E.C.G. Items

	Observed	Expected	Risk Ratio
Q-waves	5	2.40	2.1
Left axis deviation	10	9.51	1.1
ST-J depression	15	6.20	2.4
T wave changes	30	17.95	1.7
Left bundle branch block	4	0.88	4.5
Atrial fibrillation	9	3.29	2.7
Ectopic beats with sinus rhythm	7	9.93	0.7

(Definitions of E.C.G. items according to Minnesota Code[4])

Table 8.   Comparison of Stroke Victims with Others
           (standardised for age and sex)

	Non-Stroke		Stroke		
	No.	Mean*	No.	Mean*	Sig.
Height Cms.	2210	160.3	91	159.9	n.s.
Weight Kg.	2152	64.3	90	64.2	n.s.
Q.I.*	2499	25.3	91	25.4	n.s.
Q.I. age 55**	1741	25.6	61	26.7	n.s.
Q.I. age 25**	1678	23.2	63	24.1	p <0.05

*Quetelet's Index: weight divided by square of height
(Kg/m^2).

**Based on present height and reported previous weight

Table 9. Health Visitors B.P. Measurements

Age	MALES					FEMALES				
	No.	SYSTOLIC Mean	S.D.	DIASTOLIC Mean	S.D.	No.	SYSTOLIC Mean	S.D.	DIASTOLIC Mean	S.D.
65–69	377	155.7	25.9	87.7	15.6	438	161.3	25.8	92.0	15.8
70–74	300	160.1	25.5	87.3	14.8	450	166.4	26.2	91.9	16.5
75–79	184	158.7	26.0	86.1	15.7	294	168.6	25.7	91.4	16.0
80–84	101	157.6	28.9	84.9	19.0	168	169.7	27.6	91.8	16.6
85–	41	159.7	26.6	82.4	14.7	71	158.4	29.0	86.2	17.8

Table 10.   Stroke Incidence (per thousand person-months)
            According to Blood Pressure Tertiles

		SYSTOLIC TERTILE			
		1	2	3	All
	1	1.2 (19)	1.2 (9)	0.8 (2)	1.2 (30)
DIASTOLIC	2	1.5 (12)	1.7 (18)	1.8 (13)	1.6 (43)
TERTILE	3	0.0 (0)	0.9 (8)	1.2 (19)	1.0 (27)
	All	1.2 (31)	1.3 (35)	1.3 (34)	1.3 (100)

Table 11.   Stoke Incidence (per thousand person-months)
            According to Health Visitor Blood Pressure
            Measurements Excluding Respondents taking
            Medication* Modifying Blood Pressure (N=1865)

		SYSTOLIC TERTILE			
		1	2	3	All
	1	1.9 (11)	0.8 (5)	0.5 (1)	0.8 (17)
DIASTOLIC	2	1.3 (8)	1.5 (13)	1.5 (9)	1.4 (30)
TERTILE	3	0.0 (0)	0.8 (5)	1.2 (14)	1.0 (19)
	All	0.9 (19)	1.1 (23)	1.2 (24)	1.1 (66)

Figures in brackets are numbers of strokes in each group.

* Diuretics, β-blockers, hypotensives

Table 12.   Stroke Incidence (per thousand person-months)
            According to Most Recent G.P. Record of Blood
            Pressure. (Respondents with recording within
            two years of entry only included. N = 891)

		SYSTOLIC TERTILE			
		1	2	3	All
	1	1.8 (12)	1.5 (3)	1.3 (2)	1.7 (17)
DIASTOLIC	2	2.7 (9)	0.5 (2)	1.2 (3)	1.5 (14)
TERTILE	3	3.8 (2)	0.3 (1)	2.7 (14)	1.9 (17)
	All	2.2 (23)	0.7 (6)	2.1 (19)	1.7 (48)

Figures in brackets are numbers of strokes in each group.

In Table 12 we repeat the analysis for those respondents for
whom there was a general practitioner recording of blood pressure
available from their records within the two years preceding
interview.   Incidence rates in this group are higher because they
include preferentially those patients with a history of vascular
disease or raised blood pressure in the past.   Nonetheless no
relationship is observed between blood pressure level and
subsequent stroke.

Following the lead of the Chicago Stroke Study[6] we categorised
our patients into those with evidence of vascular disease or a
history of diabetes at the time of entry comprising Group S and
those respondents with no such evidence (Group H).   As expected
stroke incidence was higher in Group S than in Group H but in
neither group was a relationship between stroke and blood pressure
shown.   Table 13 shows the ratio of incidence rates for Group S
and Group H respondents according to age.   This shows that the
increased risk of stroke associated with a history of vascular
disease becomes attenuated with age and appears only to predict
stroke below the age of 80.

DISCUSSION

Our findings are surprising in the absence of a relationship
between blood pressure and stroke which has been reported from

Table 13. Stroke Incidence (per 1000 person-months)
By Age and Risk Group

| | GROUP 'H' | | | GROUP 'S' | | | INCIDENCE RATIO * |
	No.	Strokes	Incidence	No.	Strokes	Incidence	S/H
65-69	521	13	0.7	398	20	1.6	2.2
70-74	360	6	0.5	387	24	2.0	3.9
75-79	225	6	0.8	243	19	2.5	3.1
80-84	107	5	1.4	144	8	1.8	1.2
85-	42	1	0.9	49	1	0.7	0.8
All ages	1255	31	0.7	1221	76	2.0	2.7

* Computed before rounding of incidence rates.

elderly population samples in the United States[1].   There are a
number of possible explanations for this.   The first to be considered is inaccuracy in our methods of measuring blood pressure.
Our field workers were specifically trained and standardised in
the taking of blood pressure and used an instrument which removes
or obscures some forms of observer bias.   The blood pressures they
obtained were significantly correlated with general practitioner
records which also failed to show an association with stroke.
Furthermore one may point out that our methods were chosen as those
which might be expected to be used in screening programmes amont
the elderly and suggest that the pressures taken by the doctors in
research clinics as used in the American studies may not give
results that are relevant to preventive programmes in Great Britain.

        A second possibility to be considered relates to the
variability of blood pressure in the elderly.   We have shown that
among a variety of populations, including one which shows no rise
in blood pressure with age [7], that the within-individual variability
of blood pressure increases with age.   This presumably reflects
partly the decrease in accuracy and precision of homeostatic
mechanisms in the elderly but it may also indicate an increase in
the error variance with age.   A recent study from Canada[8] has shown
that the disparities between indirect and intrarterial measurements
of blood pressure are greater among the old than among the young.
The implication of this is that more readings of blood pressure
may be required accurately to categorise an older person with regard
to blood pressure than are required at younger ages: a single
blood pressure reading which is a significant predictor of vascular
morbidity at younger ages may not be so at older ages.   We are
pursuing this possibility by means of repeated blood pressure
recordings in a sub-sample of our study population.

        There is a third and particularly intriguing possibility.
The findings of our study indicate that the presence of established
vascular disease and a history of high blood pressure or of obesity
may predict stroke in old age even though present blood pressure
and present obesity do not.   This raises the possibility that
stroke in old age reflects not so much present causes of vascular
disease as causes which have been operating in the past.   Thus
present blood pressure and present obesity will only be predictors
of stroke if they accurately reflect the patient's blood pressure
and level of obesity in the past.   Probably one should be thinking
of an integration of these and other risk factors over time during
the patient's adult life.   Conceivably, therefore, the difference
between our findings and those in America could be related to the
degree of "tracking" shown by blood pressure over adult life in
American and British populations.   Various studies in the States
have shown significant degrees of tracking of blood pressure at
younger ages but studies into old age and international comparisons
are not available.

If the suggestion that the causes of vascular damage and blood pressure in old age lie many years previously one must question whether much potential remains for preventive programmes among the elderly.   Although manipulation of the causes of vascular disease might be unprofitable in old age it is still possible that intervention in the mechanisms whereby vascular damage is translated into end organ damage in the form of stroke might still be relevant. A specific possibility here lies with the anti-platelet drugs. In terms of potential side-effects and the amount of surveillance of treatment needed anti-platelet drugs are likely to prove a more attractive option in old age than are hypotensive drugs. We are carrying out some feasibility studies on the possible use of anti-platelet agents in community programmes of stroke prevention among the elderly.

ACKNOWLEDGEMENT

This work has been supported by the Department of Health and Social Security and the British Foundation for Age Research.

REFERENCES

1. J. Grimley Evans, The epidemiology of stroke, Age & Ageing 8 (Supp.):50 (1979).

2. W. B. Kannel, Blood pressure and the development of cardio-vascular disease in the aged, in: "Cardiology in Old Age," F. I. Caird, J. L. C. Dall and R. D. Kennedy, eds., Plenum, New York (1976).

3. J. Grimley Evans and W. M. G. Tunbridge, Information loss in limb-lead electrocardiograms compared with twelve-lead tracings in a population survey among the elderly, Age & Ageing 5:56 (1976).

4. G. A. Rose and H. Blackburn, Cardiovascular survey methods, WHO Monograph Series No. 56, WHO, Geneva (1968).

5. J. Grimley Evans and I. A. M. Prior, Indices of obesity derived from height and weight in two Polynesian populations, Brit.J.Prev.Soc.Med. 23:56 (1969).

6. A. M. Ostfeld, R. B. Shekelle, H. Klawans and H. M. Tufo, Epidemiology of stroke in an elderly welfare population, Amer.J.Publ.Hlth. 5:450 (1974).

7. J. Grimley Evans, (Unpublished).

8. J. D. Spence, W. J. Sibbald and R. D. Cape, Pseudohypertension in the elderly, Clin.Sci.Mol.Med. 55:399s (1978).

BRAIN DISEASE AND HYPERTENSION

J. L. C. Dall

Consultant Physician in Geriatric Medicine
Victoria Infirmary
Glasgow

For many years atherosclerosis or arteriosclerosis was thought
to be the principal underlying cause of dementia - indeed the term
arteriosclerotic dementia is still in common usage. The pathology
of dementia is now better understood and although clinical cases of
dementia may be associated with a solitary brain infarct, a few
large infarcts or multiple small infarcts[1] it is now believed that
the majority of cases are not vascular in origin[2].   Nevertheless,
focal vascular lesions causing intellectual deterioration,memory
failure and loss of social acceptance may be found in association
with hypertension. Thus intellectual changes may occur in
association with 'stroke' disease of the classical type, and
cerebral haemorrhage or infarction of large vessels with the so
called 'état lacunaire' due to disease of the smaller arterioles.

The subject is considered under four headings.

(a) Hypertension and cerebrovascular disease.
(b) Cerebral blood flow and brain disease.
(c) Hypotension and cerebrovascular disease.
(d) Treatment of hypertension.

(a) Hypertension and cerebrovascular disease. Diastolic
hypertension is uncommon in the old and very high levels of blood
pressure are rarely seen unless in association with dissecting
aneurysm of the abdominal aorta and renal infarction.   Hypertension
in the old is largely systolic hypertension which is common and has
been described as 'essential' or 'benign'. In recent years the
reports from Frammingham [3] and from Miami[4] pointed to the increased
risk of cerebrovascular disease associated with systolic hypertension and have illustrated that the term 'benign' is not justified.

Studies among the very old in a geriatric hospital[5] suggest that the
relationship between stroke and hypertension persists even into the
ninth decade and that in those without hypertension the prevalence
of stroke disease is low.   The prospect that adequate treatment of
hypertension, including systolic hypertension will prevent cerebrovascular disease, including strokes and multi-infarct dementia seems
rational and has been encouraged by the apparent success of the
Veteran's Administration Study[6] on moderate hypertension and the
improved prognosis for recurrence of strokes in the treated patient[7].
Prospective trials to establish this point continue and the outcome
of the European Working Party on Hypertension in the Elderly[8] is
awaited with interest.   Most physicians accept the need to treat
the younger patient who has "target organ damage" in the heart, the
kidney, the fundus of the eye but forget that the brain is also a
target organ and in systolic hypertension may be the most vulnerable
since E.C.G., renal and fundal changes are late or absent[4].
Unfolding of the arch of the aorta may be the only early sign of
systolic hypertension indicating the presence of vascular damage
and it is necessary to consider again the criteria for treatment of
hypertension in the elderly since to await 'target organ damage' in
the conventional sense is too late.

(b) <u>Cerebral blood flow and brain damage</u>. Perhaps the most
inhibiting aspect of treatment of systolic hypertension is the fear
of producing further cerebrovascular damage, or indeed the first
episode of cerebrovascular damage in a symptomless patient by
reducing the cerebral blood flow below a critical level.   The
pathophysiology of cerebral hypoxia and ischaemic strokes has been
reviewed by Adams[9], Brierly[10] and Graham[11] and the importance to
cerebral blood flow of the cerebral perfusion pressure has been
emphasized.   Perfusion pressure is influenced by the systolic
arterial pressure.   Autoregulation of cerebral blood flow can
maintain a relatively constant blood flow in the face of changes in
the perfusion pressure[12].   As the systolic pressure falls,
autoregulatory dilation of the cerebral arteriole maintains a fairly
constant cerebral supply despite large changes in the systolic
arterial pressure.   When, however, the cerebral vasodilation is
maximal, any further fall is systolic pressure will be accompanied
by a proportional fall in cerebral blood flow.   Autoregulation may
be impaired by the presence of pre-existing significant disease of
the cerebral vessels or of the internal carotid and vertebral
arteries, but the degree of impairment is difficult to predict
except in the presence of extensive vascular pathology[13].
Strandgaard et al[14] have suggested that when the patient is
hypertensive the level of systolic arterial pressure at which the
cerebral blood flow will fall is also higher than in normotensive
individuals but this seems more relevant to a sudden reduction in
systolic arterial pressure than to a progressive therapeutic
reduction.   While the relationship between cerebral blood flow
and mental function is still unclear, disturbed cerebral function

amounting to confusion or localised infarcts may follow cardiac
dysrhythmic episodes[15].   Ingvar[16] has shown that an increase in
cerebral blood flow in the premotor and frontal areas occurs during
problem solving mental activity - increased function stimulates
increased flow.   It has been reported that cerebral blood flow is
reduced in multi-infarct dementia[17] and in the late stages of
chronic brain syndrome[18] but it has not been established that the
circulatory change is causative in the non-vascular cases.
Nevertheless, the possibility that treatment of hypertension would
seriously disturb cerebral blood flow is of great importance.

(c) <u>Hypotension and cerebrovascular disease</u>. The experimental
work relating systolic arterial pressure and cerebral blood flow
has been conducted in a situation of an acute fall in the systolic
pressure such as is seen clinically in acute myocardial infarction,
surgical shock, major haemorrhage or postural hypotension.   In the
first three, transient or even permanent ischaemic strokes may
occur and illustrate the danger.   Jackson et al[19] illustrated the
dangers of misapplying potent hypotensive agents to systolic
hypertension by inducing severe postural hypotension with 'stroke'
consequences.   Johnson et al[20] and Lennox[21] have shown that
postural hypotension is not uncommon in elderly patients admitted
to a geriatric ward - often as a consequence of drug therapy
unrelated to blood pressure control such as the use of phenothiazines, sedatives, anti-parkinson regimes and diuretics for oedema.
Many of these patients are symptomless despite an observed fall in
systolic blood pressure and presumably are capable of normal autoregulatory function of the cerebral blood flow.   Those with
autonomic dysfunction associated with diabetes, peripheral neuro-
pathy of any aetiology or disease of the basal ganglia usually have
symptoms and are probably unable to achieve adequate cerebral
autoregulation.   This suggests that in the absence of these
associated pathologies, cerebral autoregulation can occur in the
elderly even at low levels of systolic arterial pressure and concern
for this should not prevent therapy being considered in the
hypertensive patient.

(d) <u>Treatment of hypertension in the elderly</u>.   To be acceptable
any treatment regime for use in the elderly patient, including the
symptomless elderly with systolic hypertension must not cause a
precipitate fall in blood pressure, must cause little or no postural
change and should not impair mental, renal or cardiac function.

Side-effects on renal function, uric acid secretion and glucose
tolerance of long term diuretic regimes have been fully reported
from the E.W.P.H.E. trial[22] and may be an acceptable price to pay
if there is a significant gain in reduction of cerebrovascular
disease.

Many clinicians now prefer beta-adrenergic blocking drugs as first line therapy because side effects are few and more acceptable to the patient, postural hypotension is unusual and the blood pressure can be reduced slowly over several weeks.

Twenty years ago, when ganglion blocking drugs causing severe side effects were the only treatment for hypertension no one seriously considered the treatment of symptomless elderly patients with systolic hypertension.   The therapeutic situation is now changed and there is a real opportunity to try to reduce the frequency of stroke disease and multi-infarct dementia.

REFERENCES

1. B. E. Tomlinson, G. Blessed and M. Roth, Observations on the brains of demented old people, J.Neurol. 11 (1970).
2. J. A. N. Corsellis, Ageing and the dementias in:"Greenfield's Neuropathology 3rd Edn.", W. Blackwood and J. A. N. Corsellis, eds., Arnold, London (1976).
3. W. B. Kannel, Blood pressure and development of cardiovascular disease in the aged, in: "Heart Disease in Old Age", F. I. Caird, J. L. C. Dall, R. D. Kennedy, eds., Plenum, New York (1976).
4. M. A. Colandrea, G. D. Friedman, M. Z. Nichman and C. N. Lynd, Systolic hypertension in the elderly: An epidemiologic assessment, Circulation 41:239 (1970).
5. J. L. C. Dall, Hypertension - A case for treatment, Age & Ageing Suppl. 8:36 (1979).
6. Veteran's Administration Cooperative Study Group, Effects of treatment on morbidity in hypertension, J.Amer.Med.Ass. 213:1143 (1970).
7. D. G. Beevers, M. Hamilton, M. J. Fairman and J. E. Harpur, The influence of antihypertensive treatment over the incidence of cerebral vascular disease, Postgrad.Med.J. 49:905 (1973).
8. A. Amery, P. Berthaux, C. Bulpitt, M. De Ruytterre, A. De Schaepdryver, C. Dollery, R. Faggard, F. Forette, J. Hellemans, P. Lund-Johansen, A. Mutsers and J. Tumilheto, Glucose intolerance during diuretic therapy, Lancet 1:681 (1978).
9. J. H. Adams, Ischaemic brain damage in arterial boundary zones in man, in: "Pathology of Cerebral Micro Circulation", J. Cervos-Navaro, ed., de Gruyter, New York (1974).
10. J. B. Brierly, Cerebral Hypoxia in: "Greenfield's Neuropathology 3rd Edn." W. Blackwood and J. A. N. Corsellis, Arnold, London (1976).
11. D. I. Graham, Hypoxia Ischaemia, J.Clin.Path. Supp.Royal College of Pathologists 11:170 (1977).

12. A. M. Harper,"Scientific Foundations of Neurology",
     M. Critchley, J. L. O'Leary and B. Jennett, eds., Heinemann,
     London (1972).
13. J. H. Adams, The pathology of ischaemic stroke, Age & Ageing
     Supplement 8:57 (1979).
14. S. Standgaard, J. Olesen, E. Skinhoj and N. A. Lanssen,
     Autoregulation of brain circulation in severe arterial
     hypertension, Br.Med.J. 1:507 (1973).
15. B. Livesey, Pathogenesis of brain failure in the aged,
     Age & Ageing Supplement 6:9 (1977).
16. D. H. Ingvar, Functional Landscapes of the dominant hemisphere,
     Brain Res. 107:181 (1976).
17. V. C. Hackinski, N. A. Lassen and J. Marshall, Multi infarct
     dementia: A cause of mental deterioration in the elderly,
     Lancet 2:207 (1974).
18. D. H. Ingvar and L. Gustafson, Regional cerebral flow in
     organic dementia with early onset, Acta Neurol.Scand.
     Suppl. 43:46 (1970).
19. G. Jackson, W. Mahon, T. Pierscianowski and J. Condon,
     Inappropriate antihypertensive therapy in the elderly,
     Lancet 2:1317 (1976).
20. R. H. Johnson, A. C. Smith, J. M. K. Spalding and L. Wollner,
     Effect of posture on blood pressure in elderly patients,
     Lancet 1:731.
21. I. Lennox, Postural hypotension in the elderly, pers.
     communication, British Geriatrics Society, April 1980.
22. Leading Article, Antihypertensive treatment in the elderly,
     Br.Med.J.2:1456 (1979).

ALTERED RESPONSE TO PSYCHOTROPHIC DRUGS WITH AGING

C.M. Castleden

Senior Lecturer in Geriatric Medicine
Leicester General Hospital
Leicester

It is well established that the prevalence of adverse drug
reactions is increased in the elderly particularly to psychotrophic
drugs[1,2,3].   Since these unwanted effects are largely an extension
of normal pharmacological actions[4],   the increased incidence implies
that drugs produce a greater response with aging.   There are
several possible explanations for this: firstly the elderly may
consume more drugs or produce higher tissue concentrations for a
given dose due to slow rates of drug elimination.   Secondly target
organ response may be increased with aging, and thirdly, the
prevalence of unwanted effects may merely reflect the presence of
concurrent disease in elderly patients.   The purpose of this paper
is to review each explanation with particular reference to the
central nervous system.

INCREASED DRUG CONSUMPTION

It is known that adverse drug reactions increase exponentially
with the number of drugs consumed[5],   and that drug prescription
increases with aging[6,7,8].   However, investigators have failed to
study old and young patients under the same doctor, and at the same
time.   Several studies have shown considerable geographic and
seasonal variation in the number of prescriptions issued per
patient[9], and that variations exist between members of the same
practice [10].   Thus Law and Chalmers'[11] conclusion that elderly
patients receive more drugs may be invalidated as their comparison
was made with young patients attending another general practitioner

in a different part of the British Isles.   Equally the differences
noted by other workers[6],[7],[8] could be explained by patients of
different age and sex attending different general practitioners.
An additional complication is that all the drugs given to young
patients such as oral contraceptives were not included[8].

Of more importance perhaps than absolute numbers is the type
of drug consumed.   Pearson and Havard[12] have pointed out that there
are certain areas of prescribing in which drug interactions are
particularly likely to cause dangerous effects, and in which
undesirable reactions frequently occur.   In a community based study
of old and young patients under the same general practitioner [13]
the number of elderly patients taking drugs of this type was 2.6
times higher than that of the young, and 21% of the elderly were on
drugs likely to cause CNS symptoms in toxic doses compared to only
11% of the young.   Such findings were consistent with previous
observations[9] that the elderly had different symptoms from the young
and agree with Bytheway's conclusions[14] that considerable differences
exist in the types of treatment given to patients aged 16 to 65 and
those over 65.

However such observations cannot explain why elderly patients
suffer from more unwanted reactions than young on the same
medication, nor why some drugs produce a greater pharmacological
effect in the elderly when given alone.

ALTERED PHARMACOKINETICS WITH AGING

There is firm evidence indicating that there are major problems
in drug elimination with aging[15].   Since there is no alteration in
the rate or extent of drug absorption in the old[16], a given dose may
produce higher plasma concentrations which persist for longer in an
old than in a young patient.   The pharmacological action of a drug
is related to its plasma concentration[17] which is usually closely
related to the level of drug in the tissues.   This may not be true
for the CSF because of the blood brain barrier.   There is however
very little work on the relationship between plasma and CSF drug
concentrations at any age, although a high correlation has been
shown for some e.g. chlorpromazine[18], and there is a clear association between the clinical effect of tricyclic antidepressants and
their plasma concentration[19].

The endothelial capillary cells in the CNS are tightly bound so
that compounds must pass through the cells to gain access to the CSF.
This means they must be at least partially lipid soluble[20].
In contrast non-neuronal capillaries are leaky and all molecules
move back and forth between the cells.   Penetration into the CSF
depends also on protein binding, being for some drugs inversely
related.   Since drugs must go through the cells to gain access to

the CSF, they are exposed to enzymatic traps, for example, dopa
can enter but dopamine cannot.   There are too specific carriers
for certain compounds, for example glucose, amino acids and pyruvate.
Many neurotransmitters (e.g. biogenic amines) enter the brain
minimally or not at all.   Once in the CSF drugs easily reach the
extracellular fluid in brain tissue since the two are in intimate
diffusion.   The CSF is also subjected to a constant flux, being
formed in the choroid plexus at a rate of about 530ml per 24 hours,
and reabsorbed in the arachnoid villi of the superior saggital sinus[20].

Many of the mechanisms for the passage of drugs into CSF, and
for its formation and reabsorption are active processes.   Since
there are ample examples in animals[21] and man[15] of changes in the
biochemical activity of enzymes with aging, any of these mechanisms
may become less efficient with ageing, thus allowing more drug into
the CSF or keeping it there longer.   More drug may also pass since
protein binding is frequently lower in old patients[15].

As yet this area has received virtually no research, with the
exception of work by Hewick and Shaw[22].   They injected [14]C
nitrazepam into young and old rats; the old were visibly more sedated,
but the plasma concentration at each of the 4 times measured was
similar in both groups.   This was also true of plasma clearance and
the apparent volume of distribution.   However the concentration in
the brain of the old animals was 2 to 3 times higher than that in the
young.   If these results could be extrapolated to man, they might
explain similar findings[23].   Castleden et al[23] found that the effect
of a single 10mg dose of nitrazepam was greater on psychomotor
performance in the old than in the young despite no apparent kinetic
differences between the groups.

ALTERED PHARMACODYNAMICS WITH AGING

An increased sensitivity to nitrazepam in the elderly was
proposed when the drug produced a greater effect in them in the
absence of an alternative explanation[23].   It was similarly argued
that the elderly were more sensitive to diazepam[24,25].   In contrast
other workers[26] reported a diminished responsiveness to both agonist
and antagonist beta receptor drugs with advancing years.   Hewick's
work[22] might explain the apparent discrepancy, and allow a tentative
hypothesis of decreasing target organ sensitivity with aging.   This
would fit with published work on the effects of aging on drug
receptors.   Although there is little on brain tissue in this field,
it may be possible to extrapolate from data on other tissues.
Shocken and Roth[27] measured the concentration of beta adrenergic
receptors in crude membrane fractions of mononuclear cells from human
subjects of various ages.   Receptive affinities for ^{3}H-dihydro-
aprenolol were essentially the same for all subjects.   However a
significant decrease in receptor concentration was observed in cells

from older individuals.   A mean saturation level for young subjects
24 to 48 years, (n=11) was 572 ± 58 pmols mg^{-1} protein (± SEM) whils
for subjects older than 46 years (n=12) a corresponding value was
slightly greater than half that value, 332 ± 48 pmols mg^{-1} protein
(p < 0.01).   When all data were considered, there was a significant
negative correlation of receptor concentration with age (p < 0.001).
Receptor number per cell also correlated inversely with age from
approximately 14,000 to 8,000 sites per cell in the young and old
groups respectively.   Steroid receptor concentrations are also
known to decline with age in various rodent tissues[28], and the
ability of steroid hormones to elicit physiological responses has
been shown to be diminished with age in several of these tissues[29].
The induction of acetyl cholinesterases by 17 β oestradiol in
cerebral hemispheres and cerebelli of female ovarectomised rats is
decreased to such an extent in old age that no induction occurs in
the very old[29].   Preliminary work from our laboratory agrees with
that of Shocken and Roth[27].   Cyclic AMP response in human
lymphocytes in elderly subjects was similar to that in the young at
low concentrations of isoprenaline, but markedly decreased at
higher concentrations.   These results suggest that the beta
receptors are already maximally stimulated (i.e. all sites are
occupied) at the lower concentration.

It may be valid to extrapolate the results to neuronal tissues
since most brain cells are able to produce Cyclic AMP which is a
second messenger mediating the effects of a variety of hormones, and
beta receptors are known to exist throughout the CNS[30].   However,
it is not yet clear what a decreased number of receptors with aging
means, as these are not static features.   They increase in numbers
in response to some drugs given chronically or to denervation, but
a decrease in number has not been shown previously; decreased
sensitivity being due to an increased binding of the ligand onto
the receptor site[30].

CONCOMMITANT DISEASE

The present situation therefore is that the drug concentrations
are higher in the elderly for a given dose, but end organ response
is less.   How far these two cancel each other is uncertain, but
if they do, then the increased incidence of adverse drug reactions
in elderly patients is still unexplained.   A further possibility
is an alteration in drug effect due to concommitant disease.

Multiple pathological diseases are usual in elderly patients
and have several effects on drug handling and response.
Differences between old patients and healthy young subjects are
still reported as the effects of aging despite knowledge that
disease affects drug pharmacokinetics including drug penetration

into the CSF.   An example of this is a recent report of the effect
of age on the pharmacokinetics of nitrazepam[31].   In this study the
beta elimination half life was 29 hours in the young which agreed
with previous work[23].   However, it was 40 hours in geriatric patients
whereas in the previous paper no alteration was noted with aging.
In the first study[23], the elderly subjects were all in apparently
normal health living independently in the community, whereas in the
second[31] they were ill, bed-bound and had multiple other drug
therapy.   It is clear that one study could reasonably assert to
investigate the effect of aging on the pharmacokinetics of
nitrazepam, but the other was investigating the effect of multiple
factors including disease.   This was more clearly shown by two
other studies; in 1971 O'Malley and his colleagues[32] showed that the
plasma half life of antipyrine was significantly longer in geriatric
patients compared to younger controls.   In 1975 this observation was
confirmed[33], but it was also noted that alcohol and caffeine
consumption and cigarette smoking were all lower in the older
subjects.   Both caffeine and cigarette use were positively correlated with the rate of antipyrine metabolism.   Multiple regression
analyses showed that the effect of smoking was particularly
responsible for age differences in antipyrine metabolism; smoking
explained 12% of the variance in metabolic clearance rate and age
explained only 3%.

Secondly, concommitant disease may upset compensatory mechanisms
which normally minimise an unwanted effect of a drug.   An example of
this is postural hypotension which is a particular hazard in the
elderly when certain drugs are used.   Baro-receptor activity has
been shown to decrease with aging[34], mainly due to atherosclerosis
in the carotid sinus, but also due to autonomic neuropathies.   There
is also a greater sodium loss from diuretics secondary to a decreased
number of nephrons, and hence a decreased ability to conserve sodium
ions[35].

Thirdly, disease may unmask drug effects which would pass unnoticed in healthy individuals, for example bronchospasm following
beta adrenergic neuron blockers in asthmatics.   A similar mechanism
may explain the increased incidence of extrapyramidal manifestations
after chlorpromazine therapy in elderly patients, the abnormal effect
of L-dopa in demented parkinsonium patients and the restlessness and
confusion caused by hypnotics in patients with mild chronic brain
failure[36,37].

Whether aberrant reactions to hypnotics are a reflection of
increased sensitivity to these drugs in the elderly or merely
another manifestation of additional disease, is as yet unknown.
Case report studies suggest that the elderly are particularly
susceptible to the paradoxical excitement following the use of
barbiturates[38] and of chloral hydrate[39].   However, more recent
reports suggest that the paradoxical excitement after barbiturates

is no more common in the elderly than in the young and that this is also true for flurazepam[40] and nitrazepam[3].

A final aspect of concommitant disease is dependency on other[s]. Many patients with adverse drug reactions decrease or stop therapy[1]. Elderly patients are frequently unable to make this modification as their medication is supervised by someone who may not recognise the connection between the adverse reaction and the drug.

CONCLUSIONS

The available evidence would suggest that the concept of "increased sensitivity" of the brain to drugs with aging is too simplistic. The underlying mechanism is probably a failure of compensatory manoeuvres due to disease. In addition there is frequently a greater tissue concentration of drugs which is also secondary to disease.

REFERENCES

1. N. Hurwitz, Predisposing factors in adverse reactions to drugs, Br.Med.J. 1:536 (1969).
2. Boston Collaborative Drug Surveillance Program. Clinical depression of the central nervous system due to diazepam and chlordiazeproxide in relation to cigarette smoking and age, N.Eng.J.Med.288:277 (1973).
3. D. J. Greenblatt and M. D. Allen, Toxicity of nitrazepam in the elderly : a report from the Boston Collaborative Drug Surveillance Program, Br.J.Clin.Pharma. 5:407 (1978).
4. R. J. Ogilvie and J. Ruedy, Adverse drug reactions during hospitalisation, Can.Med.Assoc.J. 97:1450 (1967).
5. J. W. Smith, L. G. Seidl and L.E. Cluff, Studies on the epidemiology of adverse drug reactions -v- clinical factors influencing susceptibility, Ann.Int.Med. 65:629 (1966).
6. M. E. J. Wadsworth, W. J. H. Bullerfield and R. Blaney, Health and Sickness. The choice of treatment, Tavistock Publications Ltd., London (1971).
7. K. Dunnell and A. Cartwright, Medicine takers, prescribers and hoarders, Routledge and Kegan Paul Limited (1972).
8. D. G. G. Skegg, R. Doll, J. Perry, Use of medicine in general practice, Br.Med.J. 1:1561 (1977).
9. D. M. Dunlop, R. S. Inch and J. Paul, A survey of prescribing in Scotland in 1951, Br.Med.J. 1:694 (1953).
10. D. J. G. Bain and A. J. Haines, A year's study of drug prescribing in general practice using computer-assisted records, J.R.Coll.Gen.Pract. 25:41 (1975).
11. R. Law and C. Chalmers, Medicines and elderly people : a general practice survey, Brit.Med.J. 1:565 (1976).

12. R. M. Pearson and C. W. H. Havard, Drug interactions, Br.J.Hosp.
     Med. 12:812 (1974).
13. C. M. Castleden, M. D. Thesis, London University (1978).
14. B. Bytheway, Prescribing in general practice. 7. Prescribing
     and the pensioner, J.Roy.Coll.Gen.Pract.Suppl.1. 26:40
     (1976).
15. F. I. Caird, Prescribing for the elderly, Br.J.Hosp.Med. 17:610
     (1977).
16. C. M. Castleden, G. N. Volans and K. Raymond, The effect of
     ageing on drug absorption from the gut, Age & Ageing, 6:138
     (1977).
17. J. Koch-Weser, Drug therapy. Serum drug concentrations as
     therapeutic guides, New Eng.J.Med. 287:227 (1972).
18. J. M. Davis, S. Erickson and H. Dekirmenjian, Plasma levels of
     antipsychotic drugs and clinical response in :"Psychopharmac-
     ology : a generation of progress", M.A. Lipton, A. Di Mascio
     and K. F. Killan, eds. Raven Press, New York (1978).
19. A. H. Glassman and J. M. Perel, Tricyclic blood levels and
     clinical outcome : A review of the art, in "Psychopharmac-
     ology : a generation of progress", M. A. Lipton, A. Di Mascio
     and K. F. Killan, eds. Raven Press, New York (1978).
20. Editorial, Clinical aspects of the blood brain barrier,
     Br.Med.J., 2:133 (1976).
21. R. Kato, P. Vassanelli, G. Frontino and E. Chiesara, Variation
     in the activity of liver microsomal drug metabolising
     enzymes in rats in relation to the age, Biochem.Pharmacol.
     13:1037 (1964).
22. D. S. Hewick and V. Shaw, Tissue distrubition of radioactivity
     after injection of   C Nitrazepam in young and old rats,
     J.Pharm.Pharmac. 30:318 (1978).
23. C. M. Castleden, C. F. George, D. Marcer and C. Hallett,
     Increased sensitivity to nitrazepam in old age, Br.Med.J.
     1:10 (1977).
24. C. M. C. Castleden, A. Houston and C. F. George, Are hypnotics
     helpful or harmful to elderly patients? J.Drug Issues 9:55
     (1979).
25. M. M. Reidenberg, M. Levy, H. Warner, C. B. Continho,
     M. A. Schwartz, G. Yu, J. Cheripko, Relationship between
     diazepam dose, plasma level, age and central nervous system
     depression, Clin.Pharmac.Ther. 23:371 (1978).
26. R. E. Vestal, A. J. J. Wood and D. G. Shand, Reduced B-adreno-
     ceptor sensitivity in the elderly, Clin.Pharmac.Ther. 26:181
     (1971).
27. D. D. Schocken and G. S. Roth, Reduced B-adrenergic receptor
     concentrations in ageing man, Nature, 267:854 (1979).
28. G. S. Roth and J. N. Livingston, Reductions in glucocorticoid
     inhibition of glucose oxidation and presumptive glucocorti-
     coid receptor content in rat adipocytes during ageing,
     Endocrinol. 99:831 (1976)

29. M. S. Kanungo, S. K. Patniak and O. Koul, Decrease in 17 B-
    oestradiol receptor in brain of ageing rats, <u>Nature</u>
    253:366 (1975).
30. S. H. Snyder, Receptors, neurotransmitters of drug responses,
    <u>New Engl.J.Med.</u> 300:465 (1979).
31. L. Kangash, E. Iisalo, J. Kanto, V. Lehtinen, S. Pynnonen,
    I. Ruika, J. Salminen, M. Sillapaa and E. Syvalahti,
    Human pharmacokinetics of nitrazepam : effect of age and
    diseases, <u>Europ.J.Clin.Pharmacol.</u> 15:163 (1979).
32. K. O'Malley, J. Crooks, E. Duke, I. H. Stevenson, Effect of
    age and sex on human drug metabolism, <u>Br.Med.J.</u> 3:607 (1971)
33. R. E. Vestel, A. H. Norris, J. D. Tobin, B. H. Cohen, N. W.
    Shock and R. Andres, Antipyrine metabolism in man:influence
    of age, alcohol, caffeine and smoking, <u>Clin.Pharmacol.Ther.</u>
    18:425 (1975).
34. B. Gribbin, T. G. Pickering, P. Sleight and R. Peto, Effect
    of age and high blood pressure on baroreflex sensitivity
    in man. <u>Circ.Res.</u> 29:424 (1971).
35. J. D. Swales, Pathophysiology of blood pressure in the elderly,
    <u>Age and Ageing</u> 8:104 (1979).
36. O. W. Sacks, C. Messeloff, W. Schartz, A. Goldford and M. Kohl,
    Effects of L-dopa in patients with dementia, <u>Lancet</u> 1:1231
    (1970).
37. W. Davison, Drug hazards in the elderly, <u>Br.J.Hosp.Med.</u> 6:83
    (1971).
38. I. I. J. M. Gibson, Barbiturate delirium, <u>The Practitioner</u>,
    197:345 (1966).
39. R. R. Miller and D. J. Greenblatt, Hypnotics <u>in</u>:"Drug effects
    in hospitalized patients", R. R. Miller and D. J. Greenblatt
    eds. John Wiley and Sons Inc. (1976).
40. D. J. Greenblatt, M. D. Allen and R. I. Shader, Toxicity of
    high-dose flurazepam in the elderly, <u>Clin.Pharmacol.Ther.</u>
    21:355 (1977).

IATROGENIC BRAIN FAILURE

W. Davison

Consultant Physician in Geriatric Medicine
Chesterton Hospital
Cambridge

There's a wicked spirit watching round you still,
And he tries to tempt you to all harm and ill.
But ye must not hear him, though tis hard for you,
To resist the evil, and the good to do.

"Jesus Calls Us" Cecil Frances Alexander
1818 - 1895

Since time immemorial people intent on doing good have
unquestionably done harm. Often the harm has been apparent at the
time of its infliction but in other instances many years have
elapsed before new knowledge and changes of opinion have caused us
to recognise that an activity which previously was regarded as
"good" is now quite clearly "bad". These general remarks apply to
all fields of human endeavour, including medicine. Originally the
term iatrogenic was introduced to describe disorders induced in the
patient by autosuggestion based on the physician's behaviour in
relation to his client. Nowadays it is applied to any ailment
occurring in a patient as the result of treatment by a physician or
a surgeon or those working with him. Presumably this definition
would include the nosocomial disorder of institutional neurosis so
well described by Barton[1]. This disease is characterised by apathy,
lack of initiative, loss of interest, especially in things of an
impersonal nature, submissiveness, apparent inability to make plans
for the future, lack of individuality and sometimes a characteristic
posture and gait. Although the production of institutional neurosis
has been common in mental hospitals for many years it is only
comparatively recently that it has been recognised to be an iatrogenic disorder which far from being inevitable is both preventable
and to a degree treatable. Presumably these cases were simply

141

accepted as "burnt out schizophrenics" or "senile dementia". The
psychiatrist failed to realise that the patient he had been treating
for melancholia or paraphrenia had developed an additional illness.
He may even have regarded the pathetic, crumpled, mute, submissive
creature as being much in need of care and attention in the
institution!

ADVERSE EFFECTS OF DRUGS ON BRAIN FUNCTION

The purpose of this paper is to review some of the main clinical
aspects of adverse drug reaction (ADR) with respect to brain
function in the elderly, drawing attention especially to those
iatrogenic syndrome of brain failure likely to impair personal inde-
independence and full engagement in family and social life.
Brain failure due to manifest drug overdose or deliberate chronic
abuse is not considered here but only those reactions due to
generally accepted therapeutic drug usage.

This will leave out much iatrogenic brain failure because
surgical attacks on the brain or on its blood supply and invasive
investigations are not considered nor is brain failure following
cardiac resuscitation or anaesthesia.   To provide a simple clinical
definition of the problem we should regard the brain as that part
of the central nervous system contained within the cranium and
'failure' as an inability to perform its tasks satisfactorily.
The brain is an organ (or a complex assembly of organs) which may
fail to a variable degree, in whole or in part, acutely or
insidiously due to some reversible or irreversible mechanism.
Thus 'brain failure' is not the same as dementia but dementia is
certainly a form of brain failure, (Table 1).

MECHANISMS OF BRAIN FAILURE

Normal brain function depends upon an intact brain at the
macroscopic, microscopic and molecular level together with both food
for thought and food for metabolism.   That is to say an
appropriate sensory input, contact with the environment and a
satisfactory blood supply.   With normal ageing, failing brain
function probably begins at the molecular level and structural
changes follow.   Similarly in senile dementia altered metabolism
appears to precede (and be much more extensive than might be
predicted from the degree of) neuronal loss[2].

Some possible mechanism of brain failure in the course of
medical treatment are listed in Table 2.   Adverse effects of drug
treatment by the very nature of things tend to occur in a setting
of poor physical and mental health so that brain failure results
from the combined efforts of nature and physician!   This makes it
difficult for the doctor to recognise the adverse drug reaction (ADR).

Table 1.   Determinants of Brain Function

Intact brain	-	structure and chemistry
Sensory input	+	emotional state
Blood supply	-	quantity and quality

for what it really is and not (as so often happens) accept it as part of the natural history of disease.   In addition to multiple diseases and drug treatments there is often coexisting some degree of sensory deprivation and emotional stress.   It is the complex interplay of all these variables which accounts for the frequency of mental disturbance in old people.   The altered mix of aetiological factors could also explain why some patients react adversely to a drug on one occasion yet fail to do so on another, (Table 2).

Table 2.   Some Iatrogenic Brain Insults

Type	Examples
Psychological	Institutions, attitudes
Haemodynamic	Diuretics, antihypertensives
Neurochemical	Antipsychotics,antiparkinsonian
Biochemical	Antidiabetics, diuretics
Thromboembolic	Oestrogens, corticosteroids
Invasive	Carotid angiogram

INCIDENCE OF IATROGENIC BRAIN FAILURE

It is not known how often medical treatment tips the patient into brain failure but it is probably a good deal more common than is generally believed.   This type of ADR is most likely to occur when an ill elderly patient is treated with multiple drugs especially those acting on the central nervous system (CNS) and of course it is least likely to be noted if the doctor is unaware of the possibility!   Lack of awareness in this area is to be expected and it is likely to continue until all health science teaching programmes give due emphasis to geriatric psychiatry and pharmac-ology[3,4,5].   Drugs acting on the CNS are very commonly prescribed both in general practice and in hospital.   For example in a two year prospective surveillance of ADR in a semirural general practice in England, Martys[6] found that drugs acting on the CNS comprised the second most commonly prescribed group after antibiotics in patients of all ages.   These drugs also gave the highest incidence of ADR and 51% of all patients who were prescribed them reported

an adverse reaction – mostly in the minimal category and usually
within a few days. Long term effects were not studied. Complaints
from all respondents (all types of drugs) included excessive
sleepiness in 12%, dizziness in 9.2%, headache in 5.6% and confusion
or hallucinations in 3%. In this study pentazocine and dihydro-
codeine were most often associated with these latter and more
serious symptoms. This study excluded the most severely ill and
those on recent multiple drug treatment. By contrast in another
English general practice, Shaw and Opit[7] found little toxicity in
elderly patients on long term treatment with drugs but noted that
about 25% were on maintenance drugs for more than one condition
especially for heart and mental illness.

In the Boston Collaborative Drug Surveillance Program (BCDSP)
according to the report of Porter and Jick[8] serious ADRs with
respect to the central nervous system are rare in general medical
wards with the notable exception of the extrapyramidal syndromes
(EPS) in patients treated with trifluoperazine and deafness in
those treated with high doses of salicylates or aminoglycosides.
Some 10% of patients on trifluoperazine (Stelazine) developed EPS
compared with only 0.3% on prochlorperazine (Stemetil). Levy et al[9]
reported from Jerusalem intensive surveillance for ADR in the
general medical wards (as part of the BCDSP) of 2499 patients.
Just over 4% were admitted because of ADR but no mention was made
of brain failure yet 40% of their cases were aged 60 years or more.
Patients admitted to geriatric departments in the UK are mostly
selected for great age and frailty compared with general medical
admissions. For example in Cambridge the mean age of admission to
geriatrics is 80 years and to general medicine 56 years. As a
result patients in geriatric wards are much more likely to have
mental disturbance due to disease or drugs. Also they are much
more liable to have iatrogenic ADR of all types than those described
in the general practice surveys being both more ill and on more
drugs. Thus in a multicentre study of 1998 newly admitted elderly
patients in geriatric departments, Williamson and Chopin[10] found
that adverse drug reaction was evident in 15% and was the sole or
contributing cause for admission in 10% of cases. Hypnotics,
sedatives, anticonvulsants and antiparkinson drugs and hypotensives
were the principal offenders in this respect and full recovery was
less likely in these patients. The implications of this are
serious in that seemingly thousands of elderly people in the UK are
being rendered mentally less capable each year by inappropriate
drug treatment.

CLASSIFICATION OF IATROGENIC PSYCHIATRIC REACTION

As McClelland[11] points out there is nothing specific about the
psychiatric reactions that drugs can cause. For example levodopa
may cause mania, delirium, mental depression, dementia or paranoid
psychosis and reserpine can give rise to insomnia, nightmares,

delirium or depression.   The classification in Table 3 is modified
from McClelland.   Two common iatrogenic syndromes seen in clinical
practice are 1) agitated confusion with or without hallucinations,
altered personality and emotional lability.   This picture is seen
especially with antiparkinsonian drugs, in the treatment of
congestive cardiac failure and with antidepressants and 2)retard-
ation and withdrawal with or without evidence of memory defect and
disorientation as seen with tranquillisers, hypotensives and
diuretics, (Table 3).

Minimal Reactions to Drugs

    Minimal reactions to drugs including drowsiness, insomnia,
nightmares and irritability are all quite common.   Many drugs
are involved viz: barbiturates, benzodiazepines and other hypnotics,
tricyclical antidepressants, beta-adrenergic blockers and antiparkinsonian agents.   Although (by definition) these ADRs are
mild they represent incipient brain failure and if the noxious
agent is not withdrawn, acute brain failure with delirium may
supervene.   Alternatively chronic brain failure with drug
dependence may develop.   Minimal reactions to drugs must not
routinely be dismissed as of no account.   Drowsiness or giddiness
are degrees of brain failure of clinical significance especially
if brain function is already on the wane.   For example chronic.
barbiturate consumption is associated in elderly people with an
increased incidence of dizziness, falls and fractures as well as
a reduction in cognitive function[12].   A disquieting feature of
this report from Nottingham is the continuing widespread use of
barbiturates and other hypnotics by the elderly despite their known
adverse effects on brain function.   Thus in 1976 51% of elderly
patients referred to the geriatric outpatient clinic in that town

Table 3.   Iatrogenic Psychiatric Reactions

          Minimal   reactions

          Delirious states

          Affective disorders

          Paranoid psychoses

          Hallucinations

          Dementia

          Extrapyramidal syndromes

were taking barbiturates, 41% taking other hypnotics and only 8%
were not taking hypnotics.   This represents considerable brain
failure in the medical profession as much as in the elderly!
It is striking that the patient's general practitioner mentioned
giddiness or falls as a reason for referral in 84% of those on
barbiturates and only in 24% of those not taking these drugs.

Delirious States

These states are often preceded by minimal reactions.   The
cardinal presentation is a fluctuating level of consciousness with
mental confusion, disorientation and maybe with anxiety and even
paranoia.   The use or sudden withdrawal of hypnotics, sedatives,
and alcohol, the use of antiparkinson and antidepressant drugs are
all common causes.   Antidepressant drugs are often used at or
near to their toxic level and a toxic confusional state may result.
Even a single evening dose of 100 mg of amitriptyline or nortriptyline can produce toxic serum levels if renal function is impaired,[13]
a state of affairs frequently found in the elderly.   Estimation
of peak rather than pre-dose plasma levels would be a better guide
to treatment especially with a once-daily regimen[14].   Levels of
plasma nortriptyline above 150 µg/l appear to inhibit recovery
from depression[15].

Lithium is being used increasingly for the treatment of
hypomania and in the prophylaxis of cyclical mood swings in affective illness.   Bipolar illness is not so rare in old age as was
previously thought[16], and there is need to reassess the value of
lithium prophylaxis in these elderly patients.   Unfortunately
lithium also has a small therapeutic ratio with toxic levels not
far above the therapeutic.   In a stabilized patient, dehydration
and salt depletion as in hot weather or use of diuretics can
precipitate gross toxicity.   Tonks[17] described a case of a 60 year
old man, stabilized on lithium, who became mentally confused,
ataxic and with slurred speech and tremor after three days dieting
and saunas at a "health farm" even though the dangers had been
spelled out to him by his medical adviser before he undertook this
hapless venture.   Thomas[18] described gross iatrogenic brain failure
in a 58 year old female stabilized on lithium for depression for
seven years who was given haloperidol 1.5 mg thrice daily for a
hypomanic swing and within two days developed severe mental
confusion with total disorientation associated with marked muscular
rigidity and orofacial dyskinesia.   Both drugs were stopped but
months later there was still evidence of severe organic brain
damage.   Curiously enough three years prior to this catastrophe
she had been prescribed haloperidol 5 mg four times in the day for
three weeks because of a hypomanic swing and on that occasion had
suffered no ill effects.

The antiparkinsonian drugs levodopa and amantadine as well as
the older anticholinergics not uncommonly produce gross mental
disorder often with hallucinations.   The latter are characteristically anthropomorphic, life size and menacing. A reduction of dose
usually relieves the psychosis.   Sudden cessation of drug treatment is contraindicated.   Anticholinergic type antiparkinson
drugs are widely used in psychiatry to prevent or control parkinsonism induced by the neuroleptics.   This practice is being called
in question as is the apparently excessive use of the neuroleptic
drugs themselves.   Elderly people with senile dementia already
have central cholinergic failure, the phenothiazine themselves are
to a variable degree anticholinergics and to add yet another
anticholinergic produces an effect to the patient's detriment.
Even anticholinergic eye drops (atropine, homatropine etc.) can
precipitate an acute confusional state.   There are well documented,
but usually forgotten dangers of impaired thermoregulation with
neuroleptic and anticholinergic drugs.   The high potency
phenothiazines suppress thalamic thermoregulation and the low
potency drugs tend more to interfere with peripheral adrenergic and
cholinergic mechanisms.   Patients on these drugs are at especial
risk in hot humid weather of heat hyperpyrexia and brain damage[19].

Digitalis glycosides are commonly thought to cause a toxic
confusional state and they are often used unnecessarily in elderly
patients.   As with the phenothiazines and barbiturates there is a
growing conviction that during the past 30 years doctors have been
over zealous in the widespread use of these drugs.   Admittedly it
is difficult to be sure in an individual case that the digitalis is
the cause of the toxic confusional state because there may be so
many other aetiological factors operating viz: hypoxia, electrolyte
imbalance and other concurrent medication.   Bellair et al[20] in a
prospective study of digitalis toxicity found no difference in
mental disturbance between his toxic and non-toxic groups of
patients.

Affective Disorders

There are many reports in the literature of mental depression
due to drugs especially in the treatment of hypertension.   The
likely rationale being depletion of, or interference with the action
of catecholamines in the brain.   Other possible explanations
include the feeling of listlessness and malaise, associated with
treatment and a fall in blood pressure, being interpreted as depression and the effect of anxiety as a precipitant of depression
or symptoms suggestive of it.   The central amine depletion effect
is seen with reserpine and methyldopa. However propranolol with its
presumed mainly peripheral beta-adrenergic blocking activity also
produces depression and it may well be that hypertensives have
depression-prone personalities.   Severe depressive reactions with

reserpine, amounting to an affective psychosis with clinical
features identical to those of primary endogenous depression, have
been reported from time to time since the 1950s.   Even doses as
low as 250 micrograms daily are not beyond suspicion with respect
to this adverse effect.   Methyldopa is less of a hazard but mild
to moderate depression is present in about 6% of cases   together
with insomnia and nightmares.   Phenothiazines which block
norepinephrine and dopamine receptors may produce depression as
may phenobarbitone, procainamide, digitalis, indomethacin and
chronic mild hypoglycaemia by different mechanisms (v.infra).

There has been a great debate over the years concerning the
alleged mental depression suffered by younger women on oral
contraceptives and mostly of mild to moderate degree.   The balance
of evidence   suggests that only a very small proportion of women
suffer adverse reactions because of the actual drugs and it is not
known if it is the oestrogen or the progesterone which is the
culprit.   The importance of a placebo effect appears to have been
underestimated in many reports.   Further evaluation of the hazards
of oestrogen therapy is of importance in the domain of geriatrics
because of the increasing use of hormone-replacement therapy (HRT)
in selected cases during and after the menopause[22].

Depression has been reported with many other drugs including
phenylbutazone and sulphonamides and in the 'let-down' period on
withdrawal of amphetamine and more especially the appetite
suppressant fenfluramine (Ponderax)[23].

Euphoria, hypomania and mania also occur due to drugs.   Minor
mood elevations are common (e.g. with levodopa) but the more extreme
mood evaluations are quite rare in the elderly presumably because,
in the presence of other potential causes of brain failure such as
structural damage, mental confusion or delirium supervenes.
Antidepressants, corticosteroids, amantadine and pentazocine
(Fortral) are all known to produce euphoria and occasionally hypomania but rarely mania.

Paranoid Psychoses

The drug-induced paranoid psychoses in all elderly people
usually occur against a background of clouding of consciousness and
disorientation and in this respect are unlike idiopathic schizophrenia.   The constellation of first-rank symptoms (auditory
hallucinations, delusions of passivity and interference, thought
alienation and delusional perception with clear consciousness and
good orientation) as seen in acute (Type 1) schizophrenia[24] is not
seen in elderly people although it characterises the amphetamine
abuse psychosis seen mostly in younger people.   Large doses are
required to produce this effect but the induction time is short

- days and even within twenty four hours smaller (medical) doses of amphetamine can exacerbate existing schizophrenia as can methyl-phenidate (Ritalin).   These drug effects simulating those of dopamine together with the undoubted antipsychotic action of dopamine receptor blockade and the postmortem evidence of a consistent increase in certain dopamine receptors in patients with schizophrenia give support to the dopamine hypothesis of molecular pathology of schizophrenia[24,25] .   However in elderly persons paranoid delusions and hallucinations due to drugs occur mostly but not exclusively in the context of an acute confusional state (q.v.).

Withdrawal Syndromes

Sudden cessation of drugs can produce an acute mental deterioration.   Many patients have become dependent upon one or more drugs.   In the elderly those likely to be implicated are barbiturates, dextropropoxyphene or a benzodiazepine but nearly all the antianxiety drugs apart from the beta-adrenergic blockers can cause dependence.   Withdrawal symptoms come on in a day or two after stopping and include tremor, agitation, headache and nausea. It seems not generally known that these psychosedative drugs in everyday use have considerable addictive potential.   Even 2 weeks of chronic use will be sufficient to cause withdrawal reactions if the drug is suddenly stopped.   Where the drug dose has been high sudden cessation may produce an acute paranoid psychosis and even convulsions and coma[26].   The antispasticity agent baclofen is a gamma-amino-butyric acid (GABA) analogue and although marketed mainly for the treatment of spinal spasticity is also used in spasticity of cerebral origin.   It is known to depress dopamine metabolism in animals and its use clinically may exacerbate pre-existing psychotic states and epilepsy.   Sudden withdrawal of baclofen has been reported to cause acute paranoid psychosis[27] but in the setting of an acute confusional state as is usually the case with drug-induced paranoid psychosis.

Dementia and Stroke

As already indicated psychosedative drugs readily produce dependence and (presumably as a result) are one of the most frequently prescribed groups of drugs.   They tend to be prescribed more often and for longer periods in older patients.   Longterm use leads to a picture of impaired mental ability, dysarthria, ataxia, poor judgement and a liability to accidents.   It is a picture which mimics arteriosclerotic dementia.   This iatrogenic disorder must be common because the numbers exposed are so large that even a small percentage incidence of brain failure in the chronic takers of psychosedatives represents a large burden of iatrogenic disability. Action - especially improved education, including appropriate motivation of doctors - is needed to reduce the total incidence

of these adverse reactions[28,29].   A bizarre syndrome of brain
failure is seen with lithium carbonate treatment, viz. mental
confusion, slurred speech, ataxia, and when toxicity increases,
psychosis, convulsions and coma.   Serious interactions can occur
with haloperidol, phenytoin and diazepam[30].

A chronic encephalopathy due to hypoglycaemia may produce a
picture of pseudodementia.   Turkington[31] described three cases of
middle aged diabetics on oral antidiabetic drugs who were so
affected to the extent that all were suspected of chronic
alcoholism by the respective employers!   The illness lasted for
some months until the diagnosis was made and appropriate reduction
of antidiabetic drug effected.   The clinical features were
intellectual impairment, memory loss, slurred speech, staggering
gait and mental confusion.   No satisfactory alternative explanatic
was found and all three remained well over the two to five year
period of subsequent observation.

Coma can occur with oral antidiabetics as with insulin.
Sometimes the precipitating factor is the introduction of another
drug (e.g. phenylbutazone) with a greater plasma binding affinity
which causes the antidiabetic agent to circulate in overdose.
Additionally coma may occur with lactic acidosis in treatment with
the biguanides especially phenformin.   This danger has undoubtedly
been underestimated in the past and it seems preferable never to
prescribe phenformin for elderly diabetics because it is impossible
to predict this very serious complication[32].

Stroke syndromes occur with a variety of drug treatments by
causing hypotension (vasodilators and antihypertensives) thrombosis
(oestrogens) or bleeding due to anticoagulants or due to acute
hypertension (pressor agents interacting with monoamine oxidase
inhibitors).   The monoamine oxidase inhibitors (MAOI) are not
generally recommended for use in the elderly because of the evident
dangers.   Nevertheless they have their advocates[33] and are said to
be especially effective in treatment of depression associated with
dementia as well as being better tolerated than the tricyclics with
that group of patients.

Abrupt reduction of systemic blood pressure is likely to
precipitate cerebral infarction.   In normal people cerebral blood
flow remains satisfactory despite the major swings in arterial
pressure which occur during the course of daily activities.   This
maintenance of cerebral perfusion is due to cerebral vascular autoregulation[34].   This regulation appears to depend in part on
autonomic function and in part on metabolic changes.   A fall in
arterial PCO_2 causing vasoconstriction and a rise causing vaso-
dilatation.   However in many elderly patients cerebral vascular
autoregulation is impaired unilaterally or bilaterally to the exter

that quite modest reductions in blood pressure put the patient at
risk of transient brain failure - giddiness, faintness, blackouts
and confusional states.   Prolonged inadequate perfusion will produce
permanent brain damage.   Wollner and his colleagues[35] were able to
demonstrate a reduction of 32 - 67% in mean cerebral blood flow in
four elderly patients with symptoms of postural hypotension with a
fall of only 9 - 33 mm Hg in mean arterial blood pressure.   By
contrast four elderly patients with no symptoms of postural
hypotension but with a similar reduction of 10 - 31 mm Hg in mean
arterial pressure showed no significant change in mean cerebral
blood flow.   Jackson et al[36] described six elderly patients with
symptomless hypertension with systolic pressures ranging from
160 - 220 and diastolic pressures from 80 - 120 mm Hg who were
prescribed treatment (methyldopa 250 mg thrice daily in five cases
and oxprenolol 160 mg thrice daily in the sixth case) by the family
doctors.   Within a week all were admitted to hospital as emergencies
on account of episodes of loss of consciousness, mental confusion and
inability to walk.   Prior to admission all had developed symptoms
of postural hypotension and had become housebound.   On cessation
of drug treatment all recovered except one who was left with a
homonymous hemianopia.   The authors wonder how many strokes world
wide are due to medical treatment.

Extrapyramidal Syndromes

    Four distinct types of drug-induced extrapyramidal syndromes
(EPS) have been recognised: 1) Parkinsonism which mimics the
idiopathic variety, 2) akathisia or uncontrollable restlessness
which mimics agitation, 3) acute dystonic reactions most often seen
in the very young and 4) the late EPS, tardive dyskinesia with a
very high incidence in elderly patients on antipsychotic drugs.
These syndromes are produced by the antipsychotic drugs to such an
extent that they constitute a major problem in the long term use of
the drugs.   Space does not allow a detailed description.   They
fall within the definition of drug-induced brain failure but their
effect is much less on intellect, affect and cognitive function
than the other ADRs which form the main body of this paper.

SUMMARY

    Iatrogenic brain failure is a common problem in sick elderly
people on drug treatment.   The precise incidence is unknown because
of difficulty in recognition against the background of a high
incidence of naturally occurring disease and the lack of thorough
surveillance in those places where ill old people are treated.
Many of the adverse reactions are predictable but doctors are so
ill grounded in geriatric psychiatry and clinical pharmacology that
they unwittingly put their patients at risk.   A high incidence of

of suspicion (together with a sound knowledge of the more likely
ADRs) is required to diagnose and treat these disorders to prevent
further and possibly irreparable loss of brain function.

REFERENCES

1. R. Barton,"Instutional Neurosis",Wright, Bristol (1976).
2. D. M. Bowen and A. N. Davison, Biochemical changes in the normal
   ageing brain and in dementia in:"Recent Advances in Geriatric
   Medicine" B. Isaacs, ed., Churchill Livingstone, Edinburgh
   (1978).
3. S. I. Finkel, Geriatric psychiatry training for the general
   psychiatric resident, Amer.J.Psychiat. 135:101 (1978).
4. R. B. Teeter, F. K. Garetz, W. R. Miller, W. F. Heiland,
   Psychiatric disturbances of aged patients in skilled nursing
   homes, Amer.J.Psychiat. 133:1430 (1976).
5. C. R. Smith, Use of drugs in the aged, Johns Hopkins Med.J.
   145:61 (1979).
6. C. R. Martys, Adverse reactions to drugs in general practice,
   Br.Med.J. 2:1194 (1979).
7. S. M. Shaw and L. J. Opit, Need for supervision of the elderly
   receiving long-term prescribed medication, Br.Med.J. 1:505
   (1976).
8. J. Porter and H. Jick, Drug-induced anaphylaxis, convulsions,
   deafness and extrapyramidal symptoms, Lancet 1:587 (1977).
9. M. Levy, M. Lipshitz and M. Eliakim, Hospital admissions due to
   adverse drug reaction, Amer.J.med.Sci. 277: 49 (1979).
10. J. Williamson and J. M. Chopin, Adverse reactions to prescribed
    drugs in the elderly. A multicentre investigation. Paper
    presented to the Spring Meeting of the British Geriatrics
    Society (1977).
11. H. A. McClelland, Psychiatric disorders in: "Textbook of
    adverse drug reactions", D. M. Davies, ed., Oxford University
    Press, Oxford (1977).
12. J. B. MacDonald and E. T. MacDonald, Nocturnal femoral fracture
    and continuing widespread use of barbiturate hypnotics,
    Br.Med.J. 2:483 (1977).
13. A. C. Carr and R. P. Hobson, High serum concentrations of
    antidepressants in elderly patients, Br.Med.J. 2:1151
    (1977).
14. I. H. Stevenson and A. A. Schiff, Plasma drug levels on once
    daily dosage, Br.Med.J. 2:579 (1977).
15. S. A. Montgomery, R. A. Braithwaite, J. C. Cramer, Routine
    nortriptyline levels in treatment of depression, Br.Med.J.
    2:166 (1977).
16. K. Shulman and F. Post, Bipolar affective disorder in old age,
    Br.J.Psychiat. 136:26 (1980).
17. C. M. Tonks, Lithium intoxication induced by dieting and saunas,
    Br.Med.J. 2:1393 (1977).

18. C. J. Thomas, Brain damage with lithium/haloperidol, Br.J.
    Psychiat. 134:552 (1979).
19. S. C. Mann and W. P. Boger, Psychotropic drugs, summer heat
    and humidity and hyperpyrexia; A danger restated,
    Amer.J.Psychiat. 135:1097 (1978).
20. G. A. Bellair, T. W. Smith, H. W. Ableman, E. Haber and
    W. B. Hood, Digitalis intoxication, New Engl.J.Med.
    284:989 (1971).
21. K. G. Granville Grossman, ed.,"Recent Advances in Clinical
    Psychiatry", Vol. 4, p.199 Churchill Livingstone, London
    (1971).
22. J. Studd, D. Oram and S. Chakravorti, eds., Management of
    the menopause, Curr.Med.Res.Opin. 3:Suppl.3 (1975).
23. J. M. Steel and M. Briggs, Withdrawal depression in obese
    patients after fenfluramine treatment, Br.Med.J. 3:26
    (1972).
24. T. J. Crow, Molecular pathology of schizophrenia: more than
    one disease? Br.Med.J. 1:66 (1980).
25. S. H. Snyder, The dopamine hypothesis of schizophrenia: Focus
    on the dopamine receptor, Amer.J.Psychiat. 133:197 (1976).
26. M. DeBard, Diazepam withdrawal syndrome: A case with psychosis,
    seizure and coma, Amer.J.Psychiat. 136:104 (1979).
27. A. J. Lees, C. R. A. Clarke and M. J. Harrison, Hallucinations
    after withdrawal of baclofen, Lancet 1:858 (1977).
28. T. H. Bewley, Drug dependence caused by medical treatment, in:
    "Drug-induced diseases", Vol. 4,L. Meyler, H.M. Peck, eds.,
    Excerpta Medica, Amsterdam (1972).
29. T. Maruta, Prescription drug-induced organic brain syndrome,
    Amer.J.Psychiat. 135:376 (1978).
30. E. M. R. Critchley, Drug-induced neurological disease,
    Br.Med.J. 1:862 (1979).
31. R. W. Turkington, Encephalopathy induced by oral hypoglycaemic
    drugs, Arch.int.Med. 137:1082 (1977).
32. E. A. M. Gale and R.B. Tattersall, Can phenformin-induced
    lactic acidosis be prevented? Br.Med.J. 2:972 (1976).
33. J. W. Ashford and C. V. Ford, Use of monoamine oxidase
    inhibitors in elderly patients, Amer.J.Psychiat. 136:1466
    (1979).
34. L. Hanson and M. Henning, eds., Negative consequences of blood
    pressure reduction, Acta med.Scand.Suppl. 628:17 (1978).
35. L. Wollner, S. J. McCarthy, N. D. W. Soper and D. J. Macy,
    Failure of cerebral autoregulation as a cause of brain
    dysfunction in the elderly, Br.Med.J. 1:117 (1979).
36. G. Jackson, T. A. Pierscianowski, W. Mahon and J. Condon,
    Inappropriate anti-hypertensive therapy in the elderly,
    Lancet 2:1317 (1976).

CONFUSIONAL STATES IN THE ELDERLY

H.M. Hodkinson

Professor of Geriatric Medicine
Royal Postgraduate Medical School
Hammersmith Hospital, London

Confusional states (here used in accordance with British geriatric practice in preference to such synonyms as toxic confusional state, acute brain syndrome, acute psycho-organic syndrome, or acute delirium) are particularly common in elderly patients.   Thus Bergmann and Eastham[1] found 16% of patients over 65 admitted to a medical department to have confusional states. Lloyd[2] had reported a 10% incidence of confusional states in patients admitted to geriatric departments in the preliminary study of a Royal College of Physicians multicentre survey. However, Hodkinson[3], reporting the main survey, showed by the use of serial mental test scoring that the true frequency was in fact far higher - of the order of 35% within the first month of admission. This finding emphasises both the high frequency of confusional states and also the tendency for the condition to be underdiagnosed. Failure to recognise confusional states is of some practical importance, for they are often the principal clinical presentation of organic illness in old people.   Indeed the delirium of a confusional state is a key symptom in this age group, often replacing pain, fever, or other symptoms more characteristic of illness in younger age groups.

PSYCHIATRIC FEATURES OF CONFUSIONAL STATES

The cardinal feature of the confusional state is clouding of consciousness. This, Lishman[4] describes as "the mildest stage of impairment of consciousness in the continuum from full alertness and awareness to coma".   It comprises a mild but global impairment of cognitive function as well as a reduction of awareness and may include drowsiness, though this is quite often absent in the

confusional states seen in elderly people.   Variability in
the degree of mental impairment is characteristic of confusional
states.   Attention is impaired and the patient is distractable.
Perception is defective so that misinterpretations, visual illusions
or, less often, visual hallucinations occur.   Thought processes are
disturbed so that there is chaotic thinking, lack of reasoning
ability, incoherence, impaired grasp and disorientation.   Memory
processes are defective in that there are difficulties of memory
recall, but also of registration so that there may be periods of
amnesia remaining after recovery.   Restlessness, sleep disturbance,
anxiousness and paranoid symptoms may also occur.   Classical
descriptions contrast the relative richness of thought processes in
confusional states with their poverty in dementia.   To quote Roth
and Myers[5]

> "psychic life in clouding is full, sometimes extravagantly so,
> but distorted: the patient lives in a private world of
> shifting experiences richly supplied with detail from the
> resources of a brain which is essentially intact...........
> In dementia, by contrast, the patient lives in the actual
> world but does so deficiently".

However, it is generally agreed that such distinctions are often
difficult to make in practice.   Hodkinson[3] found no difference in
the psychiatric symptomatology of confusional states as compared to
demented patients.   Subsequent progress of the illness may often
be a more useful guide, recovery clearly distinguishing a confusional
state from a dementing illness.   However, there may be real
difficulties where confusional states, e.g. those occurring in
association with carcinomatosis, are persistent and progressive when
they may closely mimic dementia; but, even in such circumstances they
typically have a more rapid time course.   Serial mental test scores
can certainly aid in the recognition of confusional states.

CAUSES OF CONFUSIONAL STATES

Confusional states may result from acute disorders of the brain
itself.   Thus they may result from cerebrovascular accidents, from
transient ischaemic attacks, in post-epileptic states, following
acute trauma, or in association with space occupying lesions, or
infections such as encephalitis, abscess or neurosyphilis.

Far more commonly, however, they are primarily due to extra-
cerebral disease though there may be important predisposing factors.
Thus the Royal College of Physicians Study[3] showed that confusional
states occurred in significantly older patients.   In addition those
with confusional states were far more likely to have Parkinson's
Disease or some pre-existing dementia, the incidence of confusional
states in the first month of admission being 63% and 37% respectively

However, there was no increase in the incidence of confusional states in those patients with pyramidal signs. Sensory deprivation is also believed to favour the development of confusional states. Bergmann and Eastham[1] found this for deafness in their series whilst Hodkinson[3] found that both deafness and blindness were associated with confusional states. In the same study it was also found that there was a significant association between the development of confusional states and the diagnosis of depression. Similarly, Bergmann and Eastham[1] had found an association with previous psychiatric illness and this of course included previous depression. This suggests that depression might in some way predispose to confusional states. However, there are alternative explanations for the association as depression is itself associated with severe physical illness, and furthermore treatment with antidepressant drugs might be responsible for confusional states.

Though predisposing causes are important, confusional states are very clearly associated with more severe physical disease. This can be concluded from earlier descriptive studies[6,7,8], but has been formally demonstrated by Hodkinson and Bergmann and Eastham[1]. These latter authors were able to show that physical disease was most severe in those patients who had confusional states without any pre-existing cerebral disease.

Though a wide variety of physical disease may produce a confusional state, a number of diseases are particularly likely to be responsible in the elderly. Thus significant associations between diagnostic group and confusional states were found for pneumonias, heart failure (congestive heart failure and left ventricular failure taken together), urinary infections and carcinomatosis in the Royal College Survey[3]. This confirmed findings of earlier surveys[6,7,8]. Of course many other diseases may also give rise to confusional states in this age group.

MECHANISMS OF PRODUCTION OF CONFUSIONAL STATES

The mechanisms by which physical illnesses or agents produce confusional states are poorly understood. Some effects can be plausibly explained. Thus the effects of anoxia (due to severe anaemia, respiratory anoxia or impaired cerebral perfusion) and of hypoglycaemia are readily understood as cerebral metabolism is chiefly dependent on glucose and oxygen. Deficiencies in hormones or vitamins can also be supposed to have direct effects on cerebral metabolism as for example in the "myxoedematous madness" of hypothyroidism, B12 deficiency, Wernicke's encephalopathy or Korsakov psychosis. It is also reasonable to assume that cerebral metabolism is disturbed by a variety of other metabolic abnormalities such as hypoglycaemia, electrolyte acid base, or

osmolality disturbances and dehydration.   We may postulate toxic
mechanisms in many physical illnesses - for example uraemia, hepatic
failure or hypercapnia - and drug toxicity is also of great
importance.   It is noteworthy that the drugs known to act on
neurotransmitter systems, especially on the cholinergic system, are
particularly likely to give rise to confusional states, e.g. the
anticholinergic drugs, L dopa and amantidine.   Other drugs with
central nervous system actions, such as antidepressants, sedatives,
tranquillisers, anticonvulsants and centrally acting analgesics
are also potent causes of confusional states.   However, other drugs
(e.g. digoxin) may also be responsible.   Drug withdrawal is also
relevant as in the delirium tremens of alcohol withdrawal.   Body
temperature also appears to be involved in the development of
confusional states.   Hypothermia almost invariably results in a
confusional state whilst febrile illnesses are often also associated
with them.   However, experimental fever has little effect on mental
state and toxic or other mechanisms may underlie the clinical
association between fever and confusion.   In individual disease
states, however, we are often uncertain as to how the confusional
state is produced.   For example respiratory infection might act
by a number of mechanisms, by producing anoxia, by giving rise to
hypercapnia, or by the effects of fever itself, or by the release of
bacterial toxins, or by other more subtle metabolic disturbances.

PROGNOSIS

As confusional states are often associated with severe physical
illness it is not surprising to find a high early mortality.   This
has been noted in all clinical studies, for example Hodkinson[3]
found a 25% mortality in the first month after admission compared to
$12\frac{1}{2}$% for those who were mentally normal at admission.   Furthermore
mental test score is a powerful prognostic indicator in patients
admitted to geriatric departments, low scores being associated with
higher mortality.   If, however, the precipitating physical illness
does not result in death there is usually an excellent mental
recovery within a matter of days or weeks and discharge prospects
are generally good.   Indeed, the proportion of patients discharged
within a month is only marginally lower for confusional state
patients than for the mentally normal, 47% and 53% of survivors
respectively, and these figures are markedly better than those for
demented patients of whom only 26% are discharged within a month[3].

CONCLUSIONS

Confusional states are a major feature of illness in old age.
Physical illness complicated by a confusional state probably has a
worse prognosis and, because of associated incontinence and
increased dependency, there is a considerable extra burden of care

whether this is being provided at home or in hospital.   Despite
this obvious practical importance, however, there has been little
research into confusional states.   More detailed study of the
mechanisms by which physical illnesses produce confusional states
might aid management.   Similarly, drug induced confusional states
might repay close study, particularly with regard to the changes in
neurotransmitter function.

REFERENCES

1. K. Bergmann and E. J. Eastham, Psychogeriatric ascertainment
    and assessment for treatment in an acute ward setting,
    Age & Ageing 3:174 (1974).
2. C. M. Lloyd, Royal College of Physicians Study of Mental
    Impairment in the Elderly: Report of findings of pilot survey,
    (1970).
3. H. M. Hodkinson, Mental impairment in the elderly, J.Roy.Coll.
    Physcns.Lond. 7:305 (1973).
4. W. A. Lishman, "Organic Psychiatry", Blackwell Scientific
    Publications, Oxford (1979).
5. M. M. Roth and D. H. Myers, The diagnosis of dementia,
    Brit.J.Hosp.Med. 2:70 (1969).
6. D. K. W. Kay and M. Roth, Physical accompaniments of mental
    disorder in old age, Lancet 2:740 (1955).
7. F. J. Flint and S. M. Richards, Organic basis of confusional
    states in the elderly, Br.Med.J. 2:1537 (1956).
8. J. N. Agate, Report of the 4th Congress of the International
    Association of Gerontology (Merano) 4:54 (1957).

DEMENTIA DUE TO NORMAL PRESSURE HYDROCEPHALUS

B. Guidetti and F. M. Gagliardi

Neurosurgical Institute of Rome Medical School
Rome, Italy

With the rise in average life expectancy dementia has become a serious social problem affecting some 5-6% of people over the age of 65[1], which explains the revival of interest in this humiliating and disabling disease. Unfortunately, although much more is now known about the condition, few of these patients receive adequate treatment, apart from a small group in whom the dementia is obviously related to a disturbance of the cerebrospinal fluid (CSF) circulation with consequent hydrocephalus. A simple CSF shunting procedure secures recovery of function in such patients, sometimes to a truly remarkable degree, but the selection of suitable patients is by no means easy and the operation, though simple, can have serious complications, the chief of which is a collection of CSF and blood in the subdural spaces[2,3,4]. The difficulties involved in correct selection become clear when we consider the post-operative success rates achieved by the workers who have concerned themselves with the subject and reported their results. Table 1 shows that about 60% of patients selected for shunting procedure benefit from it. Hence the need to identify as accurately as we can the clinical and laboratory criteria of selection of patients for surgical treatment.

Our aim in this paper is to assess the clinical and laboratory findings in our shunted patients in the light of the results.

PATIENTS

Between January 1967 and end-December 1977 fifty-eight patients suffering from presenile or senile dementia, which we attributed to a disturbance of the CSF circulation, were operated on for ventri-

161

Table 1.   Positive Percentage Results of Shunting Treatment in NPH

Ojemann et al.	(1969)	65%
Benson et al.	(1970)	64%
Shenkin et al.	(1973)	64%
Messert et al.	(1974)	70%
Stein et al.	(1974)	37%
Symon et al.	(1974)	50%
Belloni et al.	(1976)	63%
Jacobs et al.	(1976)	72%
Laws et al.	(1977)	50%
Guidetti et al.	(1979)	58%

Average: 59.4%

culoatrial or ventriculoperitoneal shunt by means of a Pudenz medium pressure flushing device.   One patient died 35 days after the operation and necropsy established degeneration of the brain. Three patients at varying intervals after surgery presented a large collection of blood and CSF in the subdural spaces, which was removed.   Five patients, after a good start, showed clear signs of relapse due to occlusion of the shunt but made a satisfactory functional recovery after repositioning of the device.

All but one of our patients have been followed up for anything from 2 to 12 years.   The long term results fall into three categories (Table 2).   Group 1, made up of 34 cases, comprises the patients who did well after the operation, nearly all of them being able to resume their former occupations.   Group 2 (8 cases) includes those who showed some improvement but not enough to resume their former occupations.   The 16 cases of group 3 derived no benefit and the disease took its natural course, in some cases after an initial deterioration.

ETIOLOGY

It is evident from Table 3 that shunting was most successful in patients with a history of brain disease.   Other workers[4,5,13] have found the same, the inference being that previous episodes of subarachnoid bleeding, meningitis, head injury or brain surgery in a patient presenting signs of dementia some months or years later should alert one to the possibility that the dementia stems from

Table 2. Results of Shunting in 58 Patients with NPH

	No. of Patients	Percentage
Group 1	34	(58.6%)
Group 2	8	(13.8%)
Group 3	16	(27.6%)

an upset of the cerebrospinal fluid circulation and prompt diagnostic investigations that will confirm or exclude this clinical suspicion.

The non-emergence of a causal factor does not rule out the possibility of successful treatment, but undoubtedly lessens the chances of this.

CLINICAL PATTERN

The clinical features of a patient suffering from normal pressure hydrocephalus differ little from those found in patients with other forms of dementia: loss of memory and of attention, of the capacity for calculation and judgment, disorientation in time and space, loss of will-power and so on and these signs are not reliable criteria for differential diagnosis, except for the fact

Table 3. Etiology and Results

	No. of Patients	Group 1	Group 2	Group 3
S.A.H.	13	12	–	1
Trauma	10	5	1	4
Infection	5	3	2	–
Post Surgery	7	2	3	2
	35	22	6	7
Idiopathic	23	12	2	9
	58	34	8	16

that patients with normal pressure hydrocephalus usually have less
severe dementia, in some cases confined to impairment of memory and
attention, and the course is more rapid and progressive.   But this
is far from being a rule: a case in point was a professor of
agriculture, who was so demented as to eat his own faeces. After
several investigations he was shunted and fared so well that a few
months later he was able to go back to teaching.   He had a relapse
a year later due to occlusion of the ventricular catheter, but the
dementia was not so bad as before and, once the catheter had been
cleared, functional recovery was excellent.   The patient is still
well and teaching 7 years after the operation.

If a dementia due to normal pressure hydrocephalus is difficult
or even at times impossible to differentiate from dementia of other
origin, the simultaneous presence, as in the patients of group 1,
of balance and gait disturbance is a major discriminant.   These
troubles are described by the patient or his relative as difficulty
in standing upright and in walking, reeling, fear of falling and
difficulty in standing up.   The examiner finds unsteady stance and
walking, some tendency to lurch forward and backward, difficulty
in assuming the erect position, an ataxic gait with more or less
marked cerebellar signs and rigidity. As the disease advances,
patients need support in walking and ultimately are unable to stand.

As has been pointed out by Fisher[15] among others, the balance
and walking disturbances appear early and sometimes precede the
evidence of dementia.   These findings are of the utmost importance
for the purpose of differential diagnosis because in what are called
degenerative dementias these disturbances come on late in the
disease when the dementia pattern has assumed serious proportions.

In agreement with other workers we do not attach great
importance to the presence of urinary incontinence for the purpose
of differential diagnosis[4,15,16].   Urinary incontinence, due mainly
to inattentiveness, is a late sign anyway appearing both in patients
with a favourable and in those with an unfavourable course.   The
same applies to other disturbances found in our patients: epileptic
manifestations, extrapyramidal and pyramidal deficits, and so on.

In a word, medium-grade dementia, preceded or accompanied by
difficulty in standing and walking and with a history of brain
disease should suggest a diagnosis of normal pressure hydrocephalus.

Electroencephalography. This investigation which was performed
in all 58 patients supplied only generic evidence of brain distress
and nothing useful for differential diagnosis.

CSF investigations. A manometric test and chemical and cyto-
logical examination of the CSF were carried out in all the patients,
but they furnished no data of moment.   The CSF pressure was normal

Table 4. Clinical Findings and Results

	No. of Patients	Group 1	Group 2	Group 3
Typical Syndrome	40	25	6	9
Atypical Syndrome	18	9	2	7

or below normal in all patients.   The only fact worth noting, and indeed noted by other workers[4,17,18] is a shortlived improvement in the clinical pattern after removal of CSF, found in 6 of our patients.   An interesting coincidence is that all 6 benefitted from the operation.

We have obtained, as well as other workers[4,9,19,20], discordant information from the Katzman and Hussey[21] infusion test for the measurement of CSF pressure, which we used in a small number of cases to start with and later abandoned.   Of greater value both for prognosis and treatment are the data obtained from continuous recording of the CSF pressure by means of a ventricular catheter or extradural tube.   In a high (60-70%) percentage of patients who made a good recovery after operation the records showed signs of intermittent bouts of hypertension suggesting that patients with so-called normal pressure hydrocephalus have moments of raised intracranial pressure.

Our experience of this method of pressure recording is not such that we can quantify our data and compare them with those of the literature[11,22,23,24,25],   partly because the records we obtained are not completely reliable due to artifacts produced by poor co-operation on the part of the patients.   To eliminate the snags of pressure recording by cable, we are now trying to record the signals transmitted by radio.

RADIOLOGICAL INVESTIGATIONS

Angiography. Angiographic studies were done in only 16 patients. Apart from furnishing generic information on the presence of hydrocephalus, angiography contributed nothing of importance to the differential diagnosis.   The only relevant datum is the greater tortuosity of the vessels in patients with dementia of degenerative origin than in those with CSF disturbance.   In 3 patients the investigation disclosed a megadolic basilar artery, which nicked the floor of the third ventricle, pushing it upward and giving the impression of obstructing the CSF flow.   All 3 patients benefitted from the shunting procedure, but unfortunately they had a previous

history of brain disease, which prevented confirmation of the
hypothesis of Breig et al[26], who attribute pathogenetic importance
to this vascular abnormality.

Pneumoencephalography (PEG). This investigation was done in 44
patients, great care being taken to ensure that air reached the
subarachnoid spaces.   In every case the air study highlighted the
presence of communicating hydrocephalus with more marked dilation
of the lateral ventricles, especially in their anterior portion,
which in no patient was less than 55 mm.   Without underrating the
value of the information to be gained from a study of the dimensional
ratios of the ventricular cavities and of the difference in callosal
angle between the PEGs of patients with normal pressure hydrocephalus and those with the degenerative variety, the most important
data for diagnostic and prognostic purposes are the presence or
absence of CSF block at the level of the cisterns of the base and
the presence or absence of air on the convexity.   On these criteria
the PEGs for our patients fall into three groups: Group A (22 cases)
marked by hydrocephalus and absence of air on the convexity even
some days after its injection; Group B (13 cases) with hydrocephalus
and arrest of the air in the cisterns of Sylvius or in the sub-
frontal sulci; Group C (9 cases) hydrocephalus and air in the sulci
of the convexity, widened in varying degree.

On comparing the PEG patterns with the results of treatment,
we find the best results in the patients with PEGs of group A, fair
results in those of group B and total failure in the group C cases
(Table 5).   Of some prognostic importance was the finding, already
pointed out by other workers[27,28], of temporary clinical deterioration, sometimes to the point of coma, noted in 12 patients, the
majority of whom benefitted from CSF shunting.

In short, a PEG pattern of hydrocephalus and air block in the
basal cisterns suggests normal pressure hydrocephalus, whilst
accumulation of air in the sulci of the convexity does not.   This

Table 5. PEG and Results

		No. of Patients	Group 1	Group 2	Group 3
PEG	A	22	16	3	3
PEG	B	13	8	3	2
PEG	C	9	-	-	9
		44	24	6	14

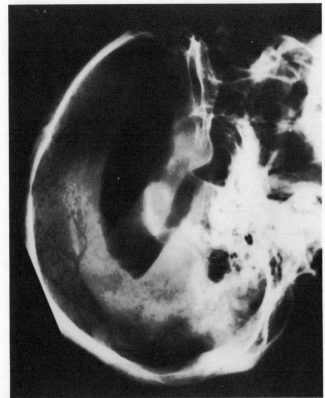

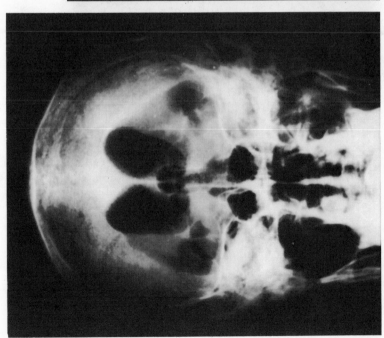

Fig. 1. (A and B)
PEG: Dilatation of ventricles and absence of air over the convexity.
Improvement after shunt.

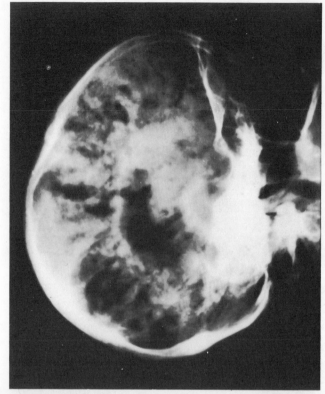

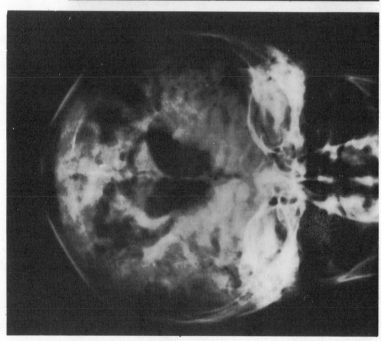

Fig. 2. (A and B)
PEG: Cortical atrophy with dilatation of the lateral ventricles.
No change after shunt.

always bearing in mind the possible errors and the risks that attend this investigation.

Radioisotope scanning of the cisterns. This diagnostic investigation, invented by De Chiro[29], less traumatic than PEG for studying the CSF circulation, was done in only 20 patients.   In all we used radioiodinated (^{131}I) human serum albumin (RIHSA) injected by lumbar route at the dose of about 200 microcuries.   As is known, the activity of the radioisotope is recorded about an hour later in the cisterna magna, 3 h later in the basal cisterns and 6 - 7 h later on the convexity.   24 h later radioactivity is recorded almost exclusively in the parasagittal areas, from which it gradually disappears around 48 h after injection.   The radioisotope diffusion differed in the 20 patients in whom this investigation was done and, as in the case of PEG, they fell into three groups: group A (17 cases) marked by early accumulation of the isotope in the ventricular cavities and none on the convexity even after 48 - 72 h; group B (1 case) with backflow of the isotope into ventricular cavities after 24 h and slow diffusion on the convexity; group C (2 cases) with a normal pattern, in which the tracer remained longer than usual in the sulci of the convexity.

As other workers have pointed out[6,30,31,32,33,34], the best results were observed in group A, although in 2 of the 17 patients the results were unsatisfactory, which suggests that one should not depend too much on this investigation (Table 6).

Computerised axial tomography(CTS).   CT scanning, done in only 10 patients by end-December 1977, revealed more or less marked hydrocephalus in all.   The sulci of the convexity were dilated in 3 patients and of normal or below-normal size in the other 7. Today CT scanning is the least traumatic investigation and the most informative in that it not only rules out an intracranial space-occupying lesion but also affords useful information on the degree of dilation of the ventricles and the width of the subarachnoid spaces of the convexity.   More detailed information can be gained by enhancing with Metrizamide by lumbar route and monitoring the diffusion and absorption.   This contrast medium was used in only 2 cases and so it is difficult to say whether it has decisive advantages over the uncontrasted investigation.

CT scanning is also of great value in postoperative follow-up for checking the course of the hydrocephalus and for detecting any collections in the subdural spaces.   As others[13,35] have already said, clinical improvement does not necessarily coincide with reduction of the ventricular cavities.   Quite the reverse, the marked reduction that occurred in some cases was accompanied by a collection of blood or CSF of varying proportions in the subdural spaces.

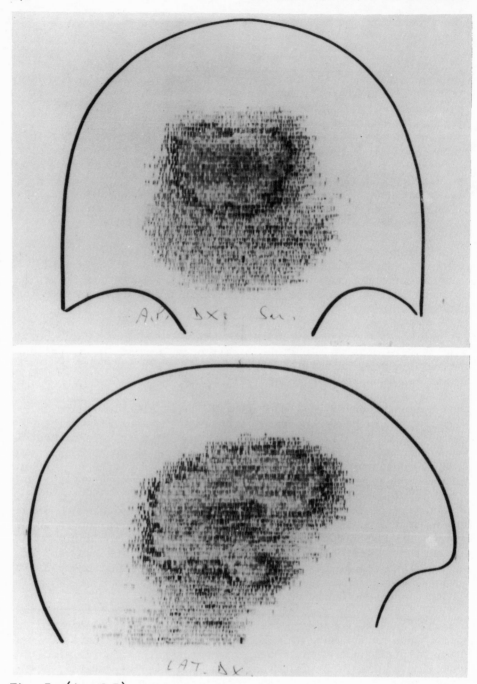

Fig. 3. (A and B)
    RIHSA Scan: Retention of the isotope in the lateral
    ventricles at 48 hours is shown. Improvement after shunt.

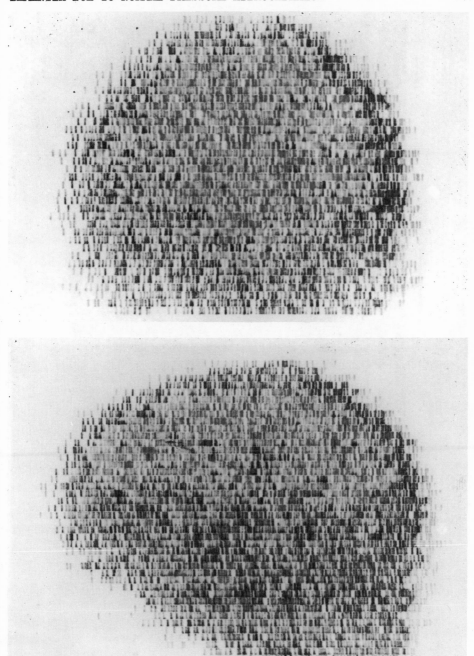

Fig. 4. (A and B)
    RIHSA Scan: Distribution of the isotope at 48 hours.
    The difficulty of resorption is shown. Moderate improvement
    after shunt.

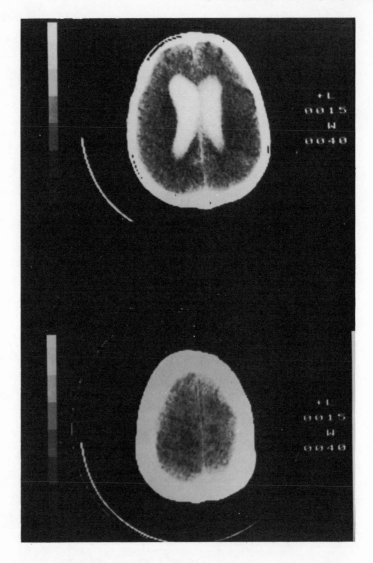

Fig. 5.   CT Scan with Metrizamide: Retention of dye in the
ventricular system is still apparent at 24 hours.
Improvement after shunt.

Cerebral blood flow. This investigation has so far béen of no
help in differentiating between hydrocephalus ex vacuo and normal
pressure hydrocephalus inasmuch as reduced blood flow was found in
both.

Table 6. RIHSA Scan and Results

	No. of Patients	Group 1	Group 2	Group 3
RIHSA Scan A	17	13	2	2
RIHSA Scan B	1	-	1	-
RIHSA Scan C	2	-	-	2
	20	13	3	4

SUMMARY AND CONCLUSIONS

We hope it is clear from what we have said that no single clinical, laboratory or instrumental investigation is capable of predicting with certainty the outcome of a shunting procedure. But several investigations, like the pieces of a jigsaw puzzle, put together, can add up with fair precision to a diagnosis of normal pressure hydrocephalus and so indicate surgery. The key features are:

1) A clinical history of brain disease in a demented patient should alert us to the possibility of normal pressure hydrocephalus. Of the 35 patients with such a history 28 (groups 1 and 2) fared well after shunting and 7 (group 3) did not. Of the 23 patients whose dementia was of unknown origin only 14 (groups 1 and 2) derived benefit from the operation (Table 3).

2) Dementia, as a rule of medium grade, accompanied or preceded by difficulty in standing and walking and sometimes by urinary incontinence are enough to warrant a suspicion of normal pressure hydrocephalus. Of the 40 patients presenting a typical clinical pattern 31 benefitted from shunting and 9 did not. Of those with an atypical clinical pattern only 11 responded to the shunting operation.

3) A PEG showing air block in the basal cisterns and no air in the sulci of the convexity (PEG AB) supplies useful guidance for treatment. Of the 35 patients presenting this PEG pattern 30 did well after surgery. On the other hand, the presence of air in widened sulci of the convexity argues for a degenerative hydrocephalus ex vacuo. None of the 9 patients with a lot of air in the sulci of the convexity benefitted from shunting.

4) The accumulation, following lumbar injection, of radioisotope in the ventricular cavities and none in the spaces of the convexity even after an interval of 48 - 72 h argue for normal pressure

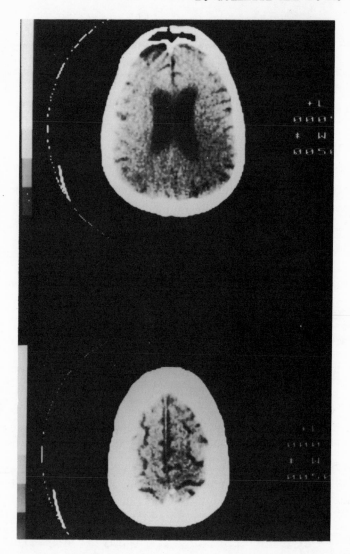

Fig. 6. CT Scan: Evident dilatation of the lateral ventricles and
         of the sulci of the cortex is shown. No change after shunt

hydrocephalus.   15 of the 17 patients with this RIHSA pattern
derived benefit from the operation.

     5) CT scanning is now the instrumental investigation of choice
in patients whose clinical picture points to normal pressure
hydrocephalus.   CT not only rules out the presence of an intracranial space-occupying lesion but also supplies useful information

on the presence or absence of hydrocephalus and on the status of the
sulci of the convexity.

6) Episodic hypertension on a continuous recording of intracranial pressure is a point in favour of normal pressure hydrocephalus.

REFERENCES

1. M. Roth, Aging of the brain and dementia: their implication in
    our society. Opening Lecture in Satellite Meeting on Aging
    of the brain and dementia, Florence (1979).
2. R. D. Illingworth, Subdural hematoma after the treatment of
    chronic hydrocephalus by ventriculocaval shunt, J.Neurol.
    Neurosurg.Psychiat. 33:95 (1970).
3. S. Samuelson, D. M. Long and S. N. Chou, Subdural hematoma as a
    complication of shunting procedures for normal pressure
    hydrocephalus, J.Neurosurg. 37:548 (1972).
4. L. Symon and T. Hinzpeter, The enigma of normal pressure
    hydrocephalus: tests to select patients for surgery and to
    predict shunt function, Clin.Neurosurg. 24:285 (1977).
5. R. G. Ojemann, C. M. Fisher, R. D. Adams, W. H. Sweet and
    P. F. J. New, Further experience with the syndrome of
    "normal" pressure hydrocephalus, J.Neurosurg. 31:279 (1969).
6. D. F. Benson, M. LeMay, D. H. Patten and A. B. Rubens,
    Diagnosis of normal-pressure hydrocephalus, N.Engl.J.Med.
    283:609 (1970).
7. H. A. Shenkin, J. Greenberg, W. F. Bouzarth, P. Gutterman and
    J. O. Morales, Ventricular shunting for relief of senile
    symptoms, J.Amer.Med.Assoc. 225:1486 (1973).
8. B. Messert and B. B. Wannamaker, Reappraisal of the adult occult
    hydrocephalus syndrome, Neurology (Minneap.) 24:224 (1974).
9. S. C. Stein and T. W. Langfitt, Normal pressure hydrocephalus.
    Predicting the results of cerebrospinal fluid shunting,
    J. Neurosurg. 41:463 (1974).
10. L. Symon and N. W. C. Dorsch, Use of long-term intracranial
    pressure measurement to assess hydrocephalic patients prior
    to shunt surgery, J.Neurosurg. 42:258 (1975).
11. G. Belloni, C. Di Rocco, C. Focacci, G. Galli, G. Maira and
    G. F. Rossi, Surgical indications in normotensive hydro-
    cephalus: a retrospective analysis of the relations of some
    diagnostic findings to the results of surgical treatment,
    Acta.Neurochir. (Wien) 33:1 (1976).
12. L. Jacobs, D. Contri, W. R. Kinkel and E. J. Manning, "Normal
    pressure" hydrocephalus: relationship of clinical and radio-
    graphic findings to improvement following shunt surgery,
    J.Amer.Med.Assoc. 235:510 (1976).
13. E. R. Laws,Jr., and B. Mokri, Occult hydrocephalus: results of
    shunting correlated with diagnostic tests,Clin.Neurosurg.
    24:316 (1977).

14. B. Guidetti and F. M. Gagliardi, Normal pressure hydrocephalus.
    Lecture in Satellite Meeting on Aging of the brain and
    dementia, Florence (1979).
15. C. M. Fisher, The clinical picture in occult hydrocephalus,
    Clin.Neurosurg. 24:270 (1977).
16. S. Jonas and J. Brown, Neurogenic bladder in normal pressure
    hydrocephalus, Urology, 5:44 (1975).
17. S. Hakim and R. D. Adams, The special clinical problems of
    symptomatic hydrocephalus with normal cerebrospinal fluid
    pressure: observations on cerebrospinal fluid hydrodynamics
    J.Neurol.Sci. 2:307 (1965).
18. J. P. Veron, Hydrocephalie a pression normale et symptomatol-
    ogie reversible, Press.Med. 77:551 (1969).
19. J. S. Wolinsky, B. D. Barnes and M. T. Margolis, Diagnostic
    tests in normal pressure hydrocephalus, Neurology (Minneap.)
    23:706 (1973).
20. S. E. Børgesen, F. Gjerris and S. C. Sørensen, The resistance
    to cerebrospinal fluid absorption in humans, Acta Neurol.
    Scandinav. 57:88 (1978).
21. R. Katzman and F. Hussey, A simple constant-infusion manometric
    test for measurement of CSF absorption. I. Rationale and
    method, Neurology (Minneap.) 20:534 (1970).
22. L. Symon, N. W. C. Dorsch and R. J. Stephens, Pressure waves
    in so-called low pressure hydrocephalus, Lancet 2:1291
    (1972).
23. J. C. Chawla, A. Hulme and R. Cooper, Intracranial pressure in
    patients with dementia and communicating hydrocephalus,
    J. Neurosurg. 40:376 (1974).
24. H. A. Crockard, K. Hanlon, E. E. Duda and J. F. Mullan,
    Hydrocephalus as a cause of dementia: evaluation by
    computerised tomography and intracranial pressure monitoring
    J.Neurol.Neurosurg.Psychiat. 40:736 (1977).
25. J. D. Pickard, G. M. Teasdale, M. Matheson and D. Wyper,
    Intraventricular pressure waves: the best predictive test
    for shunting in normal pressure hydrocephalus, Lecture at
    IV Intern. Symposium on Intracranial Pressure, Williamsburg,
    Virginia (1979).
26. A. Breig, K. Ekbom, T. Greitz and E. Kugelberg, Hydrocephalus
    due to elongated basilar artery, Lancet 1:874 (1967).
27. R. D. Adams, C. M. Fisher, S. Hakim, R. G. Ojemann and W. H.
    Sweet, Symptomatic occult hydrocephalus with "normal"
    cerebrospinal fluid pressure: a treatable syndrome.
    N.Engl.J.Med. 273:117 (1965).
28. R. L. Rovit, M. M. Schechter, B. Ortega and R. A. Brinker,
    Progressive ventricular dilatation following pneumoenceph-
    alography. A radiological sign of occult pressure
    hydrocephalus, J.Neurosurg. 36:50 (1972).
29. G. De Chiro, P. M. Reames and W. B. Matthews, RISA-ventriculo-
    graphy and RISA-cisternography, Neurology(Minneap.) 14:185
    (1964).

30. R. Bannister, E. Gilford and R. Kocen, Isotope encephalography
     in the diagnosis of dementia due to communicating hydro-
     cephalus, Lancet 2:1014 (1967).
31. F. E. Glasauer, G. J. Alker, E. V. Leslie and C. F. Nicol,
     Isotope cisternography in hydrocephalus with normal pressure,
     J.Neurosurg. 29:555 (1968).
32. M. Akerman, P. Derome and G. Guiot, Le transit radioisotopique
     dans les hydrocephalies. Application aux indications de la
     ventriculo-cisternostomie et au contrôle de son efficacité,
     Neuroshir. 16:117 (1970).
33. I. F. R. Fleming, R. H. Sheppard and V. Turner, CSF scanning
     in the evaluation of hydrocephalus: a clinical review of
     100 patients. Cisternography and Hydrocephalus.
     A Symposium, J. C. Harbert, D. C. McCullough, A. J.
     Luessenhop and G. De Chiro, eds., Charles C. Thomas,
     Springfield, III (1972).
34. B. Guidetti and F. M. Gagliardi, Normal pressure hydrocephalus,
     Acta Neurochir.(Wien) 27:1 (1972).
35. L. Jacobs and W. Kinkel, Computerized axial transverse
     tomography in normal pressure hydrocephalus, Neurology
     (Minneap.) 26:501 (1976).

CENTRAL NERVOUS SYSTEM DISTURBANCES IN MYXOEDEMA

M.G. Impallomeni

Consultant Geriatrician and Hon. Senior Lecturer
Royal Postgraduate Medical School
Hammersmith Hospital, London

Hypothyroidism may be accompanied by various forms of Central Nervous System disturbances. The purpose of this paper is to discuss the occurrence of coma in elderly patients suffering from profound myxoedema, and to describe thirteen such patients seen in a Geriatric Department in London over a three year period.

Coma in severe hypothyroidism was first described in 1880[1]; a second case was reported in 1884[2]. Both cases were reviewed in the Report of the Committee on Myxoedema of the Clinical Society of London in 1888 (Report on Myxoedema, Clinical Society of London)[3]. No further reports appeared between this date and 1953, when six new cases were reported[4,5]. Between 1953 and 1964 another 126 cases were reported[6].

Both the high prevalence of myxoedema in the elderly and the difficulty of its clinical diagnosis have been reported in the literature. It is now generally agreed that myxoedema occurs in some 2 - 3% of all admissions to Geriatric Departments in the United Kingdom[7,8]. There has however been no report of coma in these recently published series of myxoedema in the elderly.

PATIENTS AND METHOD

This is a retrospective study of thirteen elderly patients admitted to a Geriatric Department in London over a three year period. The criteria for inclusion in this study were:-
- The presence of coma not attributable to any other condition.
- The clinical and laboratory evidence of profound hypothyroidism.
- The response to thyroid hormone replacement.

During the same period of time the total number of admissions
to the Department was 4,293.   Myxoedema was diagnosed in 112
patients, which represents an incidence of 2.6%.   The Department
serves a population of around 250,000 with an estimated 31,000 over
the age of 65.

All blood tests reported in this paper were taken either while
the patients were in coma, or soon after they regained consciousnes
but always before treatment was instituted.   Blood urea, glucose,
plasma calcium and cholesterol were determined with a Technicon
Auto-analyser 'N' methodology.   Electrolytes were determined with
an EEL model 170 flame photometer.   Plasma thyroxine (T4) and
triiodothyronine (T3) resin uptake were determined with Thyopac 4
and Thyopac 3 methods respectively, as supplied by the Radiochemica
Centre, Amersham, England.   Arterial blood gases were determined
with a Radiometer BMS apparatus.   The blood gases values reported
for patients 6, 11 and 13 are corrected for the patients' low body
temperature.

Normal values were:

Sodium	135 to 145 mmol/l
Potassium	3.5 to 5.5 mmol/l
Bicarbonate	22 to 28 mmol/l
Calcium	2.25 to 2.65 mmol/l
Cholesterol	4 to 6.5 mmol/l (150 to 250 mg/dl)
PBI	270 to 600 nmol/l (3.5 to 7.5 µg/dl)
T4	75 to 180 nmol/l (5.6 to 14.3 µg/dl)
T3 resin uptake	99 to 123%, values above indicating hypothyroidism and values below indicating hyperthyroidism
T4/T3 resin uptake ratio	0.040 to 0.107, values below indicating hypothyroidism and values above indicating hyperthyroidism.
Blood pH	7.35 to 7.45
PO_2	75 to 100 mm Hg
O_2 saturation	94 to 100% of capacity
PCO_2	35 to 45 mm Hg
Bicarbonates	22 to 28 mmol/l

RESULTS

History (Tables 1 and 2)

There is a sex predilection for women 10:3.   Most cases
occurred during the winter months, but some occurred in summer as
well.   A surprising finding was that five patients had been

Table 1. History

PATIENT	AGE & SEX	MYXOEDEMA DIAGNOSED IN THE PAST	TIME OFF THYROXINE BEFORE ONSET OF COMA	MONTH OF ADMISSION
1	68 F	Yes, 2 years previously	6 months	February
2	67 F	Yes, 3 years previously	8 months	February
3	71 M	Yes, 4 months previously	6 weeks	August
4	73 F	No	–	October
5	73 F	No	–	November
6	92 F	No	–	December
7	87 F	Yes, 17 years previously	15 weeks	February
8	83 F	No	–	August
9	94 M	No	–	September
10	82 M	No	–	November
11	66 F	No	–	December
12	77 F	No	–	January
13	86 F	Yes, 3 years previously	Not known	February

Table 2. History

PATIENT	HISTORY OF HEADACHE	GENERALISED CONVULSIONS	DIAGNOSIS ON REFERRAL TO GERIATRIC DEPT.
1	+ occipital 1 week	+	CVA
2	+ frontal 1 week	0	CVA
3	0	+	CVA
4	0	+	CVA
5	+ frontal 1 year	+	Depression
6	0	0	Hypothermia
7	0	+	CVA
8	0	0	CCF
9	0	0	CVA
10	0	0	CVA
11	0	+	CVA
12	0	0	CVA
13	0	0	CVA

diagnosed as hypothyroid in the past, but their replacement therapy had been discontinued over the preceding months.

Six patients presented with generalised convulsions, of which two were in status epilepticus. Four patients were documented to have had arterial hypertension in the few weeks or months preceding the development of coma, and this was particularly severe in one. Three patients had had progressively severe frontal or occipital headaches during the few weeks preceding the development of coma. All 13 patients were admitted as emergencies. Ten were referred to the Department of Geriatrics with a diagnosis of cerebrovascular accident. Two patients had been diagnosed as suffering from depression a few weeks before admission. One of these developed a profound coma which lasted for nearly two days after taking neuroleptic drugs for a fortnight before admission.

Clinical Examination on Admission (Tables 3 and 4)

All patients were in coma at the time of arrival in the Geriatric wards. Rectal temperature was low normal in six patients. Hypothermia occurred in only three patients. A pulse of 60 or less was found in 6 patients. Hypertension occurred in four patients, and was severe in two of these. Generalised convulsions without aura occurred whilst in the wards in four patients. One patient was in severe congestive cardiac failure with anasarca.

The most consistent clinical finding was delayed relaxation of the ankle reflexes found in 11 patients. The duration of coma was longest in the most profoundly hypothermic patient. With this exception, the most prolonged coma occurred in those patients who received diazepam (Valium) 5 - 10 mg i.v. to control their convulsions.

Laboratory Investigations (Tables 5 to 8)

All patients had profoundly abnormal thyroid function tests, well in the myxoedematous range. Antithyroid antibodies, precepitin method, were positive in 50% of the patients tested.

Neutrophil leucocytosis suggestive of infection was found in seven patients, but clinical signs of one (acute bronchitis) was found in only two patients. Severe hypoglycaemia occurred in only one patient, who was found to suffer from pituitary hypothyroidism.

Hyponatraemia occurred in six patients. Blood urea was elevated over 10 mmol/l (60 mg%) in 5 patients.

Table 3. Clinical Examination during Coma

PATIENT	RECTAL TEMPERATURE	PULSE	BLOOD PRESSURE	PUFFY FACE	MALAR FLUSH
1	35.6°C (96°F)	60	110/50	+	0
2	35°C (95°F)	48	80/40	+	0
3	35.6°C (96°F)	61	200/100	0	+
4	36.6°C (97.9°F)	45	120/70	0	0
5	36.1°C (97.1°F)	70	80/40	+	+
6	27.4°C (81°F)	52	120/80	+	0
7	35°C (95°F)	52	150/90	0	0
8	36°C (96.8°F)	144	150/70	0	0
9	36.2°C (97°F)	86	140/80	0	0
10	36.8°C (98.2°F)	66	180/100	+	+
11	29.8°C (85.6°F)	56	100/50	0	0
12	35.6°C (96°F)	95	160/120	0	0
13	34.5°C (94°F)	65	160/100	+	+

Table 4. Clinical Examination during Coma

PATIENT	DELAYED RELAXATION OF TENDON REFLEXES	TRANSIENT FOCAL NEUROLOGICAL SIGNS	DURATION OF COMA
1	+	0	30 hours
2	+	Plantars extensor	52 hours
3	+	0	5 hours
4	+	Plantars extensor	2½ hours
5	+	0	24 hours
6	+	0	6 days
7	+	0	10 hours
8	0	Left hemiparesis	24 hours
9	0	0	4 hours
10	+	0	12 hours
11	+	Increased reflexes	36 hours
12	+	Left hemiparesis	3 days
13	+	0	3 hours

Table 5. Laboratory Investigations

PATIENT	T4 nmol/L (μg/dl)	T3 RESIN UPTAKE	$\dfrac{T4\ (\mu g/dl)}{T3\ resin\ uptake}$	THYROGLOBULIN ANTIBODIES (PRECIPITIN)
1	PBI = 55.1 nmol/1 (0.7 ng/dl)		Plasma cholesterol 12.9 mmol/1 (500 mg/dl)	+ 1/10
2	0.0	155	0.000	Not done
3	3.86 (0.3)	114	0.002	Negative
4	14.15 (1.1)	144	0.007	+ 1/10
5	9.00 (0.7)	127.2	0.005	Negative
6	20.20 (1.6)	125.6	0.013	Not done
7	6.50 (0.5)	124.4	0.004	Not done
8	0.0	77.8	0.000	Negative
9	0.0	86.7	0.000	Not done
10	7.0 (0.54)	131	0.004	+ 1/10 · PLASMA TSH 24.9 mU/L
11	43.0 (3.34)	95	0.035	Not done
12	0.0	135.4	0.000	+ · PLASMA TSH 24 mU/L
13	0.0	139	0.000	Negative

Table 6. Laboratory Investigations

PATIENT	Hb (g/dl)	WBC AND % NEUTROPHILS	GLUCOSE mmol/L	CALCIUM mmol/L
1	13	10.400 (78%)	4.7	2.7
2	14.1	10.600 (88%)	5.2	2.55
3	11.4	6.000 (68%)	4.0	2.6
4	14.2	11.000 (--)	5.5	2.25
5	15.3	9.400 (80%)	6.0	2.45
6	12.9	24.000 (85%)	12.5	2.2
7	12.3	8.600 (71%)	6.2	2.6
8	12.0	23.100 (98%)	4.5	2.1
9	12.9	15.500 (86%)	6.8	2.13
10	13.8	6.300 (--)	6	2.4
11	11.4	5.800 (--)	0.8	2.3
12	14.1	16.200 (92%)	6.4	2.37
13	13.2	8.800 (78%)	5.0	2.5

Table 7. Laboratory Investigations

PATIENT	UREA mmol/L	SODIUM mmol/L	POTASSIUM mmol/L	HCO3 mmol/L
1	8.0	135	3.2	34.5
2	9.0	122	2.4	28.0
3	12.0	136	3.9	26.0
4	5.8	139	2.9	29.0
5	2.0	129	4.1	24.0
6	5.1	129	4.0	29.0
7	6.0	125	4.6	29.5
8	14.0	133	4.5	21
9	49.8	136	4.6	20
10	9.8	135	3.7	29
11	13.4	132	3.6	25
12	14.0	139	3.5	25
13	7.0	129	3.7	23

Table 8.   Arterial Blood Gases

PATIENT	pH	PO2 mm Hg	PCO2 mm Hg	BICARBONATES mmol/L	BASE EXCESS	% SATURATION
1	7.45	41	46	24	+8	79
2	7.53	61	41	35	+10	94
3			N O T	D O N E		
4			N O T	D O N E		
5	7.39	55	37	22	-2	88
6	7.36	57	46	23.3	-1	94
7			N O T	D O N E		
8			N O T	D O N E		
9	7.48	75	21	14	-6	96
10			N O T	D O N E		
11	7.34	64	44	23	-3	91
12			N O T	D O N E		
13	7.39	48	37	23	-1	83

Only seven patients had their blood gases estimated.   Hypercapnia was not found among these patients.

Treatment and Progress

Six patients were treated with intravenous T3, Hydrocortisone,
antibiotics, continuous 50% oxygen 6 - 7 l/min.   Six patients who
regained consciousness soon after admission were treated with oral
Thyroxine in the usual doses.   Electrolyte balance was closely
monitored and electrolyte supplements were administered when needed.

Headache and delayed relaxation of tendon reflexes disappeared
between two and four weeks after beginning treatment.   Patients
remained slow and lethargic for 2 - 3 weeks on average.   No patient
had convulsive fits once made euthyroid, with the exception of
patient 2.   This patient had recovered from the coma but was left
with a cerebral lesion, thought to have been the consequence of the
coma.   This epileptic fit was considered secondary to the cortical
scar.

Outcome (Table 9)

Eleven patients made a full recovery from their thyroid failure.
Patient 6 bled to death from a large carcinoma of rectum 18 days
after admission.   Patient 7 died from a severe chest infection 14
days after admission.   Patient 8 died undiagnosed after five days
from admission.   Patient 10 died a sudden death from a myocardial
infarction two months after admission.   Patient 12 died suddenly
after eight days from admission, although her condition appeared
to have greatly improved on T3 treatment.   Eight patients were
finally discharged home.

DISCUSSION

Profound coma in severe hypothyroidism may be precipitated by
several factors, such as drugs, especially psychotropic, trauma,
exposure to cold, and any acute illness[9].   As the precipitating
cause improves or is withdrawn some patients may regain
consciousness even without thyroid hormone replacement[10].

Although in severe advanced hypothyroidism not complicated by
any other illness coma must necessarily precede death at some stage,
it has recently been argued that lack of thyroid hormone per se is
seldom the direct cause of depression of consciousness in severe
myxoedema.   Five factors have been implicated as the most common
precipitating causes: hypoxia, hypercapnia, hyponatraemia,
hypothermia, hypopituitarism with hypoglycaemia[10].   The majority

Table 9.   Outcome

Patient

1	Discharged
2	Discharged
3	Discharged
4	Discharged
5	Discharged
6	Died
7	Died
8	Died
9	Discharged
10	Died
11	Discharged
12	Died
13	Discharged

of the patients presented in this paper seem to have had more than
one immediate cause of loss of consciousness, but none of such a
degree to justify the coma on its own.   The coma was probably the
result of the summation of their effect.   These causes were:
hypoxia, hyponatraemia, psychotropic drugs, generalised convulsions,
all in various combinations.   Only in the patient with pituitary
myxoedema (patient 11) hypoglycaemia was severe enough to account
for the coma on its own.

Lumbar puncture was performed in only one patient while in coma
and the cerebro-spinal fluid pressure was found elevated at 40 cm of
water.   This was consistent with the clinical picture seen in these
patients and with previously reported studies in which intracranial
pressure was found to be elevated in up to 40% of untreated myxoedema
patients[11,12].

There is no clinically diagnostic finding in these patients,
except perhaps for the history of a previous diagnosis of myxoedema
followed by discontinuation of thyroxine therapy a few weeks or
months before admission.   This is seldom forthcoming at the time
the patient is seen comatose in the wards.   There are often,
however, a number of clinical signs which should arouse the suspicion
of hypothyroidism, such as delayed relaxation of tendon reflexes,
bradycardia, malar flush, E.C.G. changes etc.

The differential diagnosis of these patients is fairly complicated. One must exclude other causes of coma, such as cerebrovascular accidents; these patients may have transient signs of upper motor neurone lesion; such signs disappear once the patients are made euthyroid. The most difficult differential diagnosis is with accidental hypothermia which may mimic all the clinical signs of hypothyroidism: these signs quickly disappear on rewarming the patient. Furthermore, myxoedema accounted for only 5% of all hypothermic patients admitted to the same Geriatric Department.

This paper suggests that coma in severe myxoedema is not uncommon in elderly patients. This fact may have passed unnoticed because this condition in old age presents in a manner different from that seen in younger patients. Severe progressive headaches and generalised convulsions seem particularly common in the elderly. A proportion of these patients are sent to hospital with a diagnosis of cerebrovascular accident. This happened to ten patients in this series. They may die before the laboratory evidence of hypothyroidism becomes available.

In conclusion, thyroid function tests should be carried out as a routine in all comatose elderly patients.

SUMMARY

Thirteen patients were admitted to a Geriatric Department in London over a three year period suffering from severe impairment of consciousness and myxoedema. The majority presented with unusual clinical features - generalised convulsions were particularly common. This condition is probably not uncommon in the elderly, but can be easily missed on account of its atypical presentation.

REFERENCES

1. W. M. Ord, Cases of Myxoedema, Transactions of the Clinical Society of London, 13:15 (1880).
2. J. Allan, Myxoedema: death: necropsy, Br.Med.J. 1:267 (1884).
3. Report on Myxoedema, Supplement to Vol. XXI, Transactions of the Clinical Society of London (1888).
4. U. K. Summer, Myxoedema Coma, Br.Med.J. 2:366 (1953).
5. H. S. Lemarquand, W. Hausmann, E. H. Hemsted, Myxoedema as a cause of death, Br.Med.J. 1:704 (1953).
6. L. Leon-Sotomayor and C. Y. Bowers, Myxoedema coma, Springfield, Illinois, U.S.A. C. H. Thomas (1964).
7. P. M. Jefferys, The prevalence of thyroid disease in patients admitted to a geriatric department, Age & Ageing 1:33 (1972).
8. M. F. Green, Endocrine diseases in the elderly, Brit.J.Hosp.Med. 10:700 (1973).

9. L. J. Degroot, "Cecil Textbook of Medicine",XV edition,
        Saunders, London (1979).
10. P. C. Royce, Severely impaired consciousness in myxoedema,
        Amer.J.Med.Sci. 261:46 (1971).
11. E. C. Evans, Neurologic complications of myxoedema: convulsions,
        Annals.Intern.Med. 52:434 (1960).
12. D. Bronsky, H. Shrifter, J. de la Huerga, A. Dubin, S. S.
        Waldstein, Cerebrospinal fluid proteins in myxoedema:
        J.Clin.Endocrin.Metab. 18:470 (1958).

APHASIA IN THE ELDERLY

E. De Renzi

Neurology Clinic
University of Modena
Italy

Description of aphasia in elderly people does not require a
specific treatment, as there is little evidence that aphasic patterns
change substantially, once the process of language acquisition is
completed.   It must also be remembered that the body of knowledge
on this subject was built by neurologists predominantly investigating cerebrovascular patients i.e. patients similar to those seen
by the gerontologist with respect to age.   The present survey of
aphasia will, therefore, follow the lines of the traditional
neurological approach, which is based on the correlation between
symptoms and locus of lesion and aims at providing an anatomical
framework to understand the mechanisms of language and language
disruption.

I will first describe the three main syndromes - Wernicke's
aphasia, Broca's aphasia and global aphasia - most often occurring
in clinical practice and which provide essential information to
assess the location and size of damage.   Then I will briefly report
on the less common syndromes, which, in spite of their rarity are
of great interest, because they provide hints relevant to understanding the neural organization of language.   For a comprehensive
and up to date treatment of the topic the interested reader is
referred to the books of Goodglass and Kaplan[1] and Lecours and
Lhermitte[2] and to the article of Vignolo[3].

THE MAIN APHASIC SYNDROMES

Figure 1 displays a lateral view of the left hemisphere showing
the areas involved in language activity.   A critical role is played
by area 22 (Wernicke's area) located in the posterior portion of the

first temporal gyrus and area 44 (Broca's area), located in the
posterior third of the third frontal convolution, in front of the
motor representation of the face.   The pathway connecting the two
centres has been identified with the arcuate fasciculus.   The
language area extends further to include the posterior part of the
second temporal gyrus (area 37) and the inferior part of the
parietal lobe (area 39 and 40), but the Wernicke's and Broca's
area play a pivotal role in subserving speech function and their
lesion gives rise to the corresponding aphasic syndromes.

Wernicke's Aphasia

    Wernicke's area corresponds to the associative auditory cortex,
which borders the primary projection area of the acoustic pathways
and is on the left specialized for storing verbal sounds memory.
Its damage leaves hearing unimpaired, but prevents the patient from
understanding words because he is no longer able to reconstruct
their phonetic pattern.   Also the evocation of the acoustic model
that guides the spoken output is thwarted and consequently language
abilities are disrupted at the expressive as well as at the
comprehensive level.   In speaking, the patient shows word finding
difficulty (anomia), phonetic substitutions (phonemic paraphasias),
semantic substitutions (semantic paraphasias), or even production
of verbal patterns which present no resemblance to any word of the
language (neologisms).   In spite of these difficulties, the speech
output is marked by a certain degree of fluency, i.e. by a rate of
speaking not substantially dissimilar from that of a normal person.
The number of words in connected sentences the patient is able to
produce in a given period of time has shown to discriminate fairly

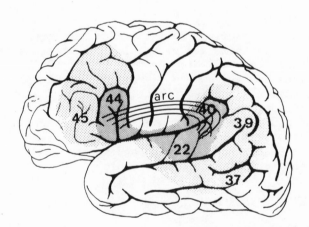

Fig. 1.   Lateral view of the left hemisphere showing the areas
          involved in language.

well Wernicke's from Broca's aphasics.   As a rule of thumb, an
aphasic output can be defined fluent if it is made up of runs of at
least four verbal units not interrupted by a pause.   There are,
however, wide variations in fluency among Wernicke's patients.
At one end you may find jargonaphasics who speak uninterruptedly in
an uncomprehensible jargon made up of neologisms, semantic and
phonemic paraphasias, completely unaware that what they are saying
is devoid of any communicative value.   At the other end there are
aphasics with an extremely reduced output because inhibition
prevails and prevents them from bridging the word finding gap with
paraphasias.   In these cases the other features marking the verbal
output of Wernicke's patients assist in making the diagnosis: lack
of effort in starting to speak, preserved articulatory agility,
normal melodic intonation, prevalance of relational words in
comparison to specific substantive words.

     One may ask why stress has been laid on the expressive features
of Wernicke's aphasics rather than on their comprehension deficit,
which is traditionally held to be the hallmark of the syndrome, as
indicated by the fact that some authors name it receptive or
sensory aphasia.   There are several reasons that justify the
emphasis on the speech output.   First, impairment in comprehension
is not specific to this type of aphasia, being equally severe in
global aphasia (see later).   Second, it does not lend itself to
qualitative distinctions, as is the case for speech.   Third, it
tends to compensate more quickly than expressive disorders, probably
because the right hemisphere has maintained a greater potential for
decoding language than for controlling the verbal output.   Evidence
directly supporting this hypothesis can be found in patients submitted to the section of the interhemispheric commissures for the
relief of epilepsy, in whom the verbal capacity of the right brain
can be tested without the aid of the hemisphere dominant for
language.   The patients were able to understand the meaning of
written stimuli projected to the left visual field and thus transmitted to the right hemisphere, though remaining unable to read
them aloud.

     The severity of comprehension impairment ranges in Wernicke's
aphasia from an almost complete inability to understand even the
simplest verbal command to subtle deficits, which escape detection
at bedside examination and show up only with tasks in which the
redundancy of the verbal message is reduced to a minimum (e.g.
the Token Test).

     From a neurological viewpoint, it is important to remember that
Wernicke's aphasics are frequently free from motor and sensory
impairment.   This lack of obvious neurological signs, together with
the preserved fluency of the verbal output, makes diagnostic blunders
easy, such as mistaking them for mental patients.   The risk is
enhanced when semantic paraphasics prevail because the speech may

mimic the schizophrenic word salad or the incoherence of confusional
states.   Benson says that the few patients who were referred to
him because of "word salad" were all eventually found to have a
fluent-paraphasic language, due to a localized brain lesion.   In
most of Wernicke's patients this misdiagnosis can be easily avoided
if visual fields are routinely tested with the confrontation method.
Contralateral (right) hemianopia is a sign freqently accompanying
Wernicke's aphasia since the optic pathways run in the white matter
of the temporal lobe and are involved if the lesion deepens to the
ventricle.

Broca's Aphasia

Broca's area corresponds to the posterior part of the third
frontal convolution and is located in front of the motor representation of the face.   It belongs to the motor association cortex and
was viewed by Broca as specialized in co-ordinating the movements
peculiar to speech.   Its lesion was held responsible for aphemia
i.e. lack of speech due to loss of memory for the movements
involved in verbal output.   Broca's aphasics have great difficulty
in choosing the right positioning of vocal tract muscles and in
regulating their rapid transition to other positions, so as to secur
a correct selection and ordering of phonemes.   They produce wrong
sounds, strive to amend them, fail and run into blocks.   As a
result, their verbal output is basically non-fluent.   There is a
wide spectrum of disabilities, depending on the stage and the size
of the lesion.   The most severe cases show a nearly complete block
of output, except for some steriotyped utterances which are
surprisingly well pronounced.   As the patient begins to recover,
his speech becomes effortfull and halting, marked by groping oral
movements and by simplifications, alliterations, transpositions of
phonemes, especially when uncommon words are pronounced.   Long
pauses occur between words and at times the patient, who is fully
aware of his difficulty, breaks down into reactions of discourage-
ment and depression.   Not only phoneme selection, but also phoneme
production is distorted to the effect that sounds are uttered which
do not correspond to the standard phonemes of the speaker's language
In milder cases, it becomes apparent the loss of the melodic line
of speech and a difficulty in organizing complex syntactic structure
with a tendency to focus on substantives and to omit relational
words (e.g. prepositions).   The corresponding pictures have been
named dysprosody and agrammatism, respectively.

Mohr[5] took exception to the thesis that a severe and persisting
Broca's aphasia is contingent upon damage to the Broca's area alone.
He convincingly argued that when the expressive symptoms are not
transient, there is a remarkably larger lesion, involving not only
the operculum but also the insula and the subjacent white matter.
It results from an infarct in the territory of supply of the upper
division of the middle cerebral artery.

Comprehension level is an important parameter for the diagnosis of Broca's aphasia and for evaluating the size of the lesion. The classical view is that comprehension is unimpaired in Broca's aphasia, unless long and complex commands are given. In this case the inability to overtly rehearse the message prevents its storage in short-term memory. As a matter of fact, most patients with persistent output reduction will usually show a mild to moderate comprehension impairment also when presented with items not particularly taxing their verbal memory. However, the discrepancy between expression and comprehension remains remarkable; if simple commands are also failed, the patient can no longer be considered a Broca's aphasic, and must be diagnosed as suffering from global aphasia.

Broca's aphasia is frequently accompanied by right hemiparesis of the oral and upper limb muscles, because the lesion tends to encroach upon the neighbouring motor area.

Global Aphasia

It is easy to misunderstand the concept of global aphasia, if it is simply defined on the basis of the contemporaneous presence of comprehension and expressive disorders. This pattern of deficit applies equally well to Wernicke's aphasia and without further qualification the distinction between the two forms is impossible. It is, therefore necessary to specify that, in addition to a remarkable deficit in comprehension, global aphasia is marked by disorders of verbal output similar to those of severe Broca's aphasia, namely extreme reduction of speech, articulatory blocks and anomia. It results from a large lesion of the left hemisphere, involving both Wernicke's area (hence, the disruption of the verbal code) and Broca's area (hence, the disruption of the articulatory control), and thwarting all channels of communication. Global aphasia frequently occurs in elderly patients with cerebrovascular disease, as a consequence of a softening in the entire territory of the Sylvian artery. Since this artery supplies most of the lateral surface of the hemisphere, also the motor and sensory cortex and the optic pathways will be damaged and global aphasics will suffer from right hemiplegia, hemianaesthesia and hemianopia.

So far only oral language has been considered. Writing and reading are equally or even more severely defective in every class in which the linguistic code is impaired. As a general rule, the less a given type of verbal activity is practised the more it is disrupted by the lesion. It is, therefore, no wonder that aphasics are at pains in decoding and encoding written symbols. The assessment of reading and writing abilities deserves, however, close examination, because there may be a discrepancy between the derangement of oral and written language, as a consequence of discrete substrates.

We shall now turn to the description of the specific aphasic syndromes, which provide cues relevant to the comprehension of the mechanisms underlying language.

SPECIFIC APHASIC SYNDROMES

In analysing these clinical pictures special attention must be paid to repetition, a linguistic activity that has been neglected in describing the main aphasic syndromes, because here its impairment strictly follows that of comprehension and expression. Repetition of a word or a sentence implies an adequate processing of the phonetic structure of the message and its transformation into a motor sequence, but does not necessarily require the understanding of its semantic value (e.g. it is possible to repeat meaningless words and babies learn to repeat words before knowing what they mean). In anatomical terms, good repetition points to the integrity of Wernicke's area, Broca's area and the pathway connecting them, but does not entail that the functioning of the semantic processor is intact. By comparing repetition and semantic comprehension in aphasics and looking for dissociation, one may gain insight into the mechanism underlying receptor language.

In Table 1 repetition is taken into account together with semantic comprehension and the qualities of speech to define the characteristics of the syndromes I am going to survey. The sign + means a relatively proficient level of functioning, the sign - a severe impairment.

Conduction aphasia is marked by a discrepancy between good or fairly good comprehension and inability to repeat. Verbal output is fluent without articulatory blocks and loss of melodic intonation, but is circumlocutory, marred by phonemic substitutions and word finding difficulties. The speech deficit is more severe in repetition than in spontaneous speech, a feature contrasting the performance of conduction aphasics to that of Broca's aphasics (the main difference remains, however, the fluency of conduction aphasics). Silent reading is good, but oral reading shows the same errors as speech. Spontaneous writing and writing to dictation are characterised by gross mis-spelling. A long-standing interpretation of conduction aphasia, dating back to Wernicke[6], envisages the disorder as due to the interruption of the arcuate fascicle linking Wernicke's area to Broca's area. The latter would consequently work deprived of the guidance provided by the acoustic model of the word. The critical lesion is thought to involve the parietal operculum.

Transcortical sensory aphasia is, in a sense, a mirror-like picture of conduction aphasia: repetition is preserved, comprehension

Table 1. Features of the Aphasic Syndromes

	Comprehension	Repetition	Expression	Speech Qualities
Wernicke's Aphasia	–	–	–	Fluent. Paraphasias and anomia
Broca's Aphasia	+	–	–	Non-fluent. Articulatory block and anomia
Global Aphasia	–	–	–	Non-fluent. Articulatory block and anomia
Conduction Aphasia	+	–	–	Phonemic paraphasias and anomia
Transcortical Sensory Aphasia	–	+	–	Semantic paraphasias
Transcortical Motor Aphasia	+	+	–	Spontaneous speech absent. Naming preserved
Anomia	+	+	–	Circomlocutions. Word-finding difficulty

is impaired. Speech is fluent, but devoid of communicative value
owing to semantic paraphasias and neologisms.   Writing and reading
are severely disrupted.   The lesion would spare Wernicke's and
Broca's areas and their connections, but would isolate Wernicke's
area from the surrounding cortex, so that the phonemic pattern,
though well analysed at the auditory level, can neither be decoded
nor evoked.   In a few patients echolalia, i.e. automatic and
compulsive repetition of the words pronounced by the examiner, is a
distinctive feature.

   Transcortical motor aphasia. The patient understands and repea
well but shows a striking reduction in spontaneous speech, which
approaches muteness in the most severe cases and is limited to a
loss of verbal fluency in the mildest ones.   Yet, no deficit in
articulation is apparent on repetition tasks and words can still be
evoked when the patient is required to name a picture or to complet
a sentence.   Reading is spared, but writing is poor.   This form
of aphasia has a fairly good predictive value with respect to the
locus of lesion.   Damage is located in the anterior or superior
region of the frontal lobe and does not encroach upon Broca's area.
The basic deficit of these patients resides in the inability to
structure information in sentences, probably because of "loss of
frontal lobe volitional influence on the speech apparatus" (Rubes[7]).

   Anomia. Word finding difficulty is an ubiquitous phenomenon in
aphasia, apparent in spontaneous speech as well as on confrontation
naming and while representing a sensitive indicator of language
deficit, it has in and of itself little diagnostic value with respec
to the definition of an aphasic picture.   A prominence of word
finding difficulty is found in the restitution stage of Wernicke's
aphasia, when comprehension and repetition have already considerably
recovered and the control exerted on the speech output is sufficien
to inhibit semantic paraphasias.   Some authors think that a
picture marked by circumlocutions in the attempt to circumvent word
finding difficulty is typical of inferior parietal lobe aphasia,
but the localizing reliability of this clinical pattern is poor.
Anomia is frequently found as a precocious symptom in tumours of
the left temporal lobe and in Alzheimer's disease, although it is
exceptional that in these cases a close examination does not
reveal also impairment of comprehension.

   I will pass over the description of so-called "pure" aphasias,
where a single output or input modality is affected (e.g. word
deafness, pure alexia, anarthria, pure agraphia), because of their
rarity, and I will briefly discuss the question whether old age has
an influence on the aphasic symptomatology.   Apart from the possibl
effect on language performance of a concomitant mental deterioration
process, it is worth considering an interesting point that has been
recently raised by Brown and Jaffe[8].   They argued that inter- and
intra- hemispheric specialization for language does not arrest at

Table 2.   Distribution of Age and Sex in 200 Aphasics

| | 50 yrs | | 51-65yrs | | 65 yrs | |
	M	F	M	F	M	F
Broca	6	7	15	6	9	6
Wernicke	6	11	30	24	24	31
Global	3	1	7	5	8	1

Factor	D.F.	chi^2	p
Age x Aphasia	4	5.46	n.s.
Sex x Aphasia	2	4.84	<.10
Age x Sex x Aphasia	4	3.93	n.s.

age 5 years, as classically thought, but evolves throughout life
up to senescence.   Accordingly, the specificity of the aphasic
patterns, produced by differently located injury of the language
area of the left hemisphere, would increase with age.   For
instance, the same lesion of Wernicke's area will result in a
picture analogous to Broca's aphasia if incurred by the child,
in anomia and phonemic paraphasias if incurred by the adult and in
jargonaphasia if incurred by the elderly.   The evidence offered
in support of this thesis was circumstantial and at variance with
some recent findings.   Woods and Teuber[9] reported that jargonaphasia can also occur in the child and Brust et al[10] that the mean
age of fluent and non-fluent aphasics drawn from a stroke population
was not significantly different.

We tried to verify what relationship, if any, existed between
the incidence of the three main aphasic syndromes and age.   Two
hundred right-handed patients affected by left hemisphere damage
following a stroke (88.5%) or a tumour (11.5%) and showing signs of
aphasia on a standard language examination were seen in our department in the period June 1975 - September 1979.   One hundred and
eight were males, 92 were females.   They were grouped under the
heading of Wernicke's aphasia (which included also transcortical
sensory aphasia), Broca's aphasia (which included also transcortical
motor aphasia) and global aphasia.   Patients of each group were
subdivided into males and females and into three classes of age:
20-50 years, 51-65 years, more than 65 years.   Table 2 presents
the distribution of the whole sample.   The statistical analysis
shows that neither the age factor nor the age by sex interaction
were significant,while the sex factor approached significance, due

to the prevalance of Wernicke's aphasia in females and of Broca's and global aphasia in males.   These findings do not lend support to the idea that old age is a variable deserving special consideration in the differentiation of aphasic symptomatology and uphold the view, shared in this paper, that the gerontologist can use the same theoretical and practical framework worked out by the neurologist in the assessment of aphasia.

REFERENCES

1. H. Goodglass and E. Kaplan, "The assessment of aphasia and related disorders", Lea and Febiger, Philadelphia (1972).
2. A. R. Lecours and F. Lhermitte, "L'Aphasie", Flammarion, Paris (1979).
3. L. A. Vignolo, Le sindromi afasiche, in: "Neuropsicologia Clinica", E. Bisiach, F. Denes, E. De Renzi, P. Faglioni, G. Gainotti, L. Pizzamiglio, H. R. Spinnler and L. A. Vignolo, F. Angeli, Milano (1977).
4. D. F. Benson, Disorders of verbal expression, in: "Psychiatric aspects of neurological disease", D. F. Benson and D. Blumer, eds., Grune and Stratton, New York (1975).
5. J. P. Mohr, Broca's area and Broca's aphasia, in: "Studies in Neurolinguistics", Vol.I, H. Whitaker and H. A. Whitaker, eds., Academic Press, New York (1976).
6. C. Wernicke, "Der Apasische Symptomencomplex", Frank U. Weigert, Breslau (1864).
7. A. B. Rubens, Transcortical motor aphasia, in: "Studies in neurolinguistics", Vol. 1, H. Whitaker and H. A. Whitaker, eds., Academic Press, New York (1976).
8. J. W. Brown and J. Jaffe, Hypothesis on cerebral dominance, Neuropsychologia 13:107 (1975).
9. B. T. Woods and H. L. Teuber, Changing patterns of childhood aphasia, Ann.Neurology 3:273 (1978).
10. C. M. Brust, S. Q. Shafer, R. W. Richter and B. Brun, Aphasia in acute stroke, Stroke 7:167 (1976).

DEPRESSION AND ITS MANAGEMENT

J. Williamson

Professor of Geriatric Medicine
City Hospital
Edinburgh

> Nessùn Maggiòr Dolòre
> Che Ricordàrsi Del Tèmpo Felìc
> Nella Misèria
>
> > Dante Alighieri (1265-1321)
> > La Divina Commedia

Old age is commonly portrayed as the period of life characterised by multiple and grevious losses. These losses include loss
of youth, of vigour, of sexuality and loss of friends and contempor-
aries. Those of us who work among the old know that this is no
better than a carricature of old age since the majority of people
as they age undergo successful adaptation to the changes in their
own minds and bodies and to altered circumstances in their life
style. Most remain as happy (or as unhappy) as they have been at
other stages of their development.

For some, however, melancholia and regret become overwhelming
and it is there that we are reminded that misery is the greater when
it is related to former happiness as in earlier life.

My task is to deal with the problem of depression in the elderly
and, perhaps, an explanation is required as to why a physician and
not a psychiatrist should undertake this important task. The
principal reason of course is that depression in old age commonly
presents in association with established physical disease and thus is

likely to come to the notice of the physician rather than the
psychiatrist. Kay et al[1] showed that even with mild depressive
disorders, half the patients had moderate or severe physical
disabilities and ordinary clinical experience confirms that patients
with long-term impairment associated with arthropathy, stroke or non-
vascular neurological conditions are particularly prone to depression
The specialist in geriatric medicine must perforce become familiar
with its manifestations and optimum management.

Depression is a common finding in the elderly and is the most
prevalent psychiatric diagnosis until age 75 after which age it is
superseded by dementia. In a detailed study of a random sample of
215 men and 272 women aged 62 to 90 years in Edinburgh[2] significant
depression was found in 5.4% of subjects and in patients referred to
the geriatric consultative clinic in that City, the diagnosis was
made in no less than 14% (Lowther et al)[3].

Since it is a common condition and since it is also an extremely
unpleasant experience, it is very important that doctors and other
health workers should appreciate its significance and be aware of
the problems in diagnosis and treatment.

Psychiatrists have an obsessional desire to classify and to
subdivide clinical conditions into neat categories and it has become
customary to refer to reactive (situational or neurotic) depression
as opposed to the endogenous (or psychotic) variety. I have always
found this distinction to be somewhat unreal and it seems probable
that the borderline, if it exists at all, is much less clear and
distinct in old age than it is in younger age groups. It is
therefore useful in elderly patients to regard depressive disorders
as encompassing a very wide spectrum. At one end of this spectrum
we find the conditions which are clearly reactions to environmental
stress as in the old lady who is overwhelmed by widowhood, loss of
her home and physical impairment. She presents with anxiety,
sleeplessness and significant and unvarying lowering of mood. At
the other end of this spectrum we encounter the old man who,
without apparent cause, becomes classically melancholic with profound disturbances of affect, severe feelings of unworthiness and
guilt and who may present with bizarre delusions and phobias.

The problem is further complicated by the fact that the mode
of presentation may significantly affect referral patterns. For
example, the milder conditions will be coped with by the general
practitioner (or often by the caring family in an essentially
instinctive fashion). The more severe disturbances, especially
those with significant somatic manifestations may be referred to
various specialists who will thereupon carry out the relevant
investigations (often at considerable expense). The physically
disabled patient who becomes depressed is likely to come the way of
the specialist in geriatric medicine while the severe "classically"

depressed patient with serious thought disorders and delusions will
be referred to the psychiatrist.   Hence different kinds of doctors
tend to acquire different views of the same group of disorders.   It
may be, also, that depression presents differently in different
cultural settings as tends to occur in individuals of different
personality (as will be discussed below).

FACTORS IN THE PRODUCTION OF DEPRESSION IN OLD AGE

Goldstein[5] has emphasised that in old age "losses are more
numerous and visible, whereas gains are fewer and less visible".
These losses are occurring at a time when the capacity of the mind
and body to adapt is significantly reduced and it is perhaps
surprising that so many ageing individuals succeed in living out
their normal span without experiencing significant depression while
only one in twenty are found to be significantly depressed.   The
most important losses are those which impair self-esteem and status
and involve widowhood, loneliness, loss of prestige and of
significant roles and the loss of function which results from
serious physical impairment due to disease.   These are all apt to
lead to preoccupation with bodily function, hypochondriasis and
that unhealthy contemplation of death which is the greatest, and
for many in our modern materialistic society, the final loss.

It is commonly asserted that modern society is prone to
exaggerate these feelings of irreparable loss because of its excessive youth orientation, its insistence upon age related retiral from
employment, its emphasis upon material success and its rejection of
the value of wisdom and experience as personified in the older
citizen.

Hereditary Factors

Kay[4] showed that there was a significant hereditary factor in
depression and that the incidence was higher in first degree
relatives of sufferers than in the general population.   The risk to
relatives seemed to be less among the relatives of probands in whom
the depression first occurred in old age.   It may therefore be
justifiable to promulgate the hypothesis that where the genetic trait
is strong, depression will tend to occur in earlier life and where
it is weak or absent, depression will either never occur or will
only supervene in old age, and then only after adverse environmental
(or biochemical or pharmacological) factors have been more or less
strongly operative.

Biochemical Factors

It has been shown that levels of certain amines are lower in the hindbrains of depressed patients. Shaw et al[6] showed low levels of 5-hydroxytryptamine (5-HT) and 5-hydroxyindoleacetic acid (5-HIAA) in the hindbrains of patients who had committed suicide while depressed. The next significant observation was that there appeared to be an increase with age in the cerebrospinal fluid levels of monoamine oxidase (MAO) in experimental animals (Bowers and Gerhade)[7]. Robinson et al[8] then demonstrated an age related increase in MAO and reduction of 5-HIAA in the hindbrains of 55 patients who had died of various causes. There was likewise a decrease with age in levels of noradrenaline.

As a result of these findings there has been an understandable temptation to assume that lowered levels of biogenic amines occurred in old age and that this led to an increased proclivity to depression Unfortunately, however, the picture is not so simple since Ashcroft et al[9] failed to find reductions in amine levels in the C.S.F. of depressed patients. To make things even more puzzling, Post et al[10] showed that, although moderately depressed patients had low C.S.F. levels of 5-HIAA and homovanillic acid (HVA), when these same patients were made to simulate the behaviour of manic patients, their C.S.F. levels of 5-HIAA and HVA rose to normal, suggesting that C.S.F. levels may be more affected by degree of psychomotor activity than by mood disturbance. This observation seems to accord with that of Bridges et al[11] that severely retarded depressed patients had low C.S.F. levels of tryptophan and tyrosine while those with agitation and anxiety had higher levels. This finding might help to explain why L-tryptophan therapy appears to be more effective in retarded depressed patients than in those presenting with agitation (Harrington et al)[12]. It has also been suggested by Jensen et al[13] that the outcome of trials of antidepressants may be significantly affected by the relative proportions of patients in the trials who present with retardation and with agitation.

THE COMMON SYNDROMES OF DEPRESSION IN OLD AGE

It is commonly (and correctly) stated that depressive illnesses do not present differently in old age compared to younger groups except that agitation rather than retardation and "severe delusional thought content, especially of a bizarrely hypochondriacal kind" are more common in the elderly depressed patients (Post)[14]. I would like, however, to try to be rather more specific and would suggest that certain syndromes associated with depression in old age present peculiar difficulties in diagnosis.

"Ordinary Depression". The classical presentation of depression
is easy to detect and ought to present few problems to the alert
physician, whether he be general practitioner, internist or
specialist in geriatric medicine.   The typical patient looks
miserable, admits to lowered mood, loss of energy, poor concentration, lack of appetite, typical sleep disturbance and may in
addition admit to feelings of being "useless" or "worthless".   The
problem in diagnosis in such patients arises from the fact that the
life situation of many old people is clearly conducive to anxiety,
agitation and unhappiness and the doctor is faced with the problem
of deciding whether the elderly widow who is clearly anxious and
who admits to feelings of "being in the way", has poor appetite and
needs a sedative to secure more than a few hours of sleep at night,
is responding normally to stress or has become depressed.   The
unfortunate truth is that many old people are exposed to considerable stress associated with widowhood, low income, poor housing,
loneliness and isolation.

Aggravation of Pre-existing Physical Disability. Old people
tend to suffer from multiple chronic and often disabling conditions
and with the development of depression the presentation may be as
an apparent aggravation of these conditions.   Thus the stroke
patient who has been making satisfactory progress ceases to respond
to rehabilitation measures and may even deteriorate.   The question
is posed "Has she had a further cerebrovascular accident?"   The
patient with rheumatoid arthritis develops depression which leads to
increased complaints of pain, reduced mobility and further erosion
of independence.   It is only too easy to conclude that she has had
an exacerbation of her rheumatoid condition but no amount of
analgesic or juggling with anti-inflammatory drugs will produce
improvement until the depressive cause of deterioration is diagnosed
and treated.   I recently saw an attractive 75 year old lady who had
known osteoarthrosis of her right hip and who was experiencing
increasing pain and disability.   Closer questioning revealed that
she had marked insomnia, recent loss of interests (her hobby had
been water colours, at which she was a remarkably competent artist)
and she admitted to feelings of "being in the way" and "a burden
to her family".   This, linked to an easily verifiable family history
of depression led to appropriate treatment with mianserin and two
months later she was found to be full of life, had started painting
again and was busily helping her daughter-in-law cope with an
illness.   She made no spontaneous comment about her hip condition
but on specific questioning she dismissed it with the words "it
still aches and there is stiffness but I can easily put up with it".
The other common condition which is apt to be similarly affected by
depression is Parkinsonism and any unaccountable deterioration in
the mobility or function of sufferers should always lead to careful
inquiry to determine whether this may be the explanation.

Hypochonriasis. The old lady who develops multiple symptoms
and complains should always be suspected of having depression.
This form of presentation is probably commoner in persons who have
always been rather meticulous and accustomed to high standards of
behaviour and self discipline.   Symptoms tend to be centred upon
headache, constipation, backache and sometimes "angina".   These
patients make themselves great nuisances, they summon relatives and
the family doctor and they are apt to receive multiple drugs and
remedies.   They become thoroughly unpopular with everyone.   It is
a good rule to suspect every such unlikeable and "manipulative"
patient of having depression.   This is, of course, especially true
where the hypochondriasis has only manifested itself in late life.

Severe localised pain. This is a most interesting and often
baffling presentation of depression.   It may occur as the only
symptom in an elderly patient who stoutly denies any mood
disturbance, loss of appetite or interference with sleep (apart from
that attributable to the pain).   This presentation was well
described by Bradley[15].   Again there is the suggestion that such
patients have been careful, obsessional and self-demanding types who
have always set themselves high standards of achievement.   When
depression occurs in such an individual it is perhaps understandable
that it should manifest in a "respectable" fashion such as a somatic
pain rather than with the patient giving way to "weakness" and
feelings of loss of energy and "giving in".   Sometimes the site of
the pain is explicable, e.g. pain in a previously injured limb or
as angina in an individual whose near relative has died of heart
disease.   These patients may be referred from doctor to doctor and
may have expensive (and unpleasant) investigations.   I recently
saw a man of 76 who presented with severe pain in the right renal
angle but with no other symptoms.   He had full urological
investigations carried out before it was discovered that the onset
of the pain followed about six months after the death of his
brother from renal carcinoma.   On treatment with antidepressants
the pain disappeared and afterwards he admitted that he had "felt
awful" quite apart from the pain, although at the time the pain was
the only symptom he would admit to.

Some cases of phantom limb pain in an amputee may have a
depressive basis and a trial of antidepressant therapy is often
warranted.

Another syndrome in which depression may be contributory is
post herpetic neuralgia.   More than 30% of cases of herpes zoster
in older women are followed by neuralgia and the virus infection
itself may help to precipitate the depression so it is always worth
enquiring carefully for other manifestations of depression and when
there is doubt a therapeutic trial may be advisable.

Association with Alcohol or other Drug Abuse. Alcoholism in old age is of increasing importance and more attention has been paid to it in recent years.   The person who has abused alcohol in younger life will continue to do so if he or she survives into old age.   The person who starts to drink excessively when old, however, may be using alcohol as a means of combating depression.   Thus the old lady who appears to have an alcohol problem may have started drinking as a means of trying to get to sleep and in such cases the possibility of underlying depression must always be considered.

Depression in the Dying Patient. It must always be borne in mind that the patient who is approaching death may become depressed and thus suffer much more in her last few months as a result.   It is too easy to conclude that dying patients must always be unhappy and depressed and to exclude the possibility that a depressive illness has supervened.   Such a patient was the 71 year old lady who had carcinoma of lung with malignant pleural effusion.   She had received treatment with intrapleural cytotoxic chemicals and the effusion only needed to be tapped every two months.   Despite this she had severe symptoms of pain and dyspnoea and, on questioning, she also admitted to being deeply depressed and just wished she could "slip away". Treatment with antidepressant therapy produced a remarkable improvement in her general condition and during the remaining year of her life she managed to see all her children and grandchildren and to lead quite a full life.   Not only did she end her days in a much happier fashion but she will be remembered much more fondly by her numerous friends and relatives (and also by her medical attendants who were previously being hard pressed to cope with her insistent and apparently unreasonable demands upon their time and energies).

Pseudodementia.   The presentation of depression as a condition akin to dementia is now well known but it still may be readily missed.   Thus the withdrawn, apathetic and retarded patient with poor concentration, poor orientation and defective memory may easily be relegated to the rag bag of "merely senile".   Where there has been insomnia, there is the extra danger that the patient will be overdosed with hypnotics or tranquillisers which may then produce confusion and thus lead to apparent confirmation of the diagnosis of dementia.

Careful assessment of all demented patients by physician, by nursing staff and by family members is therefore essential and when there is any doubt, careful use of antidepressant therapy is worthwhile.

It should be realised also that in early dementia, a depressive condition may co-exist and then cautious introduction of appropriate therapy may lead to a general improvement and the clinical progress of the dementing condition may be retarded for a time.

MANAGEMENT OF ELDERLY PATIENTS WITH DEPRESSION

As always in geriatric medicine, a complete diagnosis is essential, so a full physical assessment must be made. Any associated conditions must be adequately treated, e.g. anaemia, hypothyroidism, Parkinsonism, congestive cardiac failure and constipation associated with faecal stasis. Where anorexia has been prolonged, special attention to diet is required often with the addition of vitamin supplements.

The nurse's role here is crucial since she may use her skills to make patients feel more comfortable, to strengthen reduced self esteem and to reduce the humiliation associated with such unpleasant problems as faecal impaction and urinary incontinence.

General encouragement of the patient should be instituted with emphasis upon the good prospects for recovery and that depression, although common in old age, is generally eminently treatable. Relatives should be appropriately counselled and involved in the process of encouragement and optimism. At an early stage in recovery, patients should be deliberately reintroduced to social activities which may have been given up. For patients at home, attendence at a day hospital is very useful and for patients in hospital, staff should allocate time to speak to them and encourage them to become interested in the affairs of the other patients and in the life of the ward.

Drug Therapy

Younger doctors fail to realise what an enormous advance has been achieved in the last few decades in the drug treatment of depression. In previous generations, patients with depression had to wait for spontaneous remission when it eventually occurred and thereafter were faced with the dreaded fear of recurrence. Some remained depressed for years or died by suicide in their desperation

The drugs most commonly used in treatment of depression have been the tricyclic group of which imipramine and amitriptylline were the first and most widely used examples. Many different varieties of tricyclic drugs are now available but they are generally no more effective in their antidepressant qualities than these two. Most have been introduced in order to try to reduce the occurrence of adverse reactions which are, unfortunately, commoner in elderly patients. The principal side effects are associated with the anti-cholinergic effects of tricyclics - dry mouth, blurred vision, constipation (or aggravation of pre-existing faecal stasis), bladder atony, and perhaps most important postural hypotension and mental confusion. A baffling problem has been the widely varying plasma levels in patients on similar doses of tricyclics. Thus Carr and

Hobson[16] showed that with amitriptylline and nortriptylline in
therapeutic doses, the plasma levels varied from 60 to 600 ng/litre.
At low levels, no therapeutic response may be expected and Kragh-
Sorensen et al[17] showed that in patients who experienced very high
nortriptylline levels, the response was poor and they recommended
that the therapeutic plasma level to be sought was in the range 50
to 150 ng/litre.   It is not known why higher plasma concentrations
appear to be therapeutically ineffective, nor is it known why such
wide variations occur.   The problem was complicated by the findings
of Coppen et al[18] that in 54 patients, no important correlation was
found between plasma levels and therapeutic response.   These workers
suggested that in some patients the occurrence of unpleasant side
effects led to poorer patient compliance and hence lower plasma
levels (and non-therapeutic response).   They also suggested that
different degrees of plasma binding might explain some of the
discrepancies.   These findings have been disputed and attempts have
been made to predict plasma levels in individual patients by using a
single dose of nortriptylline followed by a 48 hour plasma concentration estimation (Montgomery et al)[19].   These authors recommended
the introduction of this procedure as a means of achieving a steady
state concentration within the therapeutic zone.   This procedure now
should be repeated to determine whether it applies in older age
groups.

     Other drugs have been introduced and used in the treatment of
depression.   Monoamineoxidase inhibitors have, however, not found
favour in older patients because of their liability to side effects
and the dangers from intake of certain incompatible food stuffs, or
simultaneous administration of some other drugs.

     The tetracyclic group have shown promising antidepressant
effects and seem to have less liability to cause anticholinergic
side effects.   Probably the best of these is mianserin, the most
widely used of these substances and it has interesting pharmacological differences from tricyclics and has only mild anticholinergic
effects.

     One interesting compound is the tricyclic nomifensine which
appears to act by blocking the re-uptake of dopamine as well as that
of noradrenaline.   It is therefore a logical drug to use in
depression occurring in patients with Parkinsonism (Hoffman)[20].

     One cautionary note should be struck in relation to the
potential cardiac toxicity of antidepressants (Coull et al)[21].   Their
effects upon the heart were thoroughly studied by Burrows et al[22]
and these workers showed that they produced an increase in heart rate
and in atrioventricular conduction.   Patients with acute overdoses
were found to have distal conduction defects and this was also seen
in some patients on therapeutic doses.   The suggested cause is a
build-up of noradrenaline in heart tissue by the blocked uptake which

these drugs produce and these authors suggested that in patients with
a history of heart disease, or with abnormal ECG (even so minor as
bundle branch block) special care is required.   A significant
finding was that the tricyclic drug doxepin seemed to be free from
blame in the production of arrhythmias or conduction defects even in
acute poisoning associated with self-poisoning overdoses.   Thus it
is logical to use this drug in depressed patients who have a history
of cardiac trouble or who currently have evidence of arrhythmia.

The drug L-tryptophan has been reported to possess useful
antidepressant qualities (Harrington et al)[12] and it seems to have
relatively few side effects apart from a tendency to produce
drowsiness in the early days of its use.   There are theoretical
reasons for suggesting its use in retarded patients while using
tricyclics or tetracyclics for those in whom agitation is the main
feature.   There has been a suggestion that combined therapy with
L-tryptophan and a tricyclic results in a synergistic effect
(Walinder et al)[23].

CONCLUSION

Depression is common in old age and causes much misery, often
by increasing the disability associated with pre-existing physical
disease.   Its presentation may be straightforward and easy to
diagnose but sometimes it presents in atypical and misleading
fashions.

Modern drug therapy offers enormous benefits for the elderly
depressed patient and doctors are recommended to get to know their
favourite antidepressant drugs and to use them carefully.

General measures directed at improving morale with support for
caring relatives are essential as is meticulous attention to the
optimum management of any associated physical illness.

Once the desired therapeutic result has been achieved the drug
must be continued for a period of months and then be withdrawn
gradually and the patient closely supervised in case of relapse.

An elderly patient who has had depression should remain
indefinitely under supervision as the chance of recurrence is
considerable and depressed patients cannot be relied upon to report
such an event.

REFERENCES

1. D. W. K. Kay, P. Beamish and M. Roth, Old age mental disorders
    in Newcastle-upon-Tyne. Part 1: A Study of Prevalence,
    Br.J.Psychiat. 110:146 (1964).
2. M. M. Maule (Unpublished data)
3. C. P. Lowther, R. D. M. McLeod and J. Williamson, Evaluation of
    early diagnostic services for the elderly, Br.Med.J. 3:275
    (1970).
4. D. W. K. Kay, Observations on natural history and genetics of
    old age psychoses; Stockholm Material 1931-1937 (abridged)
    Proc.R.Soc.Med. 52:791 (1959).
5. S. J. Goldstein, Depression in the elderly, J.Am.Ger.Soc.
    27:38 (1979).
6. D. M. Shaw, F. E. Camps and E. E. Eccleston, 5-hydroxytryptamine
    in hind brain of depressive suicides, Br.J.Psychiat. 113:1407
    (1967).
7. M. B. Bowers and F. A. Gerhade, Relationship of monoamine
    metabolites in human cerebral spinal fluid to age, Nature,
    (Lond.) 219:1256 (1968).
8. D. S. Robinson, A. Nils, J. N. Davis, W. E. Bunney, J. M. Davis,
    R. W. Colbourn, H. R. Bourne, D. M. Shaw and A. J. Coppen,
    Ageing, monoamines and monoamineoxidase levels, Lancet,
    1:290 (1972).
9. G. W. Ashcroft, I. M. Blackburn, D. Eccleston, A. I. M. Glen,
    W. Hartley, N. E. Kinloch, M. Lonergan, L. G. Murray and
    I. A. Pullar, Changes on recovery in the concentrations of
    tryptophan and the biogenic amine metabolites in the
    cerebrospinal fluid of patients with affective illness,
    Psychol.Med. 3:319 (1973).
10. R. M. Post, J. Katin, F. K. Goodwin and E. K. Gordon, Psycho-
    motor activity and cerebrospinal fluid amine metabolites in
    affective illness, Amer.J.Psychiat. 130:67 (1973).
11. P. K. Bridges, J. R. Bartlett, P. Sepping, B. D. Kantamaneni,
    and G. Curzon, Precursors and metabolites of 5-hydroxy-
    tryptamine and dopamine in the ventricular cerebrospinal
    fluid of psychiatric patients, Psychol.Med. 6:399 (1976)
12. R. N. Harrington, A. Bruce, E. C. Johnstone and M. H. Lader,
    Comparative trial of L-tryptophan and amitriptyline in
    depressive illness, Psychol.Med. 6:673 (1976).
13. F. Jensen, K. Fruensgaard, U. Ahlfors, T. A. Pihkanen,
    S. Tuomikoski, E. Ose, S. J. Dencker, D. Lindberg and A. Nagy,
    Tryptophan/imipramine in depression, Lancet, 2:920 (1975).
14. F. E. Post, Psychiatric disorders, in:"Textbook of Geriatric
    Medicine and Gerontology", J. C. Brocklehurst, ed., 2nd
    Edition, Churchill Livingston, Edinburgh, London and New
    York (1978).

15. J. J. Bradley, Severe localised pain associated with the
    depressive syndrome, Br.J.Psychiat. 109:741 (1963).
16. A. C. Carr and R. P. Hobson, High serum concentrations of
    antidepressants in elderly patients, Br.Med.J. 2:1151
    (1977).
17. P. Kragh-Sorensen, Chr. Eggert Hansen, P. C. Baastrup and
    E. F. Hvidberg, Self-inhibiting action of Nortriptylin's
    antidepressive effect at high plasma levels,
    Psychopharmacologia (Berl.) 45:305 (1976)
18. S. M. Coppen, K. Ghose, V. A. Rama Rao, J. Bailey,
    J. Christiansen, P. L. Middleson, H. M. Van Praag,
    F. Van de Poel, E. J. Minsker, V. G. Kozulha, N. Matussek,
    G. Kungkunz and A. Jorgensen, Amitriptylline plasma-
    concentration and clinical effect. A W.H.O. Collaborative
    Study, Lancet 1:63 (1978).
19. S. A. Montgomery, R. McAuley, D. B. Montgomery, R. A.
    Braithwaite and S. Dawling, Dosage adjustment from simple
    Nortriptylline spot level predictor test in depressed
    patients. Clin.Pharmacokinetics 4:129 (1979).
20. R. Hoffman, A comparative review of the Pharmacology of
    Nomifensine, Br.J.Clin.Pharmac. 4:695 (1977).
21. D. C. Coull, J. Crooks, I. Dingwall-Fordyce, A. Scott and
    R. D. Weir, Amitriptyline and cardiac disease. Risk of
    sudden death identified by monitoring system, Lancet,
    2:590 (1970).
22. G. D. Burrows, J. Vohra, D. Hunt, J. G. Sloman, B. A. Scoggins
    and B. Davies, Cardiac effects of different tricyclic
    antidepressant drugs, Br.J.Psychiat. 129:335 (1976).
23. J. Walinder, A. Scott, A. Nagy, A. Carlsson and B-E. Ross,
    Potentiation of antidepressant action of clomipramine
    by tryptophan, Lancet 1:984 (1975)

CEREBRAL BLOOD FLOW AND MOTOR NERVE CONDUCTION VELOCITY

IN THE ELDERLY

M.J. Denham, J.C.W. Crawley, H.M. Hodkinson[*] D.S.Smith[+]

Departments of Geriatrics, Radio-isotopes and
Rehabilitation
Northwick Park Hospital, Harrow

INTRODUCTION

Cerebral blood flow and motor nerve conduction velocity have
been studied separately in the elderly to determine changes with
age and dementia.   Unfortunately, the results have not been
consistent.   Little is known of changes in muscle innervation with
age, but it is possible to measure the extent of denervation by[1]
counting the number of active motor units in peripheral muscles[1].
It was, therefore, decided to make simultaneous studies of cerebral
blood flow, motor nerve conduction velocity and the number of motor
units in a peripheral muscle to see what changes occur with age and
primary (senile) dementia.   It was also hoped to see if changes in
cerebral blood flow, which reflect the function of the central
nervous system, are mirrored by changes in motor nerve conduction
velocity and the number of motor units, which reflect the function
of the peripheral nervous system.

PATIENTS AND METHODS

The patients studied were selected from those admitted to
Northwick Park Hospital geriatric unit at a convalescent stage[2] of
illness.   All patients were given a mental test questionnaire[2] when
first seen.   This test was repeated after recovery until a steady

* Now Professor H.M. Hodkinson,
      Hammersmith Hospital, London W12.

+ Now Professor D.S. Smith,
      Flinders University of South Australia,
      Bedford Park, South Australia.

mental state had been achieved.   Patients who scored more than
7/16 were generally well able to give informed consent to take part
in the study.   Because of the difficulty in obtaining fully
informed consent from those who scored less than 7, we only
proceeded where relatives were able to add their informed consent[3].

Since we wished to see how cerebral blood flow, motor nerve
conduction velocity and the number of motor units varied with
primary dementia as well as age, we studied patients with a history
of steady, progressive mental disorientation which had been present
for at least six months.   Patients with other causes of dementia
and conditions likely to affect peripheral nerve function were
excluded, e.g. those with focal neurological or upper motor
neurone signs, head trauma, tumour, Parkinsonism, diabetes,
thyroid disease, pernicious anaemia, syphilis, rheumatoid
arthritis, peripheral neuritis and those on drugs or with metabolic
disorders likely to affect peripheral nerve function.

Cerebral blood flow was measured by the Xenon-133 inhalation
technique[4,5].   The subject breathed Xenon-133 at a concentration
of 1 mCi/litre for five minutes using a spirometer with a carbon
dioxide absorber and an oxygen supply to maintain a constant volume.
The Xenon-133 was administered using a pilot's oxygen mask for
maximum patient comfort.   Detectors placed on either side of the
head measured activity in the brain and extra-cerebral tissue
whilst a third detector was used to measure the activity in expired
air.

End expired air activity was assumed to represent arterial
concentration of Xenon-133 and was used to correct the head
activity for re-circulation of the tracer[5].   Correction for extra-
cerebral activity was made by continuing cerebral recording for 40
minutes and assuming that the slow washout of tracer at the end of
the period represented extra-cerebral tissue.   The corrected
cerebral curves were analysed into two components, representing
blood flow in white (f_w) and gray matter (f_g)[6].   The gray matter
flow, the percentage weight of cortex in the field of the detector
(W_c) and the weighted mean flow (f_m) can be calculated for each of
the detectors.   The whole procedure lasts about 45 minutes and
elderly patients tolerated it well, indeed some fell asleep.   There
were practical difficulties in some patients in obtaining an air-
tight fit around the mask due to lack of teeth or gum recession,
or the shape of the nose.   Beards, too, caused problems, since
Xenon-133 tended to dissolve in the grease on the hair.

Motor nerve conduction velocity was measured in the lateral
popliteal nerve after the patient had been sitting in a room for
five minutes at a room temperature of 20°C.   The technique used
was that described by Levy and Poole[7] and Levy and his colleagues[8],
with the following differences.   The equipment was a Medelec MS6

system with an NS6 stimulator and a DF06 display and recording unit.
The NS6 produces a range of stimulus duration, but 0.1 millisecond
was used with a maximum output of 300 volts.   The stimulus
repetition frequency was 1/second.   The output impedance was less
than 300 ohms.   The techniques for stimulating and recording motor
unit activity were those described by McComas and colleagues,
except that the Medelec equipment mentioned above was used.
Tracings of motor unit activity were displayed on a cathode ray
tube and thence on to direct print paper.   The method depends on
carefully grading the strength of the electrical stimulus applied
to the motor nerve to extensor digitorum brevis to activate a single
motor unit and to recruit successive units singly.   The amplitude
produced by activating the whole muscle is then determined and
consequently the total number of units can be obtained by division.
There are a number of assumptions involved in this method, which
are fully discussed by McComas[1].

RESULTS

A total of 91 patients was studied, but not all completed the
full range of studies.   There were 33 men and 58 women, with an
average age of 78 years, and a range of 65 - 94 years.   The mental
test scores had a mean of 10 and ranged from 3 to the maximum of 16.
Analyses used were standard methods of simple and multiple
regression.   Multiple regression has the advantage of allowing
assessment of the combined contribution of two or more variables to
prediction of a dependent variable, and determining the significance
of each.

Simple regression of mental test scores against age and sex
showed both to have a significant negative correlation,
($n = 91$, $r = -0.02967$ and -0.2164, $p < 0.001$ and < 0.05 respectively).
Multiple regression against both variables showed that while age
made a significant contribution ($t = 2.41$, $p < 0.02$), sex failed to
do so in the presence of age ($t = 1.34$, not significant).

Simple regression of motor nerve conduction velocity against
mental test score and age showed no significant correlation.
Multiple regression of motor nerve conduction velocity against sex
and mental test score, however, showed both to make a significant
independent contribution, but age did not (Table 1).

Multiple regression of motor nerve conduction velocity against
cerebral blood flow, sex, age and mental test score showed that
motor nerve conduction velocity had a more powerful relationship
with cerebral blood flow (results for f_g and f_m were practically
interchangeable).   Sex and age also contributed to the regression,
but mental test score failed to do so (Table 2).

Table 1. Multiple Regression of Motor Nerve Conduction Velocity
          against Mental Test Score and Sex

Variable	regression co-efficient(b)	S.E.(b)	t	P
Sex	4.5153	1.9554	2.31	<0.025
Mental test score	0.5761	0.2658	2.17	<0.05

n = 81   R = 0.3075

Motor units showed no significant correlations with motor nerve
conduction velocity, mental test score or age.   However, there was
a significant negative correlation with cerebral blood flow
($n = 56$, f_m, $r = -0.3604$, $p < 0.01$; f_g, $r = -0.2650$, $p < 0.05$).

Cerebral blood flow was significantly correlated with mental
test score ($n = 70$; f_m, $r = 0.5317$, $p < 0.001$; f_g, $r = 0.5023$,
$p < 0.001$).
There was no correlation with age or sex.

Table 2. Multiple Regression of Motor Nerve Conduction Velocity
          against Sex, fg and Age

Variable	regression co-efficient (b)	S.E. (b)	t	P
sex	6.6245	2.2650	2.92	<0.005
f_g	0.2865	0.1242	2.31	<0.025
Age	-0.3265	0.1562	2.09	<0.05

n = 60   R = 0.4716

DISCUSSION

Increasing age is associated with deterioration of memory and
intellect in 'normal' elderly people.   There is little evidence
of widespread neuronal fall out except, however, in the cerebellum.
The finding in this study of a significant negative correlation of
mental test score with age, is not surprising, and is in line with
general experience.   The correlation of mental test score with
cerebral blood flow is as expected, the correlation being most
marked with mean blood flow and in the gray matter flow.   When,
however, cerebral blood flow is analysed against age and mental test
score, the latter is the stronger correlation, a conclusion
supported by Obrist and colleagues[26].

There is no agreement about the effect of age on motor nerve
conduction velocity.   Wagman and Less[9] , Norris and colleagues[10],

La Fratta and Canestrari[11], and Dorfman and Bosley[12] have shown a decrease with age, especially those 60 years of age[12]. Trojaborg[13] calculated the rate of decline as 2 metres/10 years. On the other hand, Levy and Poole[7] and Levy and his colleagues[8] were unable to detect any significant change. However, their age range was limited and their second study[8] did show a negative correlation which did not reach significant levels. Our study showed a negative correlation with age, which only became significant when combined with the effect of cerebral blood flow and age. This also may be because the age range is too short with no patient being younger than 65 years. Viskoper and colleagues[14] have argued that decrease in motor nerve conduction velocity could be due to hypertension, particularly if the diastolic pressure is greater than 120 mm/Hg. However, Bridgman and colleagues[15] were unable to substantiate their results and Kraft[16] argued that Viskoper and his colleagues[15] had not allowed for changes in velocity due to skin temperature or age. Johnson and Olsen[17] have demonstrated that motor nerve conduction velocity decreases by 5% for every degree Centigrade fall in temperature.

Levy and his colleagues[8] demonstrated a non-significant slowing of motor nerve conduction velocity in a group of demented patients. When the subjects were divided according to dementia score, it was found that those with values above 7 had a significantly slower motor nerve conduction velocity compared with those who scored less than 7. This compares well with this study, where there was a significant correlation of motor nerve conduction velocity with mental test score, an effect which outweighed that of sex and age. Levy and colleagues[8] argued that the degenerative process in peripheral nerves may affect only the larger fibres and might parallel changes in larger cortical neurones. The findings here of a correlation between motor nerve conduction velocity and cerebral blood flow would support this.

McComas and colleagues[1] in a study of fit subjects aged 4 - 58 years were unable to find a significant decrease in motor units with age. Our study showed a non-significant negative correlation with age. There was no correlation between the number of motor units and motor nerve conduction velocity, but the number of motor units did correlate strongly with cerebral blood flow, particularly f_g and f_m, again suggesting that changes in the peripheral nervous system mirror those of the cortical neurones. The fact that correlation is a negative one is rather surprising.

The effect of age on cerebral blood flow has been the subject of much investigation. Sokoloff[18] failed to demonstrate any significant changes in cerebral blood flow or oxygen consumption in a group of highly selected fit, elderly men up to the age of 70 years, compared with young adults. On the other hand, Kety[19] showed a fall in cerebral blood flow with age in elderly asymptomatic subjects. However, these measurements were estimates of mean

perfusion rates for the whole brain and not for the cortex itself.
Thus, small changes might occur in the cortex and not be detected.
When Wang and Busse[20] used Xenon-133 inhalation methods to study
normal aged subjects with an average age of 80 years, they found a
reduction in cerebral blood flow compared with young adults. This
agrees with this study, where there was a negative correlation of
cerebral blood flow with age, which did not reach significant levels,
perhaps because the age range was too short.

   Studies of cerebral blood flow in dementia show some discrepancies.   O'Brien and Mallett[21] and Hachinski and colleagues[22]
demonstrated a significant difference between primary and multi-
infarct dementia, but Obrist and colleagues[23] and Ingvar and
Gustafson[24] did not.   This could be due to differences in the
severity of the dementia, since Hachinski and colleagues[21] studied
earlier stages of primary dementia, while the latter studied more
severe cases.   Consequently, it appears that the end result of
primary or multi-infarct dementia is reduced cerebral blood flow.
However, O'Brien and Mallett[21] and Melamed and his colleagues[25]
failed to show any correlation between the degree of cerebral
atrophy or the size of the ventricles and cerebral blood flow,
suggesting that loss of brain substance is not an important factor
in the reduction of cerebral blood flow in dementia.

   Ingvar and Gustafson[24] and Obrist and colleagues[26] have shown a
significant correlation between regional cerebral blood flow and the
degree of mental acuity.   Wang and Busse[20] have shown such a
correlation exists within a group of elderly volunteers.   Those
with higher cerebral blood flow gave significantly better results
with a Wechsler Adult Intelligence Scale. The results of this study
confirm the correlation between mental test score and cerebral blood
flow, particularly f_m.   Furthermore, Ingvar and colleagues[27] have
shown that during activation with psychological tests, the cerebral
blood flow augmentation in association areas in demented patients is
not so marked as in demented controls.   Indeed, flow may be
diminished, and increased elsewhere (intellectual steal).

SUMMARY

   Cerebral blood flow, motor nerve conduction velocity and the
number of motor units were studied to see what changes occurred with
age and dementia.   Mental test score and motor nerve conduction
velocity deteriorated with age, and correlated with cerebral blood
flow.   The correlation between mental test scores and cerebral
blood flow was stronger than that with motor nerve conduction
velocity, suggesting that the latter may be acting as a reflection
of the former.   The number of motor units correlated only with
cerebral blood flow.

REFERENCES

1. A. J. McComas, P. R. W. Fawcett, M. J. Campbell, R. E. P. Sica,
    Electrophysiological estimation of the number of motor units
    within a human muscle, J.Neurol.Neurosurg.Psychiat. 34:121
    (1971).
2. M. J. Denham and P. M. Jefferys, Routine mental testing in the
    elderly, Modern Geriatrics 2:275 (1972).
3. M. J. Denham, A. Foster, D.A.J. Tyrell, Work of a district
    ethical committee, Br.Med.J. 2:1042 (1979).
4. B. L. Mallett, N. Veall, The measurement of regional cerebral
    clearance rates in man using Xenon-133 inhalation and
    extracranial recording, Clin.Sci. 29:197 (1965)
5. J. C. W. Crawley, Recent technical developments in the Xenon-
    133 inhalation techniques for cerebral blood flow measurement.
    in: "Progress in Stroke Research", R. M. Greenhalgh and F. C.
    Rose, eds., Pitman Medical and Scientific Publishing Co.,
    London (1979).
6. K. Høedt-Rasmussen, E. Sveinsdottir, N. A. Lassen, Regional
    cerebral blood flow in man determined by the intra-arterial
    injection of a radio-active inert gas, Circ.Res. 18:237
    (1966).
7. R. Levy, E. W. Poole, Peripheral motor nerve conduction in
    elderly demented and non-demented psychiatric patients,
    J.Neurol.Neurosurg. & Psychiat. 29:362 (1966).
8. R. Levy, A. Isaacs, G. Hawks, Neurophysiological correlates of
    senile dementia: 1. Motor and sensory nerve conduction
    velocity, Psycholog.Med. 1:40 (1970).
9. I. H. Wagman, H. Lesse, Maximum conduction velocities of motor
    fibres of ulnar nerve in human subjects of ages and sizes.
    J.Neurophysiol. 15:235 (1952).
10. A. H. Norris, N. W. Shock, H. Wagman, Age changes in the maximum
    conduction velocity of motor fibers of human ulnar nerves.
    J.Appl.Physiol. 5:589 (1953).
11. C. W. La Fratta, R. E. Canestrari, A comparison of sensory and
    motor nerve conduction velocities as related to age,
    Arch.Phys.Med.Rehabil. 47:286 (1966).
12. L. J. Dorfman and T. M. Bosley, Age-related changes in peripheral
    and central nerve conduction in man, Neurol. 29:38 (1979).
13. W. Trojaborg, Motor and sensory conduction in the musculo-
    cutaneous Nerve, J.Neurol.Neurosurg.Psych. 39:890 (1976)
14. R. J. Viskoper, J. Chaco, A. Aviram, Nerve conduction velocity
    in the assessment of hypertension, Arch.Int.Med. 128:574
    (1971).
15. J. F. Bridgman, L. Bidgood, R. Hoole, Nerve conduction velocity
    and hypertension, Br.Med.J. 3:500 (1973).
16. G. H. Kraft, Nerve conduction velocity, Arch.Int.Med. 130:447
    (1972).
17. E. W. Johnson, K. J. Olsen, Clinical value of motor nerve
    conduction velocity determinations, J.Amer.Med.Assoc.172:2030
    (1960).

18. L. Sokoloff, Cerebral circulation and metabolic changes assoc-
    iated with ageing, Res.Publ.Assoc.Res.Nerv.Ment.Dis.
    (Baltimore) 41:237 (1966).

19. S. S. Kety, Human cerebral blood flow and oxygen consumption as
    related to ageing, Res.Publ.Assoc.Res.Nerv.Ment.Dis.
    (Baltimore) 35:32 (1956).

20. H. S. Wang, F. W. Busse, Correlates of regional cerebral blood
    flow in elderly community residents, in: "Blood flow and
    metabolism in the brain", A. M. Harper, W. B. Jennett,
    J. D. Miller and J. O. Rowan, eds., Churchill Livingstone,
    London (1975).

21. M. D. O'Brien, B. L. Mallett, Cerebral cortex perfusion rates
    in dementia, J. Neurol.Neurosurg.Psychiat. 33:497 (1970).

22. V. C. Hachinski, L. D. Iliff, E. Zilhka, G. H. DuBoulay,
    V. L. McAllister, J. Marshall, R. W. R. Russell, L. Symon,
    Cerebral blood flow in dementia, Arch.Neurol. 32:632 (1975).

23. W. D. Obrist, E. Chvian, S. Cronquist, D. H. Ingvar,
    Regional cerebral blood flow in senile and presenile
    dementia, Neurol. 20:315 (1970).

24. D. H. Ingvar, L. Gustafson, Regional cerebral blood flow in
    organic dementia with early onset, Acta.Neurol.Scand.
    (Supp.43) 46:42 (1970).

25. E. Melamed, S. Lavy, F. Siew, S. Bentin, G. Cooper, Correlation
    between regional cerebral blood flow and brain atrophy in
    dementia, J.Neurol.Neurosurg. & Psychiat. 41:894 (1978).

26. W. D. Obrist, H. K. Thompson Jr., H. S. Wang, W. E. Wilkinson,
    Regional cerebral blood flow estimated by [133] Xenon
    inhalation, Stroke 6:245 (1975).

27. D. H. Ingvar, J. Risberg, M. S. Schwartz, Evidence of subnormal
    function of association cortex in presenile dementia,
    Neurology 25:964 (1975).

THE CT SCAN IN THE DIAGNOSIS OF BRAIN FAILURE

F. I. Caird

Professor of Geriatric Medicine
Southern General Hospital
Glasgow

The introduction of computerised tomography has transformed neurological investigation, particularly of the elderly, in whom it is now possible to obtain accurate anatomical information about the brain by an entirely safe and non-invasive technique. Unpleasant and indeed dangerous methods such as pneumoencephalography and angiography have been rendered largely unnecessary. This is particularly perhaps the case in the investigation of non-focal brain disease such as may result in brain failure[1,2].

In the elderly CT scanning presents only one practical problem - the need to prevent head movement in patients with brain failure. This may be easily achieved by the use of intravenous diazepam, and it is unusual to require a dose of more than 10 mg. In well over 100 cases there has been only one adverse reaction.

There are however considerable difficulties resulting from the problems of definition of the normal CT scan in old age. Anatomical studies[3] clearly show an increase with age in the extracerebral space and CT scans show an increase in the size of the ventricles, beginning at the age of about 50[5,6,7]. None of these studies has entirely overcome the formidable geometrical problems of measurement of ventricular size, which have not been satisfactorily solved even by sophisticated computer techniques[8,9,10,11] nor is there any reliable method of measurement of sulcal width[4].

The quantitation of the severity of brain failure presents further difficulties. Roberts and Caird[12] used three methods, a simple but robust clinical classification[2], a memory and information test, and the Crichton Geriatric Behavioural Rating Scale[13].

225

A statistically significant relationship between the degree of intellectual impairment as measured by these methods and the size of the ventricles as measured by a simple planimetric method was demonstrable in 66 elderly patients without mental or neurological abnormality or with brain failure of varying degree considered non-vascular in origin. Others have confirmed this finding[14]. There is however such a scatter of results, and such an overlap between patient groups that the CT scan is of little value in the individual case. The most that can be said is that if brain failure is severe and the ventricles not enlarged, an acute cause for brain failure should be sought.

The possibility of distinguishing vascular from non-vascular brain failure by CT scanning has also been investigated[15,16]. Multiple infarcts can be shown on the CT scan in only a minority of cases, presumably because they are smaller than the limits of resolution of the scan, which has difficulty in demonstrating lesions of less than about 1 cm in diameter. However, in such cases without visible infarcts, the ventricles are often considerably enlarged, more so than in non-vascular brain failure of comparable severity[15].

CT scanning cannot therefore distinguish old people with brain failure from those without, nor can it reliably decide between the two main causes of brain failure[16]. It can however exclude with confidence other conditions which may give rise to diagnostic difficulty. Intracranial tumour and subdural haematoma can be shown, but both these conditions are highly likely to produce an abnormal scintiscan, which should usually be the first investigation if they are suspected in an elderly patient[17]. Normal pressure hydrocephalus produces relatively characteristic clinical features, with striking difficulty with walking and the development of incontinence before there is severe intellectual impairment. A CT scan should always be carried out if this important though rare condition is suspected, and will show massive ventricular enlargement, to a degree much greater than is seen in either vascular or non-vascular brain failure[7].

In summary, computerised tomography has little place in the positive diagnosis of brain failure in the elderly, but is a very useful adjunct to clinical methods and other more easily available investigations in the exclusion of relatively rare though important conditions which give rise to diagnostic difficulty.

REFERENCES

1. B. Isaacs and F. I. Caird, Brain failure: a contribution to the terminology of mental abnormality in old age, _Age & Ageing_ 5:241 (1976).

2. F. I. Caird and T. G. Judge, "The Assessment of the Elderly Patient", 2nd Edition, Pitman Medical, London (1979).

3. P. J. M. Davis and E. A. Wright, A new method for measuring cranial cavity volume and its application to the assessment of cerebral atrophy at autopsy, Neuropath.Appl.Neurobiol. 3:341 (1977).

4. C. Gyldensted and M. Kosteljanetz, Measurement of the normal ventricular system with computer tomograph, Neuroradiology 10:205 (1976).

5. M. A. Roberts, F. I. Caird, K. W. Grossart and J. L. Steven, Computerised tomography in the diagnosis of cerebral atrophy, J.Neurol.Neurosurg.Psychiat. 39:909 (1976).

6. S. A. Barron, L. Jacobs and W. R. Kinkel, Changes in size of normal lateral ventricles during aging determined by computerised tomography, Neurology (Minneap.) 26:1011 (1976).

7. L. Jacobs, W. R. Kinkel, F. Painter, J. Murawski and R. R. Heffner, Computerised tomography in dementia with special reference to changes in size of normal ventricles during aging and normal pressure hydrocephalus, in: "Alzheimer's Disease: Senile Dementia and Related Disorders", Aging, Vol. 7, R. Katzman, R. D. Terry and K. L. Bick, eds., Raven Press, New York (1978).

8. K. S. Pentlow, D. A. Rottenberg and M. D. F. Deck, Partial volume summation: a simple approach to ventricular volume determination from CT, Neuroradiology 16:130 (1978).

9. W. D. Sager, G. Gell, G. Ladurner and P. W. Ascher, Calculation of cerebral tissue and cerebrospinal fluid space volumes from computer tomograms, Neuroradiology 16:176 (1978).

10. F. Brassow and K. Baumann, Volume of brain ventricles in man determined by computer tomography, Neuroradiology 16:187 (1978).

11. H. Hacker and H. Artmann, The calculation of CSF spaces in CT Neuroradiology 16:190 (1978).

12. M. A. Roberts and F. I. Caird, Computerised tomography and intellectual impairment in the elderly, J.Neurol.Neurosurg. Psychiat. 39:986 (1976).

13. R. A. Robinson, Some problems of clinical trials in elderly people, Geront.clin. 3:247 (1961).

14. M. J. de Leon, S. H. Ferris, I. Blau, A. E. George, B. Reisberg, I. I. Kricheff and S. Gershon, Correlation between computerised tomographic changes and behavioural deficits in senile dementia, Lancet 2:859 (1979).

15. M. A. Roberts, A. P. McGeorge and F. I. Caird, Electroencephalography and computerised tomography in vascular and non-vascular dementia in old age, J.Neurol.Neurosurg.Psychiat. 41:903 (1978).

16. E-W. Radue, G. H. du Boulay, M. J. G. Harrison and D. J. Thomas, Comparison of angiographic and CT findings between patients with multi-infarct dementia and those with dementia due to

primary neuronal degeneration, <u>Neuroradiology</u> 16:113 (1978).
17.   J. Macdonald and F. I. Caird, Unpublished.

COMPUTERIZED EEG IN THE STUDY OF CEREBRAL ALTERATIONS

AND BRAIN PATHOLOGY IN THE AGED

S. Giaquinto

Primario Neurologo
Ospedale Senigallia
Italy

In the second half of the human life a silent but continuous
process affects the cortico-subcortical architecture in the brain,
until either mental or neurological symptoms appear, often in a
short time.   Ageing is not strictly related to calendar years,
the amount of senile change is of greater importance.   The neurons
undergo subtle alterations with progressive loss of horizontally
orientated dendritic systems in those areas which are known to
receive specific synaptic terminals from intracortical fibre
systems.   The cortical changes are more prominent in the third
layer. The intertwined and bundled configuration was proposed as
storage sites for central programmes by Scheibel et al[1].   The loss
of dendritic arborization and the cell death damage the computing
capabilities of the nerve tissue along with mismanagement of coded
outputs.   Quality of neuropil is probably more critical that the
absolute number of dead cells.   The cortical electrical signal is
produced by the activity of a large number of vertical loops which
sweeps across the cortical surface with recruitment of the nearby
loop systems occurring during EEG synchrony.   The gross signal
taken from scalp electrodes is therefore the result of a very high
number of micro-EEGs originating in cortical columns, smaller than
800 microns.

On the other hand, the familiar picture of a large third
ventricle in old people indicates that thalamic changes occur as
well in the ageing process. The anatomy and the physiology of this
structure suggest that its cells play a role of a master control of
cortical activity for information distribution, synchronization,
interface functions: thus, defects in the latter mechanisms are
expected.   Thalamic "pacemakers" are estimated to be 200 microns
in diameter by Andersen et al[2].

It is still unclear how thalamic nuclei may control the activity of the larger cortical EEG generators; excitatory and inhibitory loops as well as post-inhibitory rebound have been proposed as models.

VALUE OF THE EEG

It follows that the study of the EEG is a useful monitor of senile changes for evaluation of both the underlying cortical activity and the remote subcortical control.   Information can be obtained on several functional characteristics, such as the mean voltage, the main frequencies, the inter-hemispheric symmetry, the maintenance of the usual spatial distribution, the reactivity, the shift with different vigilance levels, the sleep organization, the response to photic stimulation, the presence of paroxysmal activities, and changes with metabolic variations.

It therefore appears that a relatively simple, non-invasive method, at reasonable cost and practically devoid of any risk, may supply data which help in clarifying clinical patterns of elderly subjects.   In this respect the EEG provides information on dynamic processes, their time relationships and their evolution, whereas CT scan for instance, although a valuable tool in view of neuro-surgical therapies, gives as yet static and unquantified pictures, which are not necessarily related to brain malfunctions.

Compared to the study of the blood flow, EEG is a more faith-ful monitor of cortical damage in the case of a primary, metabolic brain disease when cerebral circulation is not affected.   Moreover if telemetry is used, subjects are not coupled to a machine, and they are relatively unaffected by the experimental environment. Owing to the short time, we shall deal only with the background EEG activity; evoked potentials and CNV go beyond the limits of the present lecture.

EEG CHANGES IN OLD AGE

General agreement exists on the EEG shift to slower frequencies in senescence. Obrist et al[3] found that in psychiatric patients there was a statistically significant correlation between EEG frequency measurement and circulatory variables, with the highest correlation for cerebral oxygen uptake and the proportion of slow activity.   Such a correlation was absent in healthy old subjects. Wang and Busse[4] pointed out the importance not only of the physical conditions, but also of the demographic characteristics which also affect the EEG slowing process with age.   Hughes and Cayaffa[5] demonstrated, by visual inspection only, an increase of slow and sharp waves, in addition to the positive 6 and 14/sec spikes, after the age of 70 even in subjects without evidence of organic brain disease.   Slow activity in elderly people mainly affects frontal and temporal areas and then it seems to spread to occipital regions.

Visual examination of the EEG, however, can describe the general decrease of frequency, with the major peak shifting 1 c/sec backwards.   More subtle observations require a computer analysis. Several methods have been proposed, but we shall refer to the most common one i.e. to the spectral analysis.   The proper application to the EEG requires a clear knowledge of its statistical properties, otherwise misinterpretations may arise.   The "ideal" EEG would be expressed by a mathematical model of a stationary random process with a zero mean and time-independent power spectrum.   Biological signals do not meet the requirement of a normal or Gaussian distribution, unless short epochs are taken for computer analysis under controlled conditions.   Spontaneous, non paroxysmal activity provides a signal without significant temporal changes, as in the relaxed wakefulness.   Even a mental task may disrupt normal distribution, as Elul[6] demonstrated.   Within selected epochs stationarity is conceivable, i.e. the average statistical properties do not change with time.   The process can be considered as _ergodic_ within the epochs taken under the same conditions from the same subject, when a sample is sufficient to estimate the statistical characteristics of the entire process.   In man, one sample epoch of proper length is sufficient to provide an estimate of the statistical properties of "that" subject, but intra-individual variability, as ascertained by Dumermuth[7], does not match with inter-individual variability and therefore the ergodic model does not apply.

AUTOMATED ANALYSIS

Spectral analysis moves the observer from the time-domain, the clinical EEG, to the frequency-domain, the power spectrum.   The advantage consists in the quantification of the frequencies building up a signal.   The method allows a clear description of mixed EEGs, whose components are not completely identified by visual inspection. In elderly people such an automated analysis provides therefore not only data on the spectral composition but also variations in successive recordings.   The method which we have used over 5 years at our laboratory consists in 1) analogue recording on magnetic tape; 2) off-line analysis on a 16 K digital computer by using FFT; 3) quantification of three descriptors for each of the delta, theta, alpha and beta bands.   Mathematical formulas are given in Fig.1.

The first descriptor is the "Activity", which represents the measurement of the power in each band.   The relative amount of the power is used as percentage of the band value in the whole spectrum. As a matter of fact, an absolute value can be biased by technical factors within the same recording in serial recordings at different

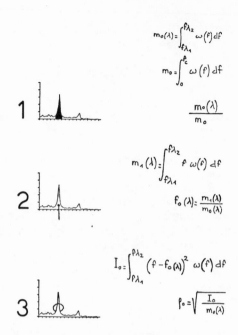

$$m_o(\lambda) = \int_{f\lambda_1}^{f\lambda_2} \omega(f)\,df$$

$$m_o = \int_0^{f_c} \omega(f)\,df$$

1

$$\frac{m_o(\lambda)}{m_o}$$

$$m_1(\lambda) = \int_{f\lambda_1}^{f\lambda_2} f\,\omega(f)\,df$$

2

$$f_o(\lambda) = \frac{m_1(\lambda)}{m_o(\lambda)}$$

$$I_o = \int_{f\lambda_1}^{f\lambda_2} \left(f - f_o(\lambda)\right)^2 \omega(f)\,df$$

3

$$\rho_o = \sqrt{\frac{I_o}{m_o(\lambda)}}$$

Fig. 1.   Activity (1) is the relative power of a band.
Barycentre(2) is the weighted mean frequency and
Radius of inertia (3) is the width of that band.

time intervals.   The sum of the relative power of the four bands
is 1.

The second descriptor is the "Barycentre" or weighted mean
frequency for each band.   This parameter allows the study of
frequency shifts as low as 0.25 c/sec.   A practical application
is particularly appropriate to the study of the process of ageing
when the regression between age and alpha frequency is calculated.

The third descriptor is the "Radius of inertia", which tells
us how wide the band is.   In the case of alpha rhythm, the value
will be higher if alpha-1 and alpha-2 (respectively slow and fast)
are together present in the signal.   On the contrary, the radius
will have a lower value in the case of an alpha consisting of a
main, steady frequency.   More detailed information on these
procedures is reported elsewhere[8].

Another parameter which should be investigated in the clinical
EEG of elderly subjects is the "Coherence".   For the computation
of this parameter the product of the amplitudes of each Fourier
components in two channels is taken.   Coherence is therefore a
further step of the cross-spectrum.   A useful application is seen
in elderly people, because figures are easily obtained on the

similarity of two homologous sources.   Mary Brazier has also shown
the usefulness of the phase information when two EEG signals are
used, especially when travelling waves from a focus are studied:
waves following a definite pathway would display high coherence and
low phase shift.

Application of mathematical methods for computerizing EEG
rhythms requires a precise knowledge of the boundaries which define
the reliability of the method itself.   None of them has unlimited
applications.   For instance, when spectral analysis is used,
transients such as spikes, sharp waves, spindles, K-complexes,
mu-rhythm are lost.   The existence of frequency components in the
spectrum does not necessarily mean that they have physiological
significance, as in the case of a spike-and-wave complex, which is
divided into different harmonics.   Moreover, slow baseline shifts
due to instrumental misadjustment, high skin resistance or sweating,
are treated as delta rhythms and the spectral composition is
therefore biased.   Spectra do not indicate how the power of a
certain band is distributed throughout the epoch; for instance, a
short spell of high voltage alpha may have the same power as a
longer but lower train at the same frequency.   Care must be taken
in the analogue-to-digital conversion;   the sampling rate according
the Nyquist criterion is twice the highest frequency in the signal.
A higher sampling rate would not increase the sample size, thereby
improving the significance as in general statistics, because, as
Cooper[10] has observed, the high sample rate provides more data
values which are not independent from their neighbours.

With a proper mathematical knowledge of the methods and when
the signal is clean, computerized analysis is certainly an improvement.   Transients may be studied, if for some reasons their
search is difficult or cumbersome, by applying mathematical models
for non-stationarities, as proposed by Lopes da Silva et al[11].

RESULTS OF EEG STUDIES

Data of major interest, obtained by using spectral analysis
in old subjects, will be reported herewith.   A study on two
populations, respectively of 50 ± 5 and 85 ± 5 years and with
a sample size $N_1 = 40$ and $N_2 = 20$, has shown typical differences
in the power of EEG bands.   All the subjects were admitted to a
neurological department for medical care, but none of them had
acute or severe brain problems.

Fig. 2 displays in the old subjects a consistent increase of
delta and theta activity associated with a marked decrease of both
alpha and beta power, as seen in the parieto-occipital region.
The highest variation, almost 10%, occurs for the alpha band, i.e.
for the most typical EEG rhythm.   Whereas in middle age slow

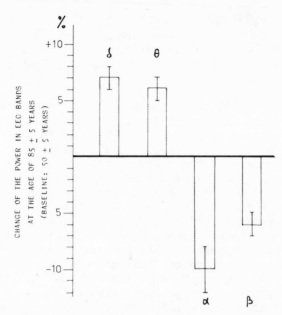

Fig. 2.   Right parieto-occipital leads. In the lifespan
          between 50 and 85 years there is an increase of
          delta and theta with a decrease of alpha and beta activity.

frequencies are often masked by the overwhelming frequency around
9-10 c/sec, in elderly people the sum of the delta and theta
relative power may override the alpha activity and the latter
parameter shows the highest variability.   Roubicek[12] found a
reduction of 7% and Matejcek[13] of 11% in the alpha power between
the younger and older age groups.   The beta frequencies behave
differently, depending on the band width.   Roubicek[12] in his
accurate study demonstrated that a non-divided beta range (12-30
c/sec) has decreased power in elderly people, like the alpha power.
A decrease is also found when the beta band is considered in the
blocks 12-16 and 16-20 c/sec.   Whereas no significant changes take
place in the 20-25 and 25-30 c/sec blocks, a real increase of
higher beta frequencies, 30-40 c/sec, is seen in the lifespan
between 50· and 92 years.   Diffuse beta rhythms have been frequently
observed by us in non-hospitalized subjects over the age of 90.
The lack of knowledge of the medical environment, a general
lowering of the emotional threshold may influence the data at this
age.   Sensitive, shy introverted persons are known to display more
beta activity; however Wang and Busse[4] reported that good health
in senescence is associated with more fast activity and less slow
activity.   It has to be said that the data on changes of the power
in EEG bands refer to a selected population of people who are
neither normal nor overtly sick; in the latter case a great variety
of EEG patterns are found, with focal or general slow waves and
paroxysms.

Whereas the spectral composition shows some variability in the ageing process, depending on pathological, psychological, genetic and demographic factors, the frequency of the alpha rhythm is less influenced.   In the same population that we studied, the barycentre shifted from a mean value of 9.25 to a 8.50 c/sec value in the older group, as shown in Fig. 3.

It was not uncommon, however, to find a frequency of 9 c/sec in 90 year old subjects.   Hubbard et al[14] found this frequency even beyond the age of 100, although without automated analysis. Roubicek[12] and Matejcek et al[13] also indicated by means of regression analysis a slowing of 1, 1.5 c/sec of the alpha frequency from the 50's to the 90's.   It follows that a barycentric frequency at or under 8 c/sec is certainly a pathological value, when collected during wakefulness with a normal blood sugar level.

Fig. 4 shows two spectra obtained by different subjects: in spite of the older age, subject LANFRE has a relative power in the alpha band of .708 against the .187 value of the subject TOMASEL, the former showing no deterioration, whereas the latter had a power of only one-quarter of that value.   One can also see the difference in the main frequency, 8.39 vs. 5.82 c/sec, and in the radius of inertia of the alpha band, which is much smaller in the undeteriorated subject.

It is interesting to observe that EEG slowing in the ageing process is not paralleled by the well-known slowing induced by hyperventilation which occurs at younger ages.   Hughes and Cayaffa[5] failed to see delta and theta rhythms under such a condition in the 70 - 80 decade;   all changes at that age referred to positive findings as focal slow waves.   One might argue that old patients may have a physical inability to hyperventilate, but Ziegler et al[15] have ruled out that hypothesis.   The photic driving response is also very poor in elderly people. Decreased response to flickering is, in our view, of the same order as the slowing of the alpha rhythm, i.e. reduced functionality of thalamo-cortical oscillatory loops.

Hypoxia has relevant effects especially on the senile EEG; automated analysis is unnecessary to check the slowing of the cortical waves due to carotid pressure.   But when autoregulation is impaired, as it can be in the case of even minimal ischemic attacks, slight shifts of the barycentre monitor the parallel changes in brain metabolism.   Freeman and Ingvar[16] have demonstrated that in such a case an increase of the cardiac output yields increased cerebral blood flow.   In the presence of a lack of oxygen the increased blood flow would restore the supply and influence EEG. Some typical patterns were observed by Isaakson et al[17] in patients with complete A-V block, at different heart rates induced by controllable pace-makers.   Elderly people also seem to be very

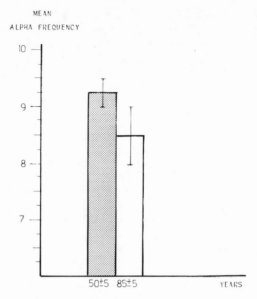

Fig. 3. Right parieto-occipital leads.   In the lifespan between
         50 and 85 years there is a shift in the alpha mean
         frequency of almost 1 c/sec.

sensitive to low blood sugar levels.   We observed complete changes
in the spectral composition with disappearance of alpha activity,
which was replaced by high quantity of theta and beta frequencies.
Unfortunately, at present sufficient data are not available for
drawing a regression line showing the influence of blood sugar
levels on EEG parameters in the elderly subjects.

        Probably the day is not far off, when automated EEG diagnosis
will be provided.   Advanced studies are in progress at the
Department of Clinical Neurophysiology of the Sahlgren Hospital in
Göteborg[18].   Since brain disorders cause displacement of the
dominant frequencies toward the slow part of the spectrum, a pathological signal is recognized when the ratio between calculated age
and real age is low; by contrast, a 1 : 1 ratio is completely
normal.   Theoretical EEG ages were calculated for each derivation
by applying linear regression to a vast amount of normal EEG
material.   At the same centre, Matousek and Petersen[19] developed
a procedure which quantifies the vigilance level according to EEG
spectra.

        Studies on drug-induced changes are helped by the availabilit
of sophisticated technological aids; on the other hand, progress in
psychopharmacology requires more refined controls on the effects at
brain level.   The concept of impaired vigilance regulation is

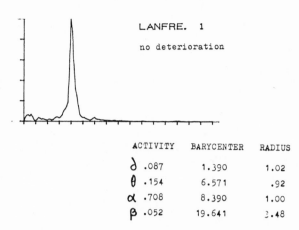

LANFRE. 1

no deterioration

ACTIVITY	BARYCENTER	RADIUS
δ .087	1.390	1.02
θ .154	6.571	.92
α .708	8.390	1.00
β .052	19.641	3.48

TOMASEL. 2

25 % deterioration

ACTIVITY	BARYCENTER	RADIUS
δ .186	2.388	1.12
θ .203	5.802	1.07
α .187	10.076	1.61
β .422	19.671	3.29

Fig. 4. Two power spectra from right parieto-occipital leads of
two subjects having different degrees of mental deterioration.   Subject 1 is 82 and subject 2 is 75 years old.
The main frequencies are, respectively, alpha and theta.
Y-axis: squared microvolts (steps of 30); X-axis:
frequencies (steps of 2 c/sec).

supposed to account for decreased integration of higher processes
in senescence.   Matejcek[20] in this sense elaborated a method,
with the main advantage of concentrating the results from various
frequency band analysis on a single curve.   The technique is based
on the priciple of sequential calculation of EEG frequency spectra.
The interval can be chosen according to the time scale of the
experiment.   The relevant parameters are then extracted from the
spectra and plotted against time.   It is therefore possible to
follow in a compressed way the time course of a drug compared to a
placebo.

SUMMARY

The intention of the present lecture was to supply a progress report on the application of computerized EEG to Geriatrics, rather than to present empirical data. Many aspects deserve further investigation and problems are far from being solved. The variability in data of the EEG analysis emphasizes that the sample must be carefully chosen. The criterion of age alone is not sufficient, because it is clear that pathological factors may worsen the EEG more than chronological age. Moreover, chronic, hospitalized patients may behave differently from similar subjects who have never had a medical check. Ethnical and genetic factors are also important and they must be kept in mind when comparing results obtained by different investigators. Within Italy itself, differences exist between regions: for instance, the region where our EEG data were taken (Le Marche) has one of the highest indices of longevity. Psychological studies help in interpreting data by displaying mental deterioration or exaggerated emotions. Circulatory as well as metabolic studies should also be made, since the EEG can be affected.

Single methods may be refined, but technology must not replace the comprehensive evaluation of an older subject. A statement of Grey Walter is pertinent to the application of EEG studies: "to adopt, to adapt and to improve".

REFERENCES

1. M. E. Scheibel, R. D. Lindsay, U. Tomiyasu and A. B. Scheibel, Progressive dendritic changes in aging human cortex, Exp.Neurol. 47:392 (1975).
2. P. Andersen, S. A. Andersson and T. Lomo, Nature of thalamo-cortical relations during spontaneous spindle activity, J.Physiol.London 192:283 (1967).
3. W. D. Obrist, L. Sokoloff, N. A. Lassen, M. H. Lane, R. N. Butler and I. Feinberg, Relation of EEG to cerebral blood flow and metabolism in old age, Electroenceph.clin. Neurophysiol. 15:610 (1963).
4. H. S. Wang and E. W. Busse, EEG of healthy old persons - a longitudinal study. I. Dominant background activity and occipital rhythm, J. Geront. 24:419 (1969).
5. J. R. Hughes and J. J. Cayaffa, The EEG in patients at different ages without organic cerebral disease, Electroenceph.clin.Neurophysiol.42:776 (1977).
6. R. Elul, The genesis of the EEG, Internat.Rev.Neurobiol. 15:237 (1972).
7. G. Dumermuth, T. Gasser and J. B. Lange, Aspects of EEG analysis in the frequency domain, in:"CEAN, Computerized EEG analysis" G. Dolce and H. Künkel, eds., Fischer Verlag, Stuttgart (197

8. F. Angeleri, S. Giaquinto, F. Marciano, G. Nolfe and C. Pierro, Quantificazione della frequenza media di banda:il baricentro". Riv.ital.EEG Neurofisiol.clin.2:437 (1979).

9. M. A. B. Brazier, Interactions of deep structures during seizures in man, in: "Synchronization of EEG activity in epilepsies", H. Petsche and M. A. B. Brazier, eds., Springer Verlag, Wien (1972).

10. R. Cooper, Measurement of time and phase relationship of the EEG, in: "CEAN, computerized EEG analysis", G. Dolce and H. Künkel, eds., Fischer Verlag, Stuttgart (1975).

11. F. H. Lopes da Silva, A. Dijk and H. Smits, Detection of nonstationarities in EEGs using the autoregressive model - an application to EEGs of epileptics, in: "CEAN, computerized EEG analysis", G. Dolce and H. Künkel, eds., Fischer Verlag, Stuttgart (1975).

12. J. Roubicek, The electroencephalogram in the middle-aged and the elderly, J.Amer.Geriat.Soc. 25:145 (1977).

13. M. Matejcek, K. Knor, P. V. Piguet and C. Weil, Electroencephalographic and clinical changes as correlated in geriatric patients treated three months with an ergot alkaloid preparation, J.Amer.Ger.Soc. 27:198 (1979).

14. O. Hubbard, D. Sunde and E. S. Goldensohn, The EEG in Centenarians, Electroenceph.clin.Neurophysiol. 40:407 (1976).

15. D. K. Ziegler, R. S. Hassanein and A. R. Dick, Effect of age and depth of hyperventilation on a quantitative electroencephalographic response, Clin.EEG 6:184 (1975).

16. J. Freeman and D. H. Ingvar, Elimination by hypoxia of cerebral blood flow autoregulation and EEG relationship, Exp.Brain Res. 5:61 (1968).

17. A. Isaakson, K. Lagergren and A. Wennberg, Visible and non-visible EEG changes demonstrated by spectral parameter analysis, Electroenceph.clin.Neurophysiol. 41:225 (1976).

18. M. Matousek and I. Petersen, Automatic evaluation of EEG background activity by means of age-dependent EEG quotients, Electroenceph.clin.Neurophysiol. 35:603 (1973).

19. M. Matousek and I. Petersen, Automatic measurement of the vigilance level and its possible application in psychopharmacology, Pharmakopsych.Neuropsychopharm. 12:148 (1979).

20. M. Matejcek, Pharmaco-Electroencephalography: the value of quantified EEG in Psychopharmacology, Pharmakopsych. Neuropsychopharm. 12:126 (1979).

PSYCHOLOGICAL ASPECTS OF THE AGING BRAIN

M. Cesa-Bianchi

Director
Institute of Psychology, Faculty of Medicine
University of Milan

Changes in the aging brain are studied by biologists, pathologists, biochemists, biophysicists and neurologists; but psychologists too cannot ignore brain changes in order to understand the factors, or some of them, which contribute in determining behavioural changes in aging people. We must look to psychology as a discipline whose trend is to integrate or to mediate between biological and social events. Therefore, psychology has to collect continuously scientific data and information derived from both these areas.

Since Broca and Wernicke, on the basis of clinical investigations, anatomical studies and experimental researches which have been carried out during many years in the framework of the contrast between "brain localizers" and "antilocalizers", and more recently by neuropsychological approaches on hemispheric lateralization, the relations between neural and psychical functioning have been scientifically examined. Today such relations are no more investigated by a one-way approach (from biological to psychical).

I trust neuroatomists and neurophysiologists for the description of the present knowledge referring to the aging brain. Let me quote only two brief considerations, related to my personal experience and to recent data by gerontological studies.

While I was, for six years, student at the Institute of Pathology of the University of Milan, I had the opportunity of analysing the morphological signs of brain aging, at that time identified as 'senile plaques' and as an hypertrophy and hyperplasia of glial cells, damaging neurons. So, I could observe by the histological examination the brain of a woman aged 102 years old

who had retained apparently normal psychic functioning until she
died as the result of an accident; no indications of brain aging
were found[1].   From these and from other observations it seemed
likely the aged brain does not necessarily show age related changes
when changes described as 'senile' appear they are to be related to
other factors, not to aging in itself.   Such changes are indeed
more frequent in old age than at any other age; but this frequency
cannot demonstrate that aging is the cause, just as a very high
correlation between two variables does not demonstrate a causal
relationship.   So, we may admit that in old age it is statistically more likely that conditions inducing brain changes will occur.
These changes, which may be morphological, biochemical or bio-
physical, are in any case revealed as neuropsychological states -
by qualitative or quantitative modifications affecting physiological or psychological activities.   But psychological activities
and especially the more complex ones which are expressed in the
various behavioural aspects, result from the interaction between
environmental influences and "personality" (the "bio-psycho-social
individuality")[2] ; the psychological disturbances may express a
brain modification and/or a disturbance or a peculiar environmental
characteristic.

The aging of the brain may be seen as the result:
a) Of a genetic programme that is strongly influenced;
b) By the quality and quantity of the environmental stimuli;
c) By life experiences;
d) By the relationship degree between individual trends and their
   potential to be realised,
e) By the health condition and by the way in which aging alters
   other biological systems.

The genetic programme, partly common to all the members of a
species and partly individually differentiated, has been demonstrat
by comparative biological and zoological studies and by studies on
heredity and human aging[3].   The nature and level of the environmental stimuli determine differences in aging of cerebral functions
according to the following principle: the rightly stimulated
functions generally persist and are sometimes bettered, the little
or not stimulated generally decay and disappear[4].

Therefore - if in general aging implies a reduction in the
number of previously utilized functions, a span narrowing for other
functions,with consequently a decay for the not acted behaviour and
a difficulty in starting the activation of new functions.   The
result is a behavioural rigidity, and a loss of plasticity.   An
environment in which stimuli are few and limited to culturally and
intellectually poor activities strongly reduces the number of still
active functions, maintaining only the psychically most elementary
ones.   In these conditions, aged people live only on a biological
dimension, while only very simple mental and social activities
rarely appear.

Life-time experiences[5] too play a basic role in aging. Such a role[6] varies according to the character tendencies of people (sensitizer versus repressor)[7] and their socio-economic status[8].

Self-realization, with the individual expression of trends represents according to some authors[9] a basic human motivation, to which all others are connected.   If this realization is achieved according to life events, aging is presumably able to consolidate the most meaningful tendencies of the genetic programme; if it does not occur, a frustration arises from which may develop a rapid decay in brain activities.

One of the most frequent examples is related to retirement;[10] the compulsory interruption of work activity, often determines, especially in industrialized societies, a sudden lessening in social role, in prestige, in economic power, in family authority; in consequence there is often engendered a sense of uselessness, a depressive and self-devaluative reaction, and a rapid decay in intellectual abilities.

In old age, diseases represent particularly frequent events, and towards them aged people react less efficiently than in previous years.   This happens because until a certain age illness is treated like an external event, a "foreign body" against which the individual reacts; in old age, on the contrary, it is lived like a personal feature, an assimilated event that one cannot eliminate, in the same way that the near death will do. This is true not-withstanding the fact that the effect induced by illness in old age[11] is not so much connected to the 'objective disease' but to the subjective feeling, to the perception of it, which express the personal attitudes towards life and death.   These attitudes are strongly influenced by an individual's socio-economic status[12]. Finally, according to its central role in the integrative processes, brain aging is influenced by functioning and aging in other biological systems.

Aging implies a reduction of sensory and psychomotor activit-ies that may be to a certain degree vitiated by the holding of other activities.   In old age, cerebral arterosclerosis and central nervous system degenerative diseases frequently occur.

As Birren[12,13] quotes referring to the former, pathological changes lead to a progressive behaviour deterioration and the progressive increase in physical symptoms; but the development is also influenced by the environment - personality interactions. These syndromes although they affect a high percentage of old people, certainly do not represent a universal phenomenon in old age.

Among the indices utilized in order to demonstrate patholog-ical decay there are the electroencephalographic ones.   Wand and

Busse[12] at Duke University examined at an interval of 3-4 years 182
volunteers, and observed that the EEC occipital frequency is strictl
related to 'physiological' aging; this relationship may be altered
by various pathological events.   It is interesting to note that
like the slowing of the alpha frequency, a behavioural slowing down
has been observed also in healthy, highly educated aged people.
Schaie and Strother[14] have studied in 1968 a group of men and women
retired from academic activity or from a correspondingly high
professional job.   The most frequent physical data include visual
and acoustic impairment, recorded in almost 90% of subjects, and
cardiovascular deterioration in about 20%.   In many cases these
authors have found a slowing down of psychomotor speed: this
slowing indicates, together with the correlations between these data
and other psychological variables, that in old age speed becomes a
general factor, presumably influencing levels of functioning in
other systems.

     Particularly interesting are the relationships observed between
blood pressure and non-verbal intelligence (Heron and Chown, quoted
by Birren and Renner[12]): a negative correlation has been found at
high, but not at average or low, levels of blood pressure.   So, we
may hypothesize that in the relationship between physical health
and behaviour in old age a mechanism similar to a homeostatic one
is operating.The same relationship has been observed between oxygen
utilisation by the brain and the quality of intellectual activity.
Therefore, the relationship between cerebral functioning and
behaviour in old age cannot be expressed as a causal, deterministic
one.

     According to Dilman[15]some alterations in this homeostatic
mechanism are fundamental to the aging process.   Such alterations,
particularly those determinant by changes in the hypothalamic
system, might induce the metabolic disorders which are more frequent
in old age: obesity, prediabetes, adult diabetes, arteriosclerosis,
from which may result impairment in cognitive processes.

     Most authors agree that it is difficult to indicate precise
limits, valid for everyone, between physiological and pathological
aging and to attribute behavioural changes to single physiological
or pathological causes.

     On the basis of all these data, we may therefore conclude that:

1.   Cerebral aging, besides being determined by the genetic
     (species specific and individual) programme, is strongly
     influenced by past and present life conditions; experience
     modifies brain function not only in the child, but also in the
     adult and aged;
2.   Cerebral aging from a neuro-psychological aspect is expressed
     in various behavioural changes, which may be influenced by
     environmental factors.

Owing to the high variability in such factors as the function of genetic programme, social class, life experiences, health, personal style, and so on we may ask if it is more meaningful to speak of brain aging, or of aging patterns particularly frequent in aging brains.

After having agreed that cerebral aging is essentially different and unrepeatable, because the life events for every person are different and unrepeatable, it is possible to indicate the trends which,also at different levels and ways, are expressed in the psychological aspects of the aging brain:

a) A reduction in the activities span, with the extinction of the less utilized ones.   The reduction may be accompanied by a bettering of the persisting activities.   The genesis of these two facts may be interpreted by the Broadbent Field Theory[16], according to which, among the many thousands of stimuli hitting every moment our sensorial receptors, only the ones having certain features allowing them to bypass the 'filter' which separates the peripheral and the central structures to reach the brain centre.   Only those stimuli will be centrally received to determine an individual reaction.

The Broadbent model was founded on a series of 'principles' that we summarize.   The first, and probably the most important, is the principle stating that the nervous system functions as a unique communication channel, with limited capacity.   Therefore, an operation is needed for the selection of channel inputs; it consists in selecting all the sensory events having common features.   The probability for some events to be selected rather than others is made greater by certain features (physical intensity, time spent since inputs of the same class have entered the channel, and so on) and by 'drives': an animal deprived of food will be more likely to select food linked stimuli ('reinforcement').

Then the probability will increase to select classes of events previously selected just before the reinforcement.   The model (see Fig. 1) provides also long term 'archives' (or memory) of the conditional probabilities for past events: the decay for the information conserved in these archives is much longer than for the memory previously considered (with a maximum time of some seconds). In the long term memory the conditional probability is conserved to select a stimulus after a previous one has been retrieved.   Thus an organism being stimulated by a drive will behave in such a way to receive those stimuli which have the highest probability to exit with reinforcement.

Therefore, the reduction in number of activities that appears with aging may be related to a progressively greater selectivity of the 'filter', which could become more and more specific;   the

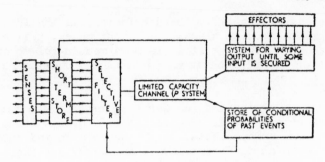

Fig. 1. Broadbent (1958)

filter would be passed by a more and more limited quality of stimuli
with the exclusion of unusual and innovative qualities.   But this
limitation could determine a progressively more acute sensibility
towards the inputs accepted by the channel, and subsequently the
ability to react to them in a more and more precise and detailed
way.

b) The difficulty in learning new tasks seems understandable
in terms of modifications in the filter which changes from an
elastic to a progressively more rigid condition.   The earlier,
juvenile condition, is therefore characterised by the ability to
receive highly differentiated inputs, and to be changed in such a
way as to receive new inputs, which had been initially rejected by
the filter.   By contrast, the old age condition seems to be
defined by the ability to receive only very similar inputs, and the
inability to adjust itself to different inputs.

c) The tendency to learn 'by doing' more than 'by remembering'.
The progressive decay in the ability for unusual inputs to be
transferred to the brain central structures determines in old age
the difficulty of behaviour modifications as a result of innovative
stimulations.   The inputs which continue to be memorized are
identical or similar to those retained for a long time; in this way,
the content of short term memory is becoming the same as that for
long term memory.   Therefore, behavioural modifications may arise
only on the basis of activities different from those habitually
carried out; in fact, these activities are self-expressing like a
new way to cope with new tasks.

d) Motor activities are carried on adequately if they are
maintained at relatively low intensity levels, but they become
maladjusted if such levels overcome a threshold which is lowered
with aging.   Thus older people are able to carry on usual
activities which are yet self-reducing in their span but they are
troubled in behaving in a customary way: for example, they walk
correctly, but do not succeed in running.

e) Motor activities are progressively reduced in speed, while their accuracy may become higher. Sensory activities, and especially the visual and acoustic ones, are progressively reduced in their efficiency; but perception is not necessarily getting worse. To understand this we must consider at least two factors; the first[17,18] concerns the strengthening by experience of perceptual modalities (i.e. in visual perception the form, size, colour and brightness) with the result that perception may correct the sensory deficit.   The second point refers to Gestalt Psychology[19,20,21], which has proposed the model of 'perceptual field', in which two kinds of operating forces are opposed: cohesive forces which are central in origin, integrative (organizing, simplifying) in function and 'perceptual' in nature, and restraining forces, which are peripheral in origin, segregative (disintegrative, autonomous) in function and 'sensory' in nature.   Perception is the end result of a resolution, a balance between the two kinds of forces, and it will appear as 'better' (that is to say, more expressive of its 'perceptual' nature) as the cohesive force influence is maximized or the restraining force strength is reduced.   Thus we may understand how the reduction of the restraining forces which is determined in aging by the decay of sensory activities may cause a shift favouring the cohesive forces in the balance between those and restraining ones.   In such a way we shall have 'better' perceptions, as are obtained by experimentally reducing the sensory inputs[22] or in visual efective children[23,24].   The changes simultaneously appearing in motor and perceptual functions may explain the behavioural features of old people: slower but often more precise behaviour, more coherent in maintaining and in improving habitual patterns of activities although less efficient in reacting to new needs, more disposed to make deeper topics already known and to find original solutions about those topics than to explore unknown areas.

f) The trend to consider diseases[25] as intrinsically connected with the person, as events which advance and announce the end of life and as conditions which cannot be overcome successfully. Illness is no longer as in the past an 'extraneous body' to expel and to fight against, and it is assimilated with own's individuality.   The 'aging brain' seems to prepare the acceptance of death as a natural fact, although determined by contingent diseases.

We should no longer speak of 'senectus ipsa morbus', as was thought in the past when the lack of preventive, therapeutic and rehabilitative procedures could perhaps justify this statement, but it is illness which when it appears acts as the announcement of death: 'morbus ipe mors in senectute'.   If the presuppositions justifying the first of the two statements could have been in great part overcome by preventive medicine, the basis from which the second statement is derived will probably be removed by a programme integrating health, psychological and social dimensions.

REFERENCES

1. M. Cesa-Bianchi, A. Degna Tommasini, Note istologiche su di un encefalo di soggetto centoduenne, Atti della Societa Italiana di Patologia 3:737 (1953).
2. P. Fraisse, J. Paiget, "Trattato di psicologia sperimentale, Vol.1: Storia e metodo, Einaudi,Torino (1972).
3. J. R. Birren, V. J. Renner,"Developments in research on the biological and behavioral aspects of aging and their applications",Institute of Medicine, National Academy of Sciences, Washington D.C. (1976).
4. M. Cesa-Bianchi, Contributo allo studio delle modificazioni psichiche in rapporto con l'età, in: "Contributi del Laboratorio di Psicologia, Vita e Pensiero, Milano (1956).
5. J. E. Birren, V. J. Renner, Research on the psychology of aging: principles and experimentation, in: "Handbook of the psychology of Aging, J.E. Birren and K. W. Schaie, eds., Van Nostrand Reinhold, New York (1977).
6. E. L. Hinkle, H. G. Wolff, Ecologic investigations of the relationship between illness, life experiences and the social environment, Amer.Med.Ass.Arch.Int.Med. 49:1373 (1958).
7. D. M. Stuart, "The rating of stressful life events by subjects of different ages", Unpublished Master's Thesis, University of Southern California (1975).
8. E. Shanas, G. L. Maddox, Aging, health and the organization of health resources, in: "Handbook of aging and social sciences", R. J. Binstock and E. Shanas, eds., Van Nostrand Reinhold, New York (1976).
9. See Carl Rogers in: M. Cesa-Bianchi, Gli aspetti teoretici della personalità, Questioni di Psicologia, a cura di L. Ancona, La Scuola Editrice, Brescia, pp.1-52 (1962).
10. M. Cesa-Bianchi, Il problema degli anziani in una società in trasformazione: aspetti psicologici, Nuova Rassegna di Legislazione, Dottrina e Giusisprudenza, 24:146 (1967).
11. R. Schmitz-Schrzer, Health, self-perceived and assessed by a physician and behavior, paper unpublished presented at the World Conference on Aging and Policy, Vichy (1977).
12. J. E. Birren, V. J. Renner, Health, behavior and aging, paper unpublished presented at the World Conference of Aging and Policy (1977).
13. M. Cesa-Bianchi (a cura di), "Psicologia della senescenza", Franco Angeli, Milano (1978).
14. K. X. Schaie, C. R. Strother, Cognitive and personality variables in college graduates of advantaged age, in: "Human Aging and Behavior", G.A. Talland, ed., Academic Press, New York (1968).
15. V. M. Dilman, The hypothalamic control of aging and age-associated pathology. The elevation mechanism of ageing, in: "Hypothalamus Pituitary and Aging", A. V. Everitt and

J. A. Burgess, eds., Charles C. Thomas, Springfield,
Illinois (1976).

16. D. E. Broadbent, "Perception and Communication", Pergamon
     Press, London (1958).

17. D. Katz, "World of Colour", Kegan, London (1933).

18. C. Musatti, "Condizioni dell'esperienza e fondazioni della
     psicologia", Universitaria, Firenze (1964).

19. K. Koffka, "Principles of Gestalt Psychology", Harount,
     New York (1935).

20. J. V. Brown, A. C. Voth, The path of seen movements as a
     function of the vector field, Amer.J.Psychol. 49:543 (1937).

21. M. Cesa-Bianchi, A. Beretta, G. Girotti, R. Luccio, P. Renzi,
     Perceptual constancies and ageing, Proceedings of 7th Int.
     Congr. of Gerontology, Wien (1966).

22. C. E. Osgodd, "Method and Theory in Experimental Psychology",
     Oxford University Press, New York (1953).

23. A. Beretta, M. Cesa-Bianchi, O. Poli Contini, Contributo
     sperimentale allo studio delle implicanze esercitate dal
     deficit visivo sulla dinamica del campo percettivo,
     Atti del XIV Congresso degli Psicologi Italiani, Napoli,
     (1963).

24. M. Cesa-Bianchi, E. Caracciolo, Ricerca sui fattori e sulle
     espressioni dell' adattamento nell' età senile, Studi e
     Ricerche di Psicologia 1:15 (1963).

25. M. Cesa-Bianchi, A. Zandomeneghi, Fattori individuali ed
     extraindividuali del disadattamento nell' età senile; la
     reazione alla malattia, Studi e Ricerche sui Problemi
     Umani del Lavoro Milano (1963).

26. M. Cesa-Bianchi, 'L' invecchiamento psichico, in:"Psicologia
     della senescenza",M. Cesa-Bianchi, ed., Angeli, Milano
     (1978).

A PSYCHOMETRIC TEST BATTERY FOR EVALUATING THE ACTION OF PSYCHOTROPIC DRUGS

A. N. Exton-Smith

Barlow Professor of Geriatric Medicine
University College Hospital Medical School
London

A comprehensive psychometric test battery is needed to solve two important problems:

1. To investigate the value of pharmacological treatment and prophylaxis against dementia. There is an expanding market in drugs claimed to improve mental function, but there is at present little objective evidence of their efficacy.

2. To investigate the way in which mental function declines in senile dementia and to ascertain if this decline differs qualitatively or quantitatively from that occurring with the natural process of ageing. It is also necessary to obtain information on the changes associated with the early development of dementia, since if a stage of depression in the function of neuronal cells exists before the permanent changes which become manifest as dementia, then it is of fundamental importance that such a stage be recognised as early as possible, because timely drug intervention may improve prognosis.

In order to investigate each of these problems serial psychometric testing is required. The assessment of the action of a drug requires repeated tests during the course of a clinical trial and the investigation of the natural history of dementia requires a longitudinal study, if possible, commencing before the onset of the dementing illness. Essential requirements for psychometric tests used in these investigations are sufficient sensitivity to detect small changes and the need for high test-retest reliability.

THE MEASUREMENT OF INTELLECTUAL CHANGES

Dementia is characterised by progressive intellectual decline
with disturbance of memory appearing among the first signs. Since
tests of memory seem particularly sensitive to change they are
included in most psychometric test batteries. Some current ideas
about the memory process and a few of the tests used in its
assessment will first be discussed. Most psychologists now accept
a multistage model of memory and the model used by Marcer[1] is shown
in Figure 1.

At the front end of the memory system there is iconic or
perceptual memory which stores a representation of the visual or
auditory stimulus for a very brief period of time. If the
information does not decay, it is available for the next stage of
processing, which occurs in primary memory.

Primary Memory

The three most important features of primary memory are its
limited capacity, its susceptibility to disruption by concomitant
activity, and its coding dimension according to the acoustic
articulatory characteristics of the stimulus. Even when information is presented visually whether it is remembered in terms of
sound or visual images depends upon how concrete or abstract the
information is. There is a limited capacity in primary memory
of about seven items - letters, figures or other discrete items of
information. As long as an individual is allowed to rehearse the
material, and provided it does not exceed the store capacity, very
little information is lost with time; but if rehearsal is prevented,
perhaps by the interpolation of a secondary task, loss of information occurs rapidly.

Secondary Memory

The next stage of the memory process is secondary memory and
information is coded on a variety of taxonomic bases, especially

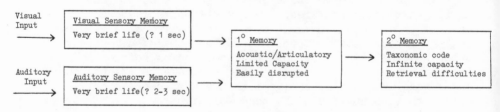

Fig. 1. Model of memory for verbal material (Marcer[1])

those of associative or semantic similarity (Marcer[1]). It is likely
that category clusters can in turn be organized to form higher
order categories.   Thus an hierarchical organization of this kind
makes it possible to process a considerable amount of information
in secondary memory; that is, a large amount of information is
stored under relatively few headings.   Secondary memory does not
require constant rehearsal in order to maintain information intact,
and breakdown in performance tends to be associated with retrieval
difficulties rather than with loss of information.

It must be emphasised, however, that memory performance is not
based on a static unitary system.   Information must be continually
transformed or reorganized and memory failure may be due to
breakdown in any one of a number of points in the information
processing chain.   Thus an older person's deficit in memory occurs
in part at the very beginning of the process with impaired
registration of stimuli and rapid decay of information from iconic
memory.   Ageing also has an effect on short-term memory (STM)
performance - but the impairment is minimal provided that the
material does not require reorganisation and is not subject to
interference.   An individual's STM span can be measured by the
length of a digit sequence that can be remembered, (Digit Span
Forwards and Backwards).   It has been found from cross sectional
studies of subjects of different ages that the digit span forwards
declines by 1 to 1.5 items between the ages of 20 and 70 years[2]
for visually presented material.   The STM performance shows a
less rapid decline in verbally presented material; this is
presumably because STM store uses acoustic/articulatory coding and
an extra transformation is required for visually presented material.
Digit span backwards declines faster with age since the individual
must maintain this material in STM while demonstrating flexibility
and agility in order to re-arrange the items.   Demented patients
show an added deficit in STM and there is evidence that this may
be due to less efficient coding[3].

The greatest degrees of impairment due to ageing and to
organic mental deterioration occur in the transference of information from STM to LTM and in the retrieving of information from
long-term memory.   Miller[4] has investigated these changes in
performance in normal control subjects (mean age 60 years) and in
patients suffering from pre-senile dementia (mean age 60.5 years),
using techniques of free-recall in which the subject listens to a
list of words and tries to recall them in any order immediately
after presentation of the whole list.   For the normal subjects
when the probability of recall of a word is plotted against its
position in the list a U-shaped curve is obtained.   The initial
elevation (the primacy effect) is due to recall from LTM, and the
terminal elevation (recency effect) is due to recall from STM.   On
the other hand, the pre-senile dementia patients remembered very
few words from the beginning of the list; that is there was almost

no initial elevation indicating reduced output from LTM due either to difficulty in recall from LTM or to impaired transfer of information from STM to LTM. The terminal curve was slightly depressed indicating reduced output from STM. Another feature of the results in demented patients is the appearance of intrusion errors, and some of these are from lists presented previously even up to a month before.

Most of the global intelligence tests can be subdivided into verbal/vocabulary and performance subtests. The verbal tests tend to concentrate on well practised and learned skills and they show little decline with age. Performance tests on the the other hand often involve memory and the ability to manipulate new relationships and they tend to show a marked deterioration with advancing age. Moreover, they are also more sensitive to the effects of organic brain damage occurring in dementing illnesses. Examples of these types of performance subtests are digit-symbol substitution and Raven's progressive matrices.

COMPREHENSIVE PSYCHOMETRIC TEST BATTERY

The psychometric tests which are included in this battery used in the Geriatric Department, University College Hospital Medical School are listed in Table 1. These are mainly performance subtests which are the most sensitive to the effects of organic brain damage and to the action of psychotropic drugs. In addition to the conventional psychometric tests there are two tests which merit further description:-

The Picture Matching Task (PMT)

This test has been described by Gedye and Miller[5] and in essence is an automated version of a paired association learning test. There are 48 stimulus pictures which are projected on to a screen in front of the subject and these are arranged hierarchically in a series of 12 filters. The patient responds to the stimulus by pressing one of the two buttons on the right or left side. Initially a simple left or right response is required to the stimulus which is on either the left or right of the screen. In the next two filters two identically matching (in shape, size or colour) stimuli are shown either to the left or the right of the centre of the screen. In the next series of filters, an extra stimulus picture similar in shape and size but not in colour is added to the two matching pictures; the task is to continue to respond to the two matched stimuli. Then pictures of objects and their names appear in the coloured oblong shapes and it is the content of the pictures rather than the colour of the surrounding shape which presents criteria for choice, e.g. two men and one

Table 1. Psychometric Test Battery

TEST	SCORING	REFERENCE
Picture Matching Test (PMT)	Computer graphed learning trajectory Peak filter and cumulative reaction time (ms).	Gedye and Miller[5]
Card Sorting Tests:- CS1 Alternately into 2 piles CS2 Red and black suites, 2 piles CS3 Each suite, 4 piles	Errors and completion time for each subtest	Crossman (modified)[10]
Digit Symbol Substitution (DIGSYM)	Number of substitutions in 90 seconds	Wechsler Adult Intelligence Scale[11]
Raven's Coloured Progressive Matrices (RAVENS)	Number of correct pattern choices before 3 consecutive errors	Raven[12]
Digit Span Forward (DSF) and Backward (DSB)	Number of digits in spans of increasing length before 2 consecutive failures	WAIS[11]
Free-recall Word-Learning List:- Immediate recall (WLI) Delayed recall (WLD)	Number of words recalled from list	Marcer & Hopkins[13]
Stockton Geriatric Rating Scale:- Physical Disability (STPD) Apathy and Inertia (STAP) Communication Failure (STCF) Socially Undesirable Behaviour (STSUB)	Score for each question 0, 1, 2 in direction of increasing deterioration	Meer and Baker[14] Taylor et al[14]
Activities of Daily Living (ADL)	% Efficiency = $\dfrac{\text{Total performance score} \times 100}{\text{Sum of max. scores for all 7 tests}}$	Keet[6]

cabbage.   The final stage of the task is the presentation of
<u>associated</u> but no longer identically matched stimuli, e.g. 'boy',
'girl' and 'knife'; 'bread', 'cheese' and 'horse'.   The computer
keeps a record of the patient's performance.   Taking each filter,
the time between each of the responses, whether right or wrong, is
recorded in graph form with filter numbers plotted against minutes
in the test.   The summary information provided in the computer
printout gives the peak filter reached and the time the patient
took to get there.

Activities of Daily Living Assessment (ADL)

This is a kit designed by Keet[6] to test functional ability.
Everyday tasks which are normally well practised and are not
dependent upon sex and cultural background are used.   The tasks
progress from fine finger movements (e.g. stacking coins), hand and
wrist movements (e.g. turning a tap, opening a Yale lock) and
shoulder movements, then displacement of the trunk and body such as
transferring from bed to chair, and walking over a timed distance.
Performance is scored on a five-point scale and timed in seconds.

PERFORMANCE IN THE PSYCHOMETRIC TESTS

The correlation between the performances in the 8 tests of the
battery have been examined in 30 elderly patients (mean age 79.8
years, age range 72 - 97 years) who were selected randomly from a
district geriatric service (Exton-Smith et al)[7].   The psychometric
testing was carried out over a period of three days.   In order to
assess the test-retest reliability of the PMT this test was given
four times, twice on day 1 and twice on day 3.   The remaining tests
were each given once in the order shown in Table 1.

Correlations Between Tests

Analysis of the performances of 30 patients has shown that the
correlation between the 4 PMTs is high (ranging from $r = 0.88$ to
$r = 0.95$).   These are illustrated in Figure 2 where the performances
for the 4 PMTs are shown as cumulative distribution curves.   It
will be noticed that there is little learning effect for those
capable of reaching the lower filters, but when the peak performance
is in the higher filters slight improvement in performance occurs
on re-testing.

Following the construction of a correlation matrix it was found
that the performances on most of the tests were highly inter-
correlated; but Digit Span Forward and Digit Span Backwards and

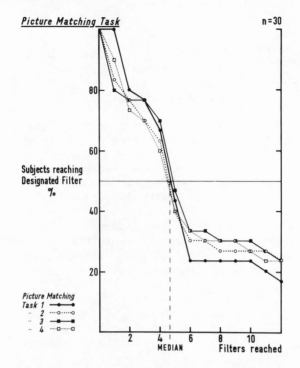

Fig. 2.   Cumulative distribution curves of performance for
           each of the four picture matching tasks.

Word Lists with immediate and delayed recall showed the lowest
correlations with other tests.

Principal Component Analysis

        The correlation matrix had 4 eigenvalues greater than unity
and the corresponding principal components explained 75% of the
total variance.

        Component I. (45% of variance) contains high loadings for:-
                Picture Matching Task (PMT)
                Card Sorting Tests (CS1, 2 and 3)
                Digit Symbol Substitution (Digitsym)
                Raven's Progressive Matrices (Raven's)
                Stockton Socially Undesirable Behaviour (STSUB)
                Activities of Daily Living (ADL)

This component could be interpreted as containing a general factor
and most of the tests measure speed and flexibility of processing

capacity.

Component II. (12% of variance) contains high loading for:-
    Digit Span Forward Test (DSF)
    Stockton Physical Disability (STPD)
    Stockton Communication Failure (STCF)

This component could be interpreted as the ability to perceive
auditory stimuli and to attend to them by immediate repetition.

Component III.(10% of variance) has high loading only for:-
    Digit Span Backward Test (DSB)
    Stockton Apathy Subtest (STAP)

The ability to hold and reprocess information may be represented
in this component and it is likely that those who have a poor
performance on the Stockton Apathy and Inertia scale are unable to
carry out the process of immediate recall and re-ordering of a
string of digits.

Component IV (7% of variance) contains a high loading only for
    Word List Immediate Recall (WLI)
    Stockton Communication Failure (STCF)

It is likely that this component tests primary verbal memory and
learning.

USE OF THE TEST BATTERY

In contrast to the large numbers of tests used in other
batteries (e.g. Birren et al)[8] this battery contains only 8
different test types which represents a great economy of time for
the subject and observer.   It would not seem justifiable to
abreviate this test battery except by the elimination of a test
which has a high loading in Component I together with low loadings
in the other components;   thus it would be possible to omit one of
the tests such as Digit Symbol Substitution.   But where one is
looking over a wide field of abilities for possible drug effects
which may alter the test loadings between components elimination
of tests which are very different from each other may result in
missing an area of change resulting from a particular therapeutic
manoeuvre.

As suggested by Birren and his colleagues[8] it is possible on
the basis of Principal Component Analysis to construct a performance profile for each individual subject.   Thus the component
scores for each subject can be computed by multiplying the
individual's standard scores on the tests by the coefficients of

the tests on the components. When serial psychometric testing is
carried out during the course of drug trials any overall change in
performance and differential change between components occurring
under the influence of the drug can be measured.

It seems likely that the Picture Matching Task gives a global
assessment of intellectual function and an indication of the way in
which the subject is able to cope with problems of everyday life.
It makes use of several abilities, especially those concerned with
non-verbal cognitive reasoning and information processing skills.
According to Bromley "Such skills enable the performer to attend
selectively to a situation, to construe certain meanings by carrying
out symbolic or imaginative transformations on perceptual data,
and to apply internalised rules of procedure - schemata - in order
to select or organise an appropriate response". He also points out
that when the connection between a perceptual display and the
required response is reasonably direct, as in the first few filters
of the PMT, no age differences in performance are evident. When,
however, the connection between the display and the responses is
made more indirect and when the subject is required to 'translate'
the situation and to use abstract reasoning, performance becomes
impaired and the number of mistakes increases. Reasoning about
abstract relationships is tested in the later filters of the PMT
and although impairment is noted in all age groups it becomes far
more pronounced in older people. It will be noted that the median
for the peak filter reached by our 30 patients occurs at the fifth
filter (see Fig. 2) and thereafter an increasing number of subjects
have difficulty in distinguishing the characteristics of criteria
classes, that is the attributes of position, colour and content of
the pictorial stimuli presented in the PMT.

The Picture Matching Task has several advantages as a psychometric instrument when included in a test-battery for drug trials.
The main advantages are:-

1. The test-retest reliability is high.
2. The PMT is sufficiently sensitive to detect small changes in
    performances.
3. The performances correlate well with those using conventional
    psychometric tests and are meaningful in terms of everyday
    living activities.
4. There is a saving in the tester's time and reduction in
    boredom of repeated delivery of routine tests.
5. The man-machine interaction reduces anxiety since the PMT is
    often seen by the subject as more of a game than a test.
6. The effect of the presence of an observer is minimised and the
    subject is unable to make use of any unconscious clue to the
    correct response which he might otherwise obtain from the
    experimenter's behaviour.

Finally, the utilisation of microprocessors instead of a large computer will facilitate the development of a more portable apparatus which will be particularly suitable for use in multi-centre drug trials.

REFERENCES

1. D. Marcer, Ageing and memory loss - role of experimental psychology, Geront.Clin. 16:118 (1974).
2. J. Botwinick and M. Storandt, "Memory, Related Functions and Age", C. C. Thomas, Springfield, Illinois (1974).
3. E. Miller, On the nature of memory disorders in presenile dementia, Neuropsychologia 9:75 (1971).
4. E. Miller, "Abnormal Ageing", John Wiley, London (1977).
5. J. L. Gedye and E. Miller, Developments in automated testing systems in:"The Psychological Assessment of the Mentally and Physically Handicapped", E. Mittler, ed., Methuen, London (1970).
6. J. Keet (unpublished)
7. A. N. Exton-Smith, J. P. Keet, A. L. Johnson and M. Lee, (1980) to be published.
8. J. E. Birren, J. Botwinick, A. D. Weiss and D. F. Morrison, Interrelations of mental and perceptual tests given to healthy elderly men, in: "Human Aging I: A Biological and Behavioural Study", J. E. Birren, R. N. Butler, S. W. Greenhouse, L. Sokoloff and M. R. Yarrow, Institute of Mental Health, DHEW, Publ.No. (ADM) 77-122 (1971).
9. D. B. Bromley, "Psychology of human ageing", Penguin, London (1974).
10. E. R. F. W. Crossman, Entropy and choice time: The effect of frequency unbalance on choice response, Quart.J.Exp.Psychol 5:41 (1953).
11. D. Wechsler,"The measurement of adult intelligence", Psychol. Corp., New York (1955).
12. J. C. Raven, "Coloured progressive matrices", H.K. Lewis, London (1962).
13. D. Marcer and S. M. Hopkins, The differential effects of memory loss in the elderly, Age and Ageing 6:123 (1977).
14. B. Meer and J. A. Baker, The Stockton geriatric rating scale, J.Gerontol. 21:392
15. H. G. Taylor and L. M. Bloom, Cross validation and methodological extension of Stockton geriatric rating scale, J.Gerontol. 29:190 (1974).

PSYCHOMETRIC EVALUATION OF COGNITIVE AND INTELLECTUAL FUNCTIONS

M. Neri, G. Feltri, G.P. Vecchi

Universita' Degli Studi di Modena
Cattedra di Gerontologia e Geriatria
Italy

INTRODUCTION

Cognitive and intellectual functions are primary topics in gerontological research.   However, owing to the lack of a definition of intelligence and also because specific tests are more useful when applied to the young rather than the old, very few data are available.

In human behaviour intelligence expresses itself by conative and cognitive activities.

The conative processes originate from motivation, intention and pulsional dynanism,   The cognitive processes are concerned with the acquisition of information from the environment, its memorization and the elaboration and the expression of an adequate answer[1,2]. A rigorous distinction between these different functions is only apparent as they interact continuously.   When evaluating intelligence, such a connection can be a determining factor.   In fact, in many conditions, excess or lack of the conative component can modify the score when testing the cognitive component.   Wisdom, which is a characteristic pattern of intellectual functions in the elderly, is derived from the integration of cognitive knowledge gained throughout life and of affective involvement[3].

Among the animal species, man has reached maximum efficiency of cognitive functions.   In the phylogenesis the onset of mental ability keeps pace with structural modifications of the brain.

In man input and output activities are elaborated by the two cerebral hemispheres in completely different ways.   The left hemisphere enables analytical evaluation, speech and permits pattern of precise and quick voluntary movement of the hands, the interpretation of symbols, and possibly, the estimation of temporal causalities.

The right hemisphere enables holistic evaluation in the 'reading' of maps and of visual spatial information.   It deals with memory and the identification of stimuli not all of which can be put into a verbal description, for example, the face.   The right hemisphere is possibly concerned with the intuition of physical laws and the organization of space[4,5].

The different performances of the cerebral hemispheres are important in every attempt at understanding the real meaning of intelligence.   In fact, psychometric evaluation is biassed by the specialization of the left hemisphere, usually being based on tests carried out in a verbal code.   Owing to the different interpretations of the intellectual functions, many theories have been proposed in order to evaluate intelligence.   The two main theoretical models of intelligence assume: 1) a general factor 'g' associated with a certain number of different specific factors; 2) several correlated primary factors which are not subordinate to any higher function[6].   In either case, the multiple components of intelligence should be examined by means of a battery of tests.   From the psychometric point of view the term intelligence denotes a function of communication rather than a function of a scientific definition[7].

In this context intelligence will include all the cognitive operations, such as abstract reasoning, the ability to synthetize many pieces of information in a single experience, the problem solving ability, the concept formation and so on[2].   We must also realise that intelligence and memory are not distinct entities. Memory is 'the sum of the effects that the past induces on present activities, the relationship between what we are doing and what we have experienced[8].   This definition is particularly relevant when considering the hypothesis that there are two types of intelligence; fluid and crystallised[9].   The first, which is related to the neurological structure, grows until the end of neuronal maturation and then decreases; whereas the second reflects cultural assimilation and, when the subject is in good health, can also grow during adulthood[10].

METHODOLOGY

The two principal methods used to collect data are cross-sectional and longitudinal ones.

That used more often than not is the cross-sectional one, but it tends to magnify or distort impairment due to ageing.   The main reason for this is the fact that different cohort subjects differ

not only for different biological age, but also for different
anthropological origin, status and development in different periods.

The longitudinal researches tend to minimize these differences,
but are more difficult for survival.

As the results referred to are those who reach the maximum age
in good health, a magnifying factor in the test scores must be
suspected.   Recently a method has been developed which tries to
combine the previous two methods[11] and it seems that with it, there
is a distinction between the ageing and cohort effect[12].

The psychometry of intelligence can be divided into two
categories of tests:[6]
1)   General capability tests, useful to measure the 'g' factor:
     typical example being Raven's Progressive Matrices.
2)   Primary Ability tests, for example, the General Aptitude Tests
     Battery, the Wechsler Bellevue test (WB) or the Wechsler Adult
     Intelligence Scale (WAIS).
Both groups of tests are limited because of the different weight
attributed to the function of each of the two hemispheres and for
the lack of information, especially in the intelligence disease
relationship.   Several authors suggest integrating the Raven test,
the WB or the WAIS one with other tests which evaluate memory and
learning, speech, motor functions[6], orientation and design and subcortical functions[13].   Other authors suggest the association of the
above with projective personality tests[14].   In general an increase
of the number of tests creates problems of availability for both the
examiners and the subjects, and limits the possibility of repeating
the tests within a short period.

On the basis of these and other objections some authors[15,16]
suggest the use of the Mental Status Questionnaire (MSQ) together
with a simple perceptual test (Face-Hand test) for practical
clinical use.   Such tests are only useful as screening diagnostic
essays.   The aim of these test batteries is to obtain a general
evaluation of intelligence.   For more specific cognitive functions
it is useful to remember the following: a) the concept formation
tests which include verbal abstraction and non verbal conceptual-
ization tests as the sorting one b) the verbal and spatial/visual
reasoning tests[2].

RESULTS

The following are the three main topics of research concerning
the modifications of the intellectual functions with age:
1) Intelligence and age are the only two variables; intelligence is
   directly modified by ageing: Intelligence and Development.
2) Also the ecology of ageing (social-economic features, environment,
   generational factors, etc.,) can influence intelligence:

Intelligence and Environment.

3) The 'disease' variable is included when considering the decline of intelligence in the ageing: Intelligence and Disease.

Intelligence and Development

Basic research is aimed towards the typical pattern of intelligence in the elderly and the possible correlations with age induced changes in the nervous system.   Even if incomplete some data is available, such as the results obtained from the WAIS tests concerning old people from different social-economic, environment and racial conditions who typically give good verbal test results, scoring less for performance[17],[18].

It is also worth noting that the decrease in score accelerated after 65-70 years of age, independently of health conditions[12].

Psycho-motor slowing. On the basis of longitudinal trials on twins[19] and on the general population[20] or of cross-sectional trials[21], many authors state that psycho-motor slowing is the most age correlated pattern[22].   Motor skill changes in the elderly have been recently described by Welford[23].   Motor skills, in the gestural rather than in the automatic form, together with speech, are also means of expressing intellectual capabilities.   The system which generates gestural expression is similar to that which generates the cognitive processes, however, this is not entirely true at the neuro-anatomical level.

Functionally it is possible to state that the measurement of the reaction time can be considered a measure of the superior processes which generate the elaboration of the stimulus and the choice and control of the adequate motor response. (See Stelmach[24] for review).

We have evaluated the age concerned correlations between motor skills and cognitive tests in a group of healthy subjects divided into classes of the following average ages: 25 years (A), 50 years (B), 68 years (C) and 80 years (D).

All these groups were tested with an intellective test (Raven), with a test of memory and verbal learning (Babcock Tale) and of memory and spatial learning (Corsi and Corsi Supra-Span, Fig. 1).

In the last test the subject is requested to memorise a sequence defined by the examiner by tapping out a series of cubes of increasing length.   The same tests were used to evaluate learning abilities by adding one or two units to the previously reached memory span until the criterion was achieved[25].   The motor skill tests were performed by Kinesthetic[26] and rotatory persuit[27] apparatuses.

Fig. 1.   The board of the Corsi test.

When using the Kinesthetic apparatus (Fig.2) the subject is
asked, in a sequence of trials, to 'learn the measurement' of a
movement and to repeat it with the apparatus.   The different
lengths of movements are defined by the examiner by holding the
subject's hand, who is blindfolded and is therefore only informed by
his own Kinesthetic perception.

When using the rotatory pursuit (Fig. 3) method, one tests the
speed and dimension of a movement necessary to reach a target placed
on the rotatory pursuit apparatus.

According to the preliminary results, the four age groups
performed as follows:
1) With regard to the cognitive and rotatory pursuit tests, the
   groups behaved differently; the scores being negatively correlated with age.
2) The four groups gave homogenous results when tested with the
   Kinesthetic apparatus.
3) In the C and D groups the rotatory pursuit was the only or the
   most discriminating tests.
4) Each group tended to learn, although at different rates, the
   rotatory pursuit.
5) In group A there were numerous correlations within cognitive
   scores and between these and motor learning ones.   During
   ageing these correlations became less numerous particularly
   within the cognitive tests.
The slowing down of psycho-memory reactions is the most typical
ageing pattern.   It is a temporal measure of an age correlated

Fig. 2. Kinesthetic apparatus
         The two boxes are equipped with a lever. The S. moves
         it from the starting position to a finish peg (stimulus),
         then he must reproduce the same distance with the free
         lever.

deteriorating mechanism of interaction amongst the perception of
the stimulus, its psychoneurological integration and the programm-
ing and execution of the motor response.

    Whatever the deteriorating mechanism is, it cannot justify the
subject's changing performances in several batteries of tests[28].
Many authors are trying to identify if ageing uniformly modifies
some or all of the mental abilities.

    Therefore, comparisons have been made between the scores of
healthy old people and those of old patients with lateral or widespread cerebral disorders.

    Overall[29] and Goldstein[30] conclude that the normal performance
of the old differs from that of patients with widespread cerebral
disorders.   Klisz[14] suggests that the right hemisphere functions
deteriorate more than those of the left hemisphere.

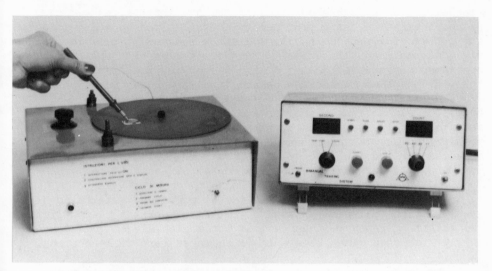

Fig. 3.   Rotatory pursuit, detecting both contact time and
          number of contacts.

He quotes Semmes[31] who suggests the existence of widespread
psychoneurological organization in the right hemisphere versus a
more focalised organization in the left hemisphere.

A more detailed analysis on psychometric results and their
relationships with cerebral anatomical pathology, or on the
difference between the psychological and pathological ageing
processes would be lengthy and inconclusive.

The correlations between test scores and life expectancy are
obvious.   Much data confirm that the results of the cognitive tests
are generally well correlated with age and health conditions[14,21,32]
they seem to reflect life expectancy rather than past life[19].
Unfortunately, more precise practical data on survival are lacking
at present[12].

The following cognitive tests of specific intellective functions
related to daily behaviour, are worth mentioning: Problem solving
ability tests, memory tests and learning tests.

Problem solving ability. Many studies have proved that a good
problem solving ability is correlated with good results in
intelligence tests, (for instance the Raven) and vice-versa.

Such ability requires good memory and the use of certain
strategies which allow the S. to codify the information and to
process effectively the data necessary for the solution of the

problem.   The aged S shows lack of ability in this task[33,34,35,36,]
either because he has a poor memory and a limited capacity to detect
any information processing strategies or because he prefers to adopt
rigid stereotyped strategies.   Such strategies although complex and
spontaneously created cannot adapt to changing problems.   To verify
this, S. has been trained to behave in a particular way in a test,
his performance being examined in a subsequent test where the same
behaviour was no longer appropriate.   The results of such studies
seem to support the second hypothesis.   Alternatively, the above
problem can be studied by using real life tests where the S is not
expected to find a unique solution, but, while taking probability
into account, must choose one of a number of possible answers.

Results show that when given the alternatives, risk plus high
gratification and safety plus low gratification, the aged S prefers
the latter.   However, when choices contain an element of risk
anyway, the aged S may behave as if he takes more risky chances.

The reasons for such contradictory behaviour lies in the
subjects possible decreased capacity to predict the optimum balance
between risk and gratification[37].

Memory and learning.   M and L analysis have proved the importance of memory in the cognitive behaviour of the aged S.   Such
studies, especially in the cross-sectional ones, have resulted in
more flexible data than those obtained in other kinds of analysis,
as they reveal if and how the S. acquires and retains the information[38].   In the latter case, which relates to a shorter life span,
the whole experience is completely forgotten.   Finally objective
tests have proved that the aged S does not have a good memory for
remote events[41,42].

Intelligence and Environment

Many articles report the importance of factors other than
ageing in intelligence tests.   In fact, to think that ageing alone
can influence the results, would be an over simplification.   Other
factors are stressed i.e. fatigue, level of education, social-
economic condition and the continuation of a working activity etc.
There is some controversy among authors concerning the influence of
environmental conditions and the fact that scores and institutional-
ization are negatively correlated[12,15, 43].

With respect to this, it is important to remember the guidelines
and preliminary data obtained in studies applying Baltes' method.
This author[3] has recently stressed the inadequacy of studies regarding the psychometry intelligence in the aged S.   In fact, the
studies of developmental intelligence rely on experiences drawn
from young people and simply apply as they were.

The aim is not to define abstract references based on mean
values drawn from cross-sectional validations, but to limit values
of intraindividual variability.   This could be obtained through the
interaction of the longitudinal and cross-sectional methods which
permit quantifying the interaction of factors connected with age
and other determinants; cultural, economics etc.

Preliminary data reporting known models can be summarized as
follows:-
a) In subjects of 70 years or less, chronological age is only
   partially responsible for intelligence variations, whereas after
   70 it becomes significant.
b) Cohort differences in subjects up to 70 are as important as
   chronological age, the same applies to education, status and
   health.
c) There are important differences in the changes of the two types
   of intelligence (crystallized and fluid) due to age and cohort
   factors.
It follows that if to the age of 70, age alone holds little importance, a quantitative or qualitative modification in intellectual
behaviour will be caused by other factors (health, environmental
conditions etc.).

Intelligence and Disease

A great many studies analyse the relationship between intelligence and disease.   It has already been mentioned that intelligence
test scores diminish during disease.   Obviously negative effects
are greater when the C.N.S. is affected by an acute or chronic
ailment.   The final result will be one syndrome with many names:
intellectual impairment, cognitive disorders or dementia.   The
medical, social and economic importance of this syndrome derives
from the following:-
The percentage of aged population is steadily increasing. According
to different authors cognitive disturbance in SS over 65 vary from
1% to 7.2% for severe dementia, from 2.6% to 15% for mild dementia
and from 10% to 18% for all various kinds of intellectual disturbances[13].   Until recently, the majority of these SS were superficially defined as 'arteriosclerotic'.   At present vascular
determinants are deemed less important, though many intriguing
cases are still undefined and a lot more work is necessary to noso-
graphically classify them; therefore a multidimensional approach to
the problem is necessary[44,45,46,47,48].

Psychometric evaluation should be accounted for in the same
context.   Two crucial points should be taken into consideration:
1) An early diagnosis of dementia, which like cancer is a life
    shortening disease, is necessary. In both cases it is still
    impossible to intervene on the primary causes, however, it is

more useful and effective to intervene as early as possible.
2) A differential diagnosis which is important to distinguish
   between organic and non-organic disturbances. In the former
   case it would be useful to discover the probable origin of the
   disturbances through the identification of characteristic
   patterns of impairment.

The personal experience we would like to quote is a study
on 106 subjects over 70 years with at least 3 years of schooling,
no focal neurologic deficiencies, and not being treated for neurotic
or psychotic manifestations. The subjects were divided into 3
groups: A) Hospitalized for C.N.S. extraneous diseases,
R) Institutionalized for social reasons, D) Suffering from cognitive
disturbances. The subjects underwent intellectual tests
(W-B verbal scale, Raven Coloured Matrices) spatial and verbal
memory and learning tests (Corsi, Corsi Supra-Span, Babcock tale
and word list). The aim was to define possible differences among
the groups and discover which tests were more discriminating. Each
of the three groups gave a wide range of scores. In intellective
tests A and R groups showed almost equal mean standardized scores
while in D group the Raven score was remarkably lower than the
W-B score, analogy and vocabulary in W-B being the lowest.Primary
memory data differ greatly between group A-R and D, the latter
lacking memorising ability, especially in spatial and tale tasks,
but showing some ability in the digit verbal task. With regard to
learning A-R and D groups behaved differently and showed patterns
similar to the previous ones.

The results are summarized in Fig. 4;by means of discriminant
analysis we have obtained a map with three areas each pertaining to
a group. The most discriminant tests, intellective, spatial
memory and partly verbal, restrict each subject to an area. With
this method groups A and R are undistinguishable (false positive
from 24% to 38%). However, both groups differ remarkably from
group D, excluding the possibility to assign a subject of group D
to either group A or R.

CONCLUSION

At present cognitive psychometry is somewhat limited as to the
solution of the two problems described. As previously stated it
offers a static image of a moment in S's life. Therefore, it is
often not possible to utilise it in analysing S's past and it is of
little help in monitoring the evolution of the syndrome itself.

It is difficult, if not impossible, to apply cognitive
psychometry to SS with cognitive impairment (aphasia-apraxia, etc.)
and it does not always discriminate between organic and functional
forms of impairment[6,13,15,50].   These being the negative connot-

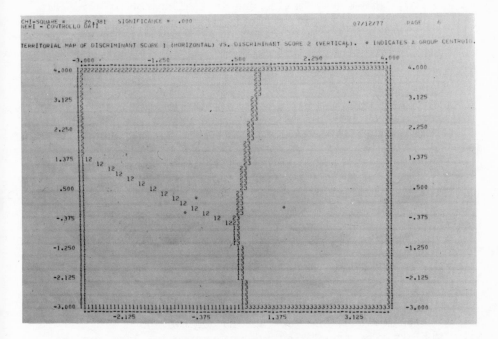

Fig. 4.   Discriminant map

ations there are also some positive values.   It is harmless to the
slightest health changes.   There are some, as yet unreliable
studies, that show the possibility of obtaining different patterns
of impairment owing to different causes and of utilising non-tiring
tests suitable for the longitudinal monitoring of intellectual
impairment[14,51,52,53,54,55].   Psychometry is therefore still of
great importance in evaluating both healthy and sick aged subjects,
however, a great deal of research is still to be carried out.

REFERENCES

1. J. de Ajuriaguerra, A. Rego, J. Richard, R. Tissot - Psychologie
        et Psychométrie du vieillard, <u>Confrontations Psychiatriques</u>
        5:27 (1970).
2. M. D. Lezak,"Neuropsychological assessment", Oxford University
        Press, New York (1976).
3. P. B. Baltes, S. L. Willis, The critical importance of approp-
        riate methodology in the study of aging: The sample case of
        psychometric intelligence <u>in</u>: "Brain function in old age",
        F. Hoffmeister, C. Müller, eds., Springer Verlag,Berlin (1979).
4. R. Levi-Montalcini, Le basi structturali del messaggie nervose
        <u>in</u>: "Il Messaggio Nervoso", Rizzoli, Roma (1975).

5. J. Levy, Human cognitation and lateralisation of cerebral
    function, T.I.N.S. 2:222 (1979).
6. G. Gainotti, I disturbi dell'intelligenza in: "Neuropsicologia
    Clinica", Franco Angeli, Milano (1977).
7. P. Orelon, Pour un depassement du concept d'intelligence,
    Int.Rev.App.Psychol. 24:107 (1975).
8. H. Spindler, R. Sterzi, G. Wallar, "Le amnesie", Franco
    Angeli, Milano (1977).
9. R. B. Cattel, Theory of fluid and crystallised intelligence:
    a critical experiment, J.Educ.Psychol. 54:1 (1963).
10. R. W. Cunningham, V. Clayton, G. Owerton, Fluid and crystallised
    intelligence in young adulthood and old age, J. Gerontol.
    30:53 (1975).
11. P. B. Baltes, Longitudinal and cross-sectional sequences in the
    study of age and generation effects, Hum.Develop. 11:145
    (1968).
12. I. Botwinick, Intellectual abilities in: "Handbook of the
    psychology of aging", J. E. Birren, K. W. Schaie, eds.,
    Van Nostrand Reinhold, New York (1977).
13. C. Eisdorfer, D. Cohen, The cognitive impaired elderly:
    differential diagnosis, in: "The clinical psychology of
    aging", M. Storandt, I. C. Siegler, M. F. Elias, eds.,
    Plenum Press, New York and London (1978).
14. D. Klisz, Neuropsychological evaluation in older persons, in:
    "The clinical psychology of aging", M. Storandt, I.C.
    Siegler, M. F. Elias, eds., Plenum Press, New York and
    London (1978).
15. R. L. Kahn, N. E. Miller, Assessment of altered brain function
    in the aged, in: "The clinical psychology of aging.
    M. Storandt, I. C. Siegler, M. F. Elias, eds., Plenum Press,
    New York and London (1978).
16. R. Kazman, T. B. Karasu, Differential diagnosis of dementias,
    in: "Neurological and sensory disorders in the elderly",
    W. S. Fields, ed., Stratton Inter.Med. Book, New York (1975).
17. C. Eisdorfer, C. W. Busse, L. D. Cohn, The W.A.I.S. performance
    of an aged sample : The relationship between verbal and
    performance I. Q. J. Gerontol. 14:197 (1959).
18. L. F. Jarvik, C. Eisdorfer, J. E. Blum, "Intellectual function-
    ing in adults", Springer, New York (1973).
19. L. F. Jarvik, Thoughts on the psychology of aging, Am.Psychol.
    30:576(1975).
20. F. J. Mathey, Psychomotor performance and reaction speed in old
    age in: "Patterns of aging", H. Thomae, ed., S. Karger,
    Basel (1976).
21. R. N. Butler, The National Institutes of Mental Health Study,
    in: "Alzheimer's Disease: Senile dementia and related
    disorders", R. Katzman, R. Terry, K. L. Bick, eds., Raven
    Press, New York (1977).

22. J. Birren, Speed of behaviour as an indicator of age changes
    and the integrity of the nervous system, in: "Brain function
    in old age", F. Hoffmeister, C. Müller, eds., Springer
    Verlag, Berlin (1979).
23. A. T. Welford, Motor performance in: "Hand book of the psycho-
    logy of aging", J. E. Birren, K. W. Schaie, eds., Van
    Nostrand Reinhold, New York (1977).
24. G. E. Stelmach,"Motor control - Issues and trends",Academic
    Press, New York (1976).
25. E. de Renzi, Le amnesie, in:"Neuropsicologia clinica",
    Franco Angeli, ed., Milano (1977).
26. M. Posner, Components of skilled performance, Science 152:1712
    (1966).
27. S. Corkin, Acquisition of motor skill. After bilateral medial
    temporal lobe excision, Neuropsychologia 6:255 (1968).
28. M. Storandt, Age, ability level and method of administering
    and scoring the WAIS, J. Gerontol. 32:175 (1977).
29. J. E. Overall, D. R. Gorham, Organicity versus old age in
    objective and projective test performance, J.Consult. &
    Clin.Psychol. 39:98 (1972).
30. G. Goldstein, C. H. Shelly, Similarities and differences
    between psychological deficit in aging and brain damage,
    J. Gerontol. 30:448 (1975).
31. J. Semmes, Hemispheric specialisation : a possible clue to
    mechanism, Neuropsychol. 6:11 (1968).
32. V. A. Kral, Benign senescent forgetfulness in:"Alzheimer's
    disease : senile dementia and related disorders",
    R. Katman, R. D. Terry, K. L. Bick, eds., Raven Press,
    New York (1978).
33. E. Simon, Depth and elaboration of processing in relation to
    age, J. of Exp.Psychol. Hum.Learn.& Mem. 5:115 (1979).
34. C. A. Furry, Baltes, P.B., The effects of age differences in
    ability-extraneous performance variable on the assessment
    of intelligence in children, adults and the elderly,
    J. Gerontol. 28:73 (1973).
35. F. I. M. Craik, Age difference in human memory in: "Handbook
    of the psychology of aging", Van Nostrand Reinhold,
    New York (1977).
36. E. Robertson-Tchabo, D. Arenberg, P. T. Costa, Temperamental
    predictors of longitudinal changes in performance on
    Benton revised visual retention test among seventy year
    old man : an exploratory study in: "Brain function in old
    old age", F. Hoffmeister, L. Müller, eds., Springer
    Verlag, Berlin (1979).
37. P. Rabbitt, Changes in problem solving ability in old age in:
    "Handbook of the psychology of aging", J. Birren, K. W.
    Schaie, eds., Van Nostrand Reinhold, New York (1977).

38. K. W. Schaie, E. Zelinski, Psychometric assessment of
    dysfunction in learning and memory, in: "Brain function in
    old age", Springer Verlag, Berlin (1979).
39. J. L. Gilbert, H. W. Lewee, Patterns of declining memory,
    J. Gerontol. 26:70 (1971).
40. S. Gordon, L. Crawford, W. Clark. Adult age differences in
    word and nonsense sillabe recognition memory and response
    criterion, J. Gerontol. 29:659 (1974).
41. E. K. Warrington, H. T. Sanders, The fate of old memories,
    Exp.Psychol. 23:432 (1971).
42. L. R. Square, Remote memory as affected by aging, Neuropsychol.
    12:429 (1974).
43. M. Neri, G. Feltri, G. Ciotti, L. Belloi, R. Lugli, Studio
    delle funzioni intellettive in soggetti anziani sani e
    mentalmente deteriorati, Giorn.Gerontol. 27:33 (1979).
44. W. S. Fields, Neurological and sensory disorders in the
    elderly, in: "Stratton Med.Book, New York (1975).
45. R. Katzman, R. Terry, K. L. Bick,"Alzheimer's disease :
    Senile dementia and related disorders", Raven Press,
    New York (1978).
46. M. Storandt, I. L. Siegler, M. F. Elias, "The clinical
    psychology of aging", Plenum Press, New York (1978).
47. K. Nandy, Senile dementia. A biomedical approach,
    Elsevier-North, Holland, New York (1978).
48. F. Hoffmeister, C. Müller, "Brain function in old age",
    Springer Verlag, Berlin (1979).
49. M. Neri, G. Ciotti, G. Feltri, V. Baldelli, M. Gasparini,
    Studio delle funzioni mnesiche in soggetti anziani sani
    e mentalmente deteriorati, Giorn.Geront.27:287 (1979).
50. M. Roth, Diagnosis of senile dementia and related forms of
    dementia, in: "Alzheimer's disease, senile dementia and
    related disorders", R. Katzman, R. Terry, K. L. Bick,
    eds., Raven Press, New York (1977).
51. J. Brull, J. Wertheimer, E. Haller, Evolutive profiles in
    senile dementia, in: "Brain function in old age", Springer
    Verlag, Berlin (1979).
52. P. A. Fuld, Psychological testing in the differential diagnosis
    of the dementias, in: "Alzheimer's disease, senile dementia
    and related disorders", R. Katzman, R. Terry, K. L. Bick,
    eds., Raven Press, New York (1978).
53. H. J. Neville, M. F. Folstein, Performance in three cognitive
    tasks in patients with dementia, depression or Korsakoff
    syndrome, Gerontology 25:285 (1979).
54. F. Perez, A. Gray, R. Taylor, Wais performance of neurologic-
    ally impaired aged, Psychol.Rep. 37:1043 (1975).
55. G. Gainotti, C. Caltagirone, C. Masullo, G. Miceli, Patterns
    of neuropsychological impairment in various diagnostic types
    of dementia, Communication held at the meeting "Aging of
    the brain and dementia",Florence - August 1979.

PRINCIPLES AND METHODS OF EVALUATING MENTAL DISORDERS IN THE AGED AND THE MODIFICATIONS FOLLOWING DRUG ADMINISTRATION

M. Passeri and D. Cucinotta*

Department of Geriatrics
University of Parma
Italy

Ageing is characterised by a reduced capacity of adaption and fewer possibilities of keeping homeostasis constant when faced with internal and external stimuli[1]. This is due to a deterioration in the self-regulatory systems which is revealed when some type of stress or even minor illness triggers off a chain of irreversible events.

This impaired homeostatic balance is expressed at brain level in elderly patients by nervous and psychic manifestations derived from alterations in the brain's capacity to adapt, since the brain is an organ which controls psychological processes, the capacity to communicate, affectionate behaviour, motor activity and vegetative activities, etc. Thus the capacity to recover and react, which makes the brain a true compensatory organ is undermined.

The manifestations involve the personality and behaviour of the old subject, and produce various clinical disorders, ranging from modifications in the sleep-wakefulness rhythm to slow brain functioning, lack of personal care, an inability to adapt to surroundings, a lack of interest and motivation and a reduced psychic-intellectual activity.

There is no consistent correspondence between the important anatomical and pathological alterations and the above modifications. In fact there is good evidence from recent research[2,3] that the brain is relatively well preserved in healthy old people. The

*With the collaboration of:
M. Mancini, M. D. - G. Stuard Hospital, Parma.
A. Sirianni - Department of Geriatrics, Parma.
M. G. Michelotti, M. D. - Institute of Neuropsychiatry, Parma.

brain suffers little loss of weight except in extreme old age, without there being a high loss of neurones with ageing.

In the elderly human brain some morphological changes are seen in neurones, particularly in the neocortex and hippocampus. These include neurofibrillary degeneration, granulovacuolar degeneration and the accumulation of insoluble lipofuscin granules[4], but they do not lead to any serious deficit in overall functioning.

A characteristic alteration in the normal ageing brain is a moderate increase in microglia[4] which is not to be neglected, however, since it can produce some modifications in cerebral permeability and in the functioning of the vessel-glia-nerve cell complex on which the latter feeds.

Indeed metabolic alterations characteristic of the ageing brain are being discussed in this symposium; one has only to remember Carlsson et al[5] who showed a reduction in the level of 3 metossitiramine in the ageing brain; McNamara et al[6] and Kohn[7] have even gone so far as to say that the functional depression of the cerebral cholinergic system may well be an important factor in the general lack of resistance to stress in old age.

Various studies have confirmed neurotransmitter modifications as a result of ageing: i.e. tyrosine-hydroxylases, DOPA decarboxy-lase[8,9] choline acetyltransferase and acetylcholinesterase[10], glutamate-decarboxylase and GABA-binding to specific receptors[11] and monoamine oxidase[2].

The behaviour of the serotoninergic system too, of the various polypeptides (somatostatine, neurotensine, TRH, gastrin, angiotensin - Vale et al[12]), of the polyamines and nucleic acids[13], of the qualitative and quantitative variations of the cerebral lipid content[14,15,16] created interest in the possible importance they may have both on ageing and as a cause of some disorders of mental function in old age.

The variations in the brain enzymatic activities[17] connected with oxygen uptake and glucose consumption in the elderly have been widely studied, because they are fundamental in maintaining both cerebral homeostasis and functioning.

Senile functional modifications in the vessel-glia-neuron unity can be the result of associated diseases, which are very common in old age.

The widespread vascular alterations, represented by manifest damage or only functional defects, must be considered. Among these are thrombosis, haemorrhage, aneurysm, atherosclerosis in general, as well as the worsening of zonal circulation, the

permeability variations in the microvascular tissue complex, the
alterations in blood supply due to extracranial vascular defects or
to a general circulatory malfunctioning etc.

We must also consider possible damage to the brain from
endogenous or exogenous toxic agents or from hypo and hypernutrition
or from modifications in plasma constituents or in red cell
composition[18,19].   In all these cases, all too frequent in old age,
there are changes in brain metabolism, which determine local
endorphine production, neurotransmitter release, reduction in glucose
and oxygen uptake, fundamental in brain cell life[2,3].

One only need mention observations from biochemical studies on
the brain, e.g. neurotransmitter deficit not only in particular
illnesses such as Parkinson's disease, but also in many other
pathological states in old age[2,3,8,9,20,54].

It is only today that we fully understand the brain function
alterations which result partly from ageing (perhaps premature
ageing), and partly from vascular and metabolic factors conditioning
first functional defects and then organic damage which may reach
cell necrosis.   Consequently anatomical and pathological pictures
more serious than normal simple brain ageing are derived; these can
be seen in degenerative phenomena and the results of ischaemic
necrosis, variously combined.

From a clinical point of view, the very nature of these
situations leads to a series of extremely common pathological
manifestations, ranging from motor function defects to psychic
alterations. When considering the mental alterations present in
these situations, it is obviously difficult to form a classification
and a methodology which would evaluate the extent of the deficit
or alterations present; it is even more difficult to estimate and
quantify possible modifications of the morbid pictures following,
for example, drug intake.

We can distinguish two different clinical forms of organic
mental impairment:-

1. Organic brain syndrome (O.B.S.) also called chronic brain
    failure[21] or C.C.V.I. (cerebral chronic vascular insufficiency),
    a long and progressive illness, with disastrous social
    consequences.
2. Acute mental disorders, in which the brain failure is caused by
    general illnesses[21].   The diagnosis must take into account many
    objectives, such as recognition and prevention of psychological
    stress which can disturb the delicate and unstable homeostasis
    of the ageing brain, and the treatment of associated morbid
    processes.

In addition to clinical examination and to the psychological and social history certain tests must be performed (EEG, ECG, Chest X ray, laboratory tests), including the assessment of vision and hearing, since very often impairment of the special senses is responsible for disorientation, mood change and psychic and behavioural disorders in the elderly.

For some years now our research group has been experimenting on various methods of examination in order to single out with greater precision the extent of the patient's disorder and, if possible, to estimate the effects of various therapies, so as to be able to institute appropriate treatment.

PSYCHOMETRIC TESTS

In order to have some element of evaluation regarding the intellectual and cognitive functioning of the patient it is best to test his space-time orientation, his ability to recognise people and his long and short-term memory.

Expressions which may seem threatening and worrying or which may frighten the patient into feeling that failing the test may have negative results in his relationship with the doctor and therefore the possibility of being treated, must at all costs be avoided: everyone must always appear friendly and sympathetic.

One of the commonest and oldest tests used is the Kahn and Goldfarb Questionnaire on Mental Status (M.S.Q.)[22]. However, the data, owing to its extreme simplicity, may be of little help especially in a sensitive estimation of the changes caused by therapy, since the results are notably influenced by the cultural level of the patient and his social environmental conditions.

The widely used Face-Hand Test[23] is predominantly a measure of organic disease per se, and will detect the sequelae of severe strokes or of intracranial operations even when the patient exhibits no confusion or disorientation.

The essential characteristic of geriatric OBS is the reduced ability to manipulate new information. OBS has an effect on memory, which involves three phases of information processing: the first phase is sensory memory, which can maintain a literal copy of a complex stimulus for up to two seconds; the second phase is short-term memory, which requires attention and the ability to retain certain aspects of a stimulus for a brief period, estimated to range from 30 seconds to half an hour; the third phase is long-term memory, a separate information processing system which must be employed if the subject is to retain a stimulus or concept for a longer period[23].

Thus, a valid instrument for assessing the symptoms of organic brain syndrome in geriatric patients should focus on short-term information processing, rather than on sensory or long-term memory.

In order to test current short-term memory, the examiner must present a stimulus for which the patient will be tested shortly afterwards, but not before several seconds have erased the original sensory memory trace.   Furthermore, in order to differentiate confusion from normal decrements, the tests should measure recognition rather than recall[23].   For this purpose various tests have been proposed and used to examine the memory deterioration rate[24,25,26], but we know that OBS also determines failure in many other psychophysical performances for which complex test batteries have been used at times; these are partly from the Wechsler scale, partly from other psychometric scales which are quite represent- ative of elderly behaviour, e.g. Kugler et al[27].   It would be even easier if we limited ourselves to considering vigilance, attention, memory[28,29,30] . Thus we are able to make a quantitative analysis of the parameters studied in the various experimental situations.

Various test batteries sufficiently precise and selective were recently employed to evaluate the performances most commonly influenced by the different types of drug treatment, used mainly to study the side effects of anti-epileptic drugs[31,32].   In spite of the various positive aspects of this methodology we do not think it suitable for older patients where one must consider their varying personalities.   We also know the sophisticated tachystoscopic methods or similar ones based on the use of very precise, refined and expensive apparatus but we have found great difficulty in test- ing elderly patients with complicated machinery (which alarms them and distorts the response).   This is also true with specialized and complicated psychometric tests which not only tire elderly patients but also require the presence of a very experienced psychologist.

Thus there is a need for simple, convenient, inexpensive instruments which can be administered and used by persons without extensive training in psychometry.   This is why we tried to find new, very simple tests, which are objective and easy to use. We are still standardizing these methods and we cannot anticipate preliminary results which could be invalidated by successive tests. But we wish to mention briefly the possibility of evaluating motor ability with a "tremblingmeter", which examines movement with a point or ring along various pre-established lines, connected to an electronic device which counts the number of mistakes, the time taken and therefore the presence or absence of trembling.   We have tested another electronic apparatus which explores together and singly three neuropsychic functions: vigilance, attention and memory.   For a pre-established and variable time, the patient observes through circular windows, in a quiet semi-obscure room,

stimuli of coloured light (red, green, white, yellow, blue) singly
or two together.   In the latter case it is possible to give 25
different sequences.   Having a timer and a stop button the patient
stops the circuit when he wants to answer.

Evaluation of the cognitive functions, vigilance and attention,
can be obtained using the instrument, which presents tasks relative
to visual-motor reaction time:

a) "Simple reaction time" (the patient must press the button as
   soon as the light stimulus appears), to study vigilance.
b) "Choice reaction time" (the patient must press the button when
   a pre-determined light stimulus from 5 possibilities appears),
   for attention.
c) "Stimulus recognition test", for short-term memory, as a test of
   visual discrimination and recognition (the patient is asked to
   recognise as quickly as possible from a random series of light
   stimulus sequences, a sequence shown one minute before).

We have also employed a recency test which we prepared following
Umilta and Ladavas [23] methodology, but which we have simplified and
adopted to elderly patients*.   We prepared this test in two
equivalent forms so as to be able to use it as a test and as a re-
test, both after some time and following therapy.   The patient is
shown 3 series of 8 photographs each with abstract pictures, one
at a time with no time limit; he is then given another two, both of
which he may have already seen, or one or neither of them.   The
patient must recognise whether he has already seen the images and,
if possible, indicate the order in which they were presented.
With a recency test we can study separately 2 functions: short-term
memory and attention.

Another very simple test in use in our department is the
crossing-out test [24,34] ; but it has been slightly modified and
simplified with respect to the original.   The patient is given a
list of 50 concrete names.   Twenty of these are very easy animal
names which he has to cross out as quickly as possible.   It tests
attention and strategy, requiring little intellect or cultural
knowledge and resisting senile deterioration, but it explores the
patient's vigilance, and allows us to discover any semantic or
memory disorders.

Other more common methodology in the study of psychobehavioural
alterations in the elderly, is based on the contemporary manifestation of the extent of some target symptoms, either neurological or

---

*We thank Professors Ladavas and Umilta who kindly showed us their
 recency test and allowed us to make the necessary modifications
 and simplifications.

psychic, of some rating scales, and also on psychometric subtests from the Wechsler scale.   For this purpose an experimental protocol widely used by us in the past years to study drug activity in the organic brain syndrome, requires the use of symptomatological features, rating scales and psychometric tests (see Table 1).   In spite of the uncertainties arising from these tests due to a subjective interpretation, we thought it suitable to examine the action in the human being of certain so-called vascular-metabolic drugs, capable of acting more or less evidently on the circulation (above all the micro-circulation) and/or on neuronal metabolism. There are many of these drugs today (see Table 2) used more and more frequently although often with disappointing results.   In fact all these substances have particular actions (not always known) which reflect on some brain structure activities, with mechanisms elucidated in vitro in the experimental animal.

Our data is intended to be an attempt at correlating the changes in brain metabolism behaviour following the use of drugs with the clinical picture of the patient; they are intended to form a logical basis for the management of OBS.

PATIENTS AND METHODS

In this research our methodology was as follows:
1. Selection of a group of patients with similar characteristics from a general anamnestic clinical point of view and covering the majority of symptoms present in OBS.

Table 1.   Target symptoms, Rating Scales and Psychometric Tests
           Used for the Evaluation of Activity of the Drugs

Target symptoms: NEUROLOGICAL   PSYCHICAL
                 Confusion   Apprehension
                 Headache   Irritability
                 Dizziness   Scarce cooperation
                 Poor vigilance   Impulsivity
                 Tinnitus   Emotional lability
                 Trembling   Poor sociability
                 Physical weakness   Depression

Bronx State Hospital Geriatric Scale (Plutchick et al.)

Psychometric tests: Wechsler Memory Scale:
                    - Information
                    - Orientation
                    Wechsler- Bellevue Adult Intelligence
                    scale
                    - Comprehension   - Digit Span
                    - Arithmetical reasoning - Block design

Table 2.   Main Drugs at Present in Use in Therapy of Organic
           Brain Syndromes

1. Drugs with a mainly eumetabolic action: Cyticholine,
   naphthydrofuril, phospholipids, piracetam, piritiossine.

2. Drugs with a vasoactive and eumetabolic action: Cinnarizine,
   (-)eburnamonine,   3-GS,   pyribedil,   pyridinol-carbamate,
   suloctidil, vincamine, viquidil, xantinol-nicotinate.

3. Drugs with a mainly alphalitic action: Ergot alkaloids,
   nicergoline, pipratecol, raubasine.

4. Drugs with a mainly betafacilitating action: Buphenine,
   cetiedil, isoxuprine.

5. Drugs with a mainly myolitic action: Bamethane, bencyclan,
   nifedipine, papaverine, papaveroline, propaxoline.

2. Other psychotropic therapy was strictly forbidden during the
   research period, while other therapies (digitalis, diuretics,
   etc.) to which these patients had already been submitted for a
   long time were retained[35].
3. Groups of patients were homogenous for age, school level, male
   or female, duration of OBS.
4. No patients suffering from brain damage due to trauma, infection
   or tumour were included; neither were patients with serious
   disabling organic illness, nor psychosis or those who needed
   continuous treatment with hypotensive drugs.   We also excluded
   all those patients who before or during the experiment had
   undergone some psychic-traumatic event.
5. The patients chosen were affected by mild OBS and had at least
   two target symptoms, neurological or psychic (according to
   Table 1), and they were able to carry out the tests of the rating
   scales chosen.   Patients with sequelae of stabilized strokes
   which did not incur serious physical or psychic deterioration
   (without aphasia, agnosia or apraxia), were also admitted[35].

     In the field of the vast group of the active drugs (at least
experimentally on animals) or symptoms of OBS (see Table 2), we
have up to now made the following comparison of their activity.
- Suloctidil (S) versus Placebo 1st (P_I)
- (-) Eburnamonine (E) versus Cinnarizine (C)
- (-) Eburnamonine (E) versus Vincamine (V)
- (-) Eburnamonine (E) versus Placebo 2nd (P_{II})
- Vincamine (V) versus Placebo 2nd (P_{II})
- Bamethane (B) versus Placebo 3rd (P_{III})

     The drugs were always administered orally; Table 3 shows the
dosage and length of treatment.   A random doubleblind protocol was
always followed.

Each subject underwent the above-mentioned test battery (WMS,
WBAIS, BSHGS) , and the recording of neurological and psychic
symptoms, were were evaluated according to the following scores:
absent: 0; slight: 1; bad: 2; very serious: 3.   An overall clinical
judgement on the symptomatological variations induced by treatment
was given, taking into account both our own and the patient's
impressions, (see Table 3).

General behaviour aspects (daily living activities, orientation,
etc.) were evaluated on the GRS36, considered specific for elderly
patients; it consists of 31 items evaluated on a three point scale
(0, 1, 2). For a statistical evaluation we took into account the
total score reached.

The evaluation of the intellectual abilities of the patients
was performed using some standardized psychometric tests and retests
from WMS and WBAIS, such as information, orientation, comprehension,
digit span, arithmethic reasoning, block design.   With this group
of subtests (chosen among those which for the most part record
senile brain failure), we were able to make a good evaluation of
the principal intellectual abilities of the elderly subjects; they
were neither difficult nor long, which avoided tiring the elderly
patients.

RESULTS AND DISCUSSION

In view of the homogeneity of the various groups of patients,
we have analysed our results to assess the general average
variation with regards to the base situation and their possible
statistical significance rather than give an analytical evaluation
of the performance of the individual subject when comparing it to
therapeutic activity.

The equal distribution of the sexes in our research allowed us
to avoid keeping the two groups separate in our evaluation of the
tests.   In fact, although the existence of a differentiation
between men and women is quite possible in the field of language
and spatial functions, perhaps as a consequence of cultural and/or
biological factors37, we admitted patients at random, and our
interest was directed not to a comparison between patients but to
an eventual modification in symptomatology.

Table 3 shows the various activities compared and completed
up to now, the number of patients admitted, the length of therapy,
and daily administration of drugs as well as a synthetic clinical
judgement.   An evaluation of the data must necessarily take into
account the facts that the various comparisons made were carried
out in successive periods, on different groups of subjects and with
different length of therapy.

Table 3. Dosage, Length of Treatment and Overall Clinical Results

TREATMENT	NUMBERS OF PATIENTS	DAYS OF TREATMENT	DOSE FOR ONE DAY	VERY GOOD	GOOD	FAIR	NIL
SULOCTIDIL	10	84	300 mg	2	4	2	2
PLACEBO 1st	10	84	======	1	1	1	7
(-)EBURNAMONINE	55	30	60 mg	6	27	13	9
CINNARIZINE	51	30	150 mg	2	28	11	10
(-)EBURNAMONINE	28	28	60 mg	4	16	5	5
VINCAMINE	30	30	60 mg	2	18	4	6
PLACEBO 2nd	30	30	======	1	3	6	20
BAMETHANE	36	60	37,5 mg	not yet performed			
PLACEBO 3rd	37	60	======	not yet performed			

All the same, we will try to single out the possible specific major or minor activity of each drug examined for a given symptom or group of symptoms or intellectual ability. At the time of writing the present work the data relative to the Bamethane-Placebo 3rd comparison, have not been completely evaluated or statistically analysed; therefore the tables and figures only show results available up to now.

We can see from Table 3 that the various placebos used did not give any satisfactory clinical result, while certain drugs often induced favourable modifications in the overall clinical picture, which throws a positive light on the activity as a whole.

The average values of the total score of the neurological and psychic symptomatology (Table 4) are notably lower after drug treatment, but with greater statistical significance in the groups treated with E and C.

Table 4.   Results of the Rating Scales Total Scores ($\pm$ SE)
(Before and After Treatment)

	DAYS OF TREAT.	BASELINE	NEUROGICAL SYMPTOMS	PSICHIC SYMPTOMS	B.S.H. GERIATRIC RATING SCALE	WMS/WBAIS (Total scores)
SULOCTI-DIL	84	Baseline	$10,06 \pm 0,82$	$12,04 \pm 0,58$	$28,62 \pm 1,09$	$20,79 \pm 1,03$
		After treat	$5,92 \pm 0,36$	$6,82 \pm 0,49$	$23,04 \pm 1,18$	$24,58 \pm 1,29$
		n	10	10	10	10
		p	0,001	0,05	0,01	0,01
PLACEBO 1st	84	Baseline	$10,22 \pm 0,68$	$9,84 \pm 0,55$	$24,83 \pm 1,44$	$24,28 \pm 1,05$
		After treat	$8,84 \pm 0,44$	$8,85 \pm 0,50$	$23,12 \pm 1,17$	$26,51 \pm 1,19$
		n	10	10	10	10
		p	N. S.	N. S.	N. S.	N. S.
(-) EBURNA-MONINE	3C	Baseline	$9,06 \pm 0,60$	$7,81 \pm 0,75$	$51,60 \pm 1,63$	$21,67 \pm 1,98$
		After treat	$4,90 \pm 0,50$	$4,94 \pm 0,56$	$44,85 \pm 1,47$	$26,18 \pm 1,84$
		n	55	55	52	34
		p	0,001	0,001	0,001	0,01
CINNARI-ZINE	30	Baseline	$10,04 \pm 0,67$	$9,06 \pm 0,80$	$52,90 \pm 1,51$	$24,43 \pm 1,65$
		After treat	$4,84 \pm 0,62$	$5,90 \pm 0,79$	$46,30 \pm 1,57$	$26,26 \pm 1,38$
		n	51	51	50	35
		p	0,001	0,001	0,001	N. S.
(-) EBUR-NAMONINE	30	Baseline	$10,22 \pm 0,68$	$11,38 \pm 0,55$	$26,64 \pm 1,12$	$20,84 \pm 1,08$
		After treat	$5,93 \pm 0,54$	$6,69 \pm 0,66$	$20,14 \pm 1,68$	$28,80 \pm 1,96$
		n	28	28	28	28
		p	0,001	0,001	0,05	0,01
VINCAMINE	30	Baseline	$9,82 \pm 0,82$	$12,24 \pm 0,48$	$23,65 \pm 1,27$	$23,17 \pm 1,36$
		After treat	$6,84 \pm 0,56$	$7,29 \pm 0,52$	$20,18 \pm 1,53$	$25,68 \pm 1,77$
		n	30	30	30	30
		p	0,01	0,01	0,01	0,05
PLACEBO 2nd	30	Baseline	$9,08 \pm 0,58$	$10,84 \pm 0,39$	$24,86 \pm 1,08$	$22,64 \pm 1,92$
		After treat	$8,01 \pm 0,36$	$9,28 \pm 0,62$	$25,02 \pm 1,84$	$23,84 \pm 1,86$
		n	30	30	30	30
		p	N. S.	N. S.	N. S.	N. S.
BAMETHANE	60	Baseline	$8,8 \pm 0,4$	$9,0 \pm 0,4$	$14,9 \pm 1,0$	$27,5 \pm 1,7$
		After treat	$6,3 \pm 0,6$	$6,1 \pm 0,6$	$9,9 \pm 1,0$	$26,0 \pm 2,0$
		n	36	36	36	36
		p	N. S.	N. S.	N. S.	N. S.
PLACEBO 3rd	60	Baseline	$9,9 \pm 0,4$	$9,9 \pm 0,4$	$17,4 \pm 1,0$	$23,0 \pm 1,1$
		After treat	$7,3 \pm 0,6$	$8,0 \pm 0,7$	$13,5 \pm 1,5$	$23,1 \pm 1,8$
		n	37	37	37	37
		p	N. S.	N. S.	N. S.	N. S.

In fact, S would appear to be less efficient on the psychic symptoms rather than on the neurological disorders, while V acts analogously on both groups, even if less evidently than other drugs.   B and placebos did not induce any significant modification even on the total score of the items of the Plutcick GRS or of the WMS/WBAIS.   We must note that there always exist behavioural differences between the patients treated with the various drugs which suggest that S and E have a more pronounced activity.   An examination of the average changes in the various neurological symptoms (Fig.1) with respect to base values, leads to some interesting observations, in spite of the limitations incurred.

The symptom 'confusion' seems more sensitive to C and E; headache is reduced considerably by C, as well as by E and V; dizziness by C; poor vigilance is less evident after therapy with S while V (as also E) has a marked effect on tinnitus.

Trembling present in just a few patients, and with slight intensity, does not seem to be influenced in any clear way by the therapies carried out, even though V, S and C have always lowered the intensity of the symptom.

Physical weakness appeared to be less after administration of nearly all the drugs examined, but especially after therapy with S.

With regards to the neurological symptomatology, we feel it is important to emphasize the modest or absent placebo activity; from this we can deduce that the reduction in the degree of symptomatology obtained with drug treatment was not by chance.

There was also a general tendency for a reduction in psychic symptoms (see Table 5), but for some (apprehension, impulsivity) the placebos were discretely efficient.   Therefore, it is most probable that a part of the psychic symptomatology of OBS may be influenced not only by drug treatment, but also with substances having a placebo effect which may determine a different psychic response on the part of the patient.   However, for the other disorders C noticeably relieved irritability (evidently due to the sedative effect of the drug), V impulsivity and emotional lability, E impaired cooperation and depression and both E and V almost analogously poor sociability.   The modifications in symptomatology and differences in the activity of the various groups of drugs and comparisons in therapy carried out, do not, however, often reach any degree of statistical significance.

The subtests of the psychometric WMS and WBAIS scales have further shown no particular significance after treatment with placebo, while S showed very significant modifications in information and comprehension, E mainly information, but also comprehension and digit span and, even, but to a lesser degree, orientation and block design.

Table 5. Scores ($\pm$ SE) of the subtests of the WMS/WBAIS before and After some Therapies (B=baseline; A.T. = After Treatment)

		INFORMATION	ORIENTATION	COMPREHENSION	DIGIT SPAN	ARITHMETIC REASONING	BLOCK DESIGN
SULOCTIDIL	B.	2,89 $\pm$ 0,18	3,89 $\pm$ 0,20	5,33 $\pm$ 0,15	3,00 $\pm$ 0,11	1,67 $\pm$ 0,22	3,01 $\pm$ 0,08
	A.T.	3,44 $\pm$ 0,22	4,00 $\pm$ 0,28	7,55 $\pm$ 0,38	4,02 $\pm$ 0,14	2,33 $\pm$ 0,18	3,11 $\pm$ 0,18
	p	0,001	N.S.	0,001	0,01	0,01	N.S.
PLACEBO 1st	B.	2,22 $\pm$ 0,36	4,33 $\pm$ 0,18	5,03 $\pm$ 0,58	4,14 $\pm$ 0,21	3,95 $\pm$ 0,46	4,35 $\pm$ 0,26
	A.T.	3,22 $\pm$ 0,4	4,22 $\pm$ 0,33	6,22 $\pm$ 0,44	3,86 $\pm$ 0,19	4,01 $\pm$ 0,49	4,08 $\pm$ 0,31
	p	0,001	N.S.	N.S.	N.S.	N.S.	N.S.
(-)EBURNAMONINE 1st	B.	2,78 $\pm$ 0,15	3,90 $\pm$ 0,24	5,75 $\pm$ 0,38	2,23 $\pm$ 0,42	3,35 $\pm$ 0,58	3,03 $\pm$ 0,22
	A.T.	3,46 $\pm$ 0,18	4,29 $\pm$ 0,16	6,96 $\pm$ 0,41	3,11 $\pm$ 0,58	4,00 $\pm$ 0,48	3,78 $\pm$ 0,18
	p	0,001	0,05	0,01	0,01	N.S.	0,05
CINNARIZINE	B.	3,14 $\pm$ 0,21	4,18 $\pm$ 0,35	6,42 $\pm$ 0,46	3,51 $\pm$ 0,28	4,06 $\pm$ 0,42	3,63 $\pm$ 0,28
	A.T.	3,40 $\pm$ 0,19	4,57 $\pm$ 0,41	6,78 $\pm$ 0,49	4,03 $\pm$ 0,34	4,45 $\pm$ 0,46	3,93 $\pm$ 0,30
	p	N.S.	N.S.	N.S.	N.S.	N.S.	N.S.
(-)EBURNAMONINE 2nd	B.	2,89 $\pm$ 0,26	3,94 $\pm$ 0,28	5,39 $\pm$ 0,16	3,76 $\pm$ 0,18	3,01 $\pm$ 0,28	3,11 $\pm$ 0,20
	A.T.	3,29 $\pm$ 0,18	4,36 $\pm$ 0,25	6,18 $\pm$ 0,22	4,39 $\pm$ 0,38	3,44 $\pm$ 0,31	3,84 $\pm$ 0,16
	p	0,01	N.S.	0,001	0,05	N.S.	0,01
VINCAMINE	B.	3,22 $\pm$ 0,09	4,32 $\pm$ 0,11	5,44 $\pm$ 0,14	4,09 $\pm$ 0,20	2,88 $\pm$ 0,36	3,28 $\pm$ 0,17
	A.T.	3,48 $\pm$ 0,14	4,28 $\pm$ 0,23	6,08 $\pm$ 0,22	4,14 $\pm$ 0,09	3,55 $\pm$ 0,42	3,95 $\pm$ 0,18
	p	N.S.	N.S.	0,01	N.S.	N.S.	0,05
PLACEBO 2nd	B.	2,68 $\pm$ 0,25	4,02 $\pm$ 0,20	6,35 $\pm$ 0,16	2,38 $\pm$ 0,30	3,50 $\pm$ 0,18	3,88 $\pm$ 0,24
	A.T.	3,09 $\pm$ 0,19	4,14 $\pm$ 0,15	6,02 $\pm$ 0,22	3,11 $\pm$ 0,11	3,36 $\pm$ 0,27	4,05 $\pm$ 0,44
	p	N.S.	N.S.	N.S.	0,05	N.S.	N.S.

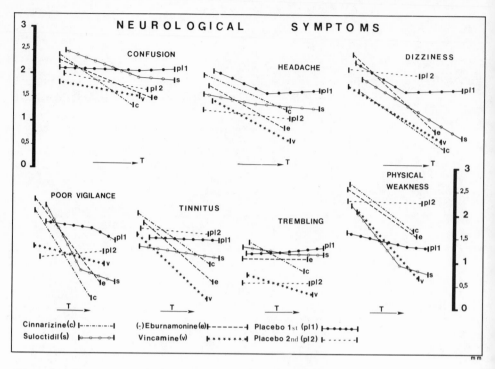

Fig. 1.   Mean scores of neurological symptoms after some therapies.

On the other hand, C did not reveal any noticeable action and
V induced significant variations only in comprehension and block
design.

From this picture we can begin to see the particular action
the various drugs have on some intellectual abilities, especially
those which worsen with the deterioration of brain functioning in
old age.

CONCLUSIONS

Our methodology can be used as an approach to the problem of
diagnosis in OBS, and it can be useful in the clinical evaluation
of drug treatment.   As we have already mentioned, these methods are
still not completely satisfactory and are justifiably criticised.
The complexity of the disorders to be examined and the difficulty
in distinguishing between the different psychic functions are
obvious.   However as numerous authors have asserted[26,32,35,38],
our methodology can supply some satisfactory judgements especially
when results are not considered separately, but evaluated globally
in a general clinical perspective of the patient.

A clinical evaluation must in fact precede any type of technical examination; it must guide it. It is from a general clinical evaluation of the patient that the same application of the various tests can be made. None of these will be of any value unless performed on patients whose overall clinical conditions allow them to be carried out.

Our work shows that some subtests of WMS and WBAIS and the GRS can be applied to the elderly patient and that we can make some judgements from them regarding the mental state of the subject and possible variations obtainable with suitable treatment.

We have, however, also proposed some methods of examination (tremblingmeter, recency test, crossing-out test, and the use of an electronic detector for V.A.M.) which can certainly be applied to the elderly patient and prove useful in evaluating the most frequent symptoms.

With regards to the other methods of examination we think that the results obtained with the administration of certain drugs show both the usefulness of the method, and their value in a study of the action of particular substances, the characteristics of which indicate their use in OBS therapy.

No definitive judgement can be given on the basis of this data and with such a small group of drugs. We must remember that many other pharmacological substances in daily use have given proof of some therapeutic activity in the symptomatology of OBS (cyticholine, piracetam, viquidil, etc.) especially if used on certain symptoms or groups of symptoms.

But we think that we can, on the whole, affirm that the drugs we considered have showed therapeutical effects which may be referred to the pronounced vascular and/or metabolic action they exert on the ageing brain. We hope that future research will amplify our results with more certain data, of greater use in the differentiation of therapies for the various clinical manifestations of geriatric OBS.

The therapeutic importance of environment in relation to brain functioning must not be forgotten. There are some interesting experimental data on animals: different environmental situations (light stimulation and remaining in the dark) are related to variations in the opposite direction to do esterasic activity, thus indicating a biological turn to environment[20]. With this in mind, the acquisition of a consistent cultural knowledge enriched by continual stimulation may allow alternative solutions and a blocking of the involutional process.

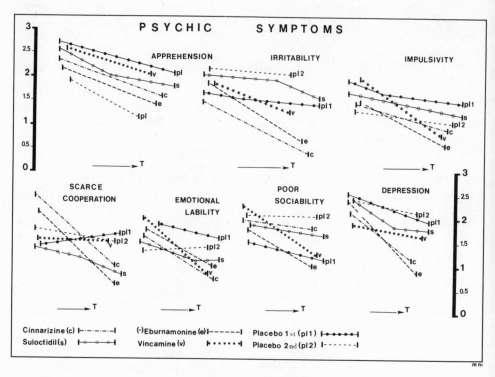

Fig. 2. Mean scores of psychic symptoms after some therapies.

This extremely difficult problem must be faced and programmed long before old age. Therefore we must make every effort to prevent and cure symptoms of deterioration in the psycho-physical functions since this always presents a serious clinical picture and often leads to a lack of self-sufficiency in the patient, which is inevitable without the timely use of drugs together with all the psychological, social, and rehabilitative means available.

An indication to use these substances might be of help to the medical practitioner in that it permits a more correct therapeutic approach to the morbid forms in question. With the development of research in this field, it will perhaps be possible to use the drugs more correctly and make a very important contribution not only to the therapy of chronic geriatric OBS, but also to the knowledge of molecular mechanisms which regulate the various manifestations of the whole neuropsychic functioning of the human brain; from a study of symptomatologic variations, and the biochemical actions seen in vitro, we might perhaps begin to recognize the mechanisms involved in the various brain function activities.

REFERENCES

1. E. Musco and F. M. Antonini, "Adattamento e scompenso psicolo-
   gico nell'anziano", Monografia Sandoz (1976).
2. D. M. Bowen and A. N. Davison, Biochemical changes in the normal
   ageing brain and in dementia, in: "Recent Advances in
   Geriatric Medicine I", Bernard Isaacs, ed., Churchill
   Livingstone (1978).
3. D. M. Bowen, C. B. Smith and A. N. Davison, Molecular changes
   in senile dementia, Brain 96:849 (1973).
4. B. E. Tomlinson and G. Henderson, Some quantitative cerebral
   findings in normal and demented old people, in: "Neurobiology
   of Aging" (Aging v.3), S. Gershon and R. D. Terry, eds.,
   Raven Press, New York (1976).
5. A. Carlsson and Winblad, J.Neurol.Neurotransmission 38:271
   (1976).
6. M. C. McNamara, A. T. Miller, V. A. Benignus and J. N. Davis,
   Brain Res. 131:313 (1977).
7. R. R. Kohn, "Principles of Mammalian Ageing", Englewood Cliffs,
   Prentice Hill (1971).
8. K. G. Lloyd and O. Hornykiewicz, Occurrence and distribution
   of aromatic L-aminoacid (L-Dopa) decarboxylase in the human
   brain, J.Neurochem. 19:1549 (1972).
9. K. G. Lloyd, L. Shemin and O. Hornykiewicz, GABA binding in
   human brain: specific alterations in substantia nigra of
   Parkinsonian patients, Abstract 6th Annual Meeting of the
   Society for Neurosciences, Toronto (1976).
10. E. G. McGeer and P. L. McGeer, Age changes in the human for
    some enzymes associated with metabolism of catecholamines.
    GABA and acetilcholine, in: "Neurobiology of Ageing",
    J.M. Ordy, K. R. Brizzee, eds., Plenum Press, New York (1975).
11. S. Fahn, in: "GABA in nervous system function", E. Roberts,
    T.N. Chase and D. B. Tower, eds., Raven Press, New York
    (1976).
12. W. Vale, C. Rivier and M. Brown, Ann.Rev.Physiol. 39:473 (1977).
13. N. Seiler and T. Schmidt-Glenewinkel, Regional distribution of
    Putrescine, Spermidine and Spermine in relation to the
    distribution of RNA and DNA in the rat nervous system,
    J.Neurochem. 24:791 (1975).
14. A. Merat and J.W.T. Dickerson, The effect of development on
    the gangliosides of rat and pig brain, J.Neurochem. 20:873
    (1973).
15. G. Price and T.Makinodan, Gerontologia 19:58 (1973).
16. A. Rosenberg, "Inborn disorders of sphingolipid metabolism",
    Pergamon Press, New York (1976).
17. G. Benzi, E. Arrigoni, F. Dagani, F. Marzatico, A. Manzini,
    O. Pastoris and R. F. Villa, Drugs action on cerebral energy
    state during and after various hypoxic conditions, Ach.Int.
    Pharm.Ther. 236:234 (1978).

18. P. Ernst, D. Badash and P. Beran, Incidence of mental illness
    in the aged: unmasking the effects of a diagnosin of chronic
    brain syndrome, J.Am.Geriatric Soc. 25:371 (1977).
19. E. H. Liston, Occult presenile dementia, J.Nerv.&Ment.Dis.
    164:263 (1977).
20. L. Amaducci, La demenza senile, Il Polso 3:20 (1979).
21. J. C. Brocklehurst and T. Hanley, "Geriatric Medicine for
    Students", Churchill Livingstone (1976).
22. R. L. Kahn, A. I. Goldfarb and M. Pollack, Brief objective
    measures for the determination of mental status in the
    aged, Am.J.Psychiat. 117:326 (1960).
23. T. L. Brink, J. Bryant, M. L. Catalano, C. Janakes and
    C. Oliveira, Senile Confusion: Assessment with a New
    Stimulus Recognition Test, J.Amer.Geriatric Soc. 126:3
    (1979).
24. E. de Renzi, Le amnesie, in: "Neuropsicologia clinica",
    F. Angeli, ed., Milano (1977).
25. C. de Risio and F. Rossi, "La memoria negli alcoolisti.
    Alterazioni nella memorizzazione temporale di elementi
    visivi", Unpublished Ph.D. thesis, Medicine School,
    University of Parma (1977).
26. M. Neri, G. Ciotti, G. Feltri, V. Baldelli, M. Gasparini
    Casari, Studio delle funzioni mnesiche in soggetti anziani
    sani e mentalmente deteriorati, Giorn.Geront. 27:287 (1979).
27. J. Kugler, Dtsch.Med.Wsch. 103:456 (1978).
28. A. L. Benton, "Problemi di Neuropsicologia", Giunti-Barbera,
    ed., Firenze (1966).
29. B. Milner, Interhemispheric differences in the localization
    of psychological processes in man, Brit.Med.Bull. 27:272
    (1971).
30. B. Milner, Psychological aspects of epilepsy and its neuro-
    surgical management, in: "Advances in Neurology",
    D. P. Purpura and P.I.K. Walter, eds., Raven Press, New
    York (1975).
31. C. B. Dodrill, A neuropsychological battery for epilepsy,
    Epilepsia 19:611 (1978).
32. G. F. Marchesi, O. Scarpino, M. Signorino and P. Fuà,
    Prestazioni visuomotorie nell'epilessia "Piccolo male":
    studio con una prova di discriminazione visiva, in:
    "Progressi in Epilettologia", F. Angelesi and R. Canger,
    eds., Boll.Lega It.Contro Epilessia, 22:89 (1978).
33. C. Umiltà and E. Ladavas, Personal communication (1979).
34. P. Fraisse, Les niveaux d'efficience in: "Manuel Pratique de
    Psychologie Experiementale",Presses Universitaires de
    France.
35. M. Passeri, Therapy of chronic consequences of brain ischemia,
    Eur.Neurol. (suppl.1) 17:150 (1978).
36. R. Plutchik, H. Conte, H. Lieberman, M. Bakur, J. Grossman
    and N. Lehrman, Reliability and validity of a scale for

assessing the functioning of geriatric patients, <u>J.Am.
Geriatric Soc.</u> 18:491 (1970).

37. S. B. Jacobson, Geriatric psychiatry today, <u>Bull.N.Y.Acad.Med.</u>
54:568 (1978).

38. J. Pearce and E. Miller, "Clinical aspects of dementia",
Bailliere and Tindal, London (1973).

PROBLEMS WITH THE PSYCHOMETRIC AND NEUROPSYCHOLOGICAL

EVALUATION OF AGING

N. Martucci,* M. Fioravanti,** S. Mearelli,*A. Agnoli*

*Neurological Clinic
** Institute of Psychology
School of Medicine, L'Aquila, Italy

The aging process is defined in diverse ways with each author stressing specific aspects of the process, e.g. biological, environmental, and/or social aspects. Birren and Renner[1] have proposed a more global and comprehensive definition of aging. They identify this process as the regular changes which occur in a mature organism living in its natural environment as it progresses in chronological age.

The concept of age is more complex than its mere chronological aspect. Biological age, for example, is defined by the life span of the organism. Psychological age, on the other hand, is determined by the organism's capability to adapt to change within its environment. Social age is a consequence of the attributions by other people and refers to the extent to which the organism meets society expectations characteristic of its particular chronological age. Statements such as "he acts older than his age" or "he acts younger than his age" reflect this social evaluation.

As a consequence of these conceptual difficulties, the psychological study of the process of aging has encountered a series of methodological problems which must be identified and minimized before reliable research results can be expected.

There has been an increased interest in the ageing processes in both medical and social disciplines. Research focusing on the biological, psychological and/or sociological aspects of aging, however, offers results which are not always consistent and, often conflicting.

In the sphere of clinical application, this branch of psychology is confronted by problems of defining the range of normality to employ as an evaluation criterion and the need to account for socio-cultural variables related to the characteristic behaviour of the aged person.

The scope of the gerontopsychologist should be that of offering the clinician valid instruments capable of identifying, first of all, diagnostic criteria to justify and orient eventual therapeutic intervention and then identifying those criteria which permit evaluation of therapeutic outcome.

It should be noted that psychological assessment instruments developed for adults are often inappropriate for the aged and create problems which can invalidate subsequent results. These problems can be related to sensory-motor deficits, to greater susceptibility to physical and mental fatigue, to longer reaction and performance times, and to cultural background differences in the normal aged person and result in altered scores on tests developed for the normal adult.

At present, the diagnostic instruments used in clinical practice can be grouped as follows:

1) Clinical Interview
2) Rating Scales
3) Intellectual level and Performance Scales
4) Self-description Scales
5) Projective Techniques

1) The purpose of the interview is to identify the presence, the intensity, and the quality of eventual disabling symptoms. The interviewer must also evaluate the patient's contact with reality, his or her emotional state, reasoning and problem solving capabilities, and ability to develop and maintain satisfactory interpersonal relationships. Interviewing aged patients presents specific problems for the interviewer. He or she must be specially trained to reduce the length of the interview in order to avoid unnecessary fatigue, to conduct the session in optimal environmental conditions, and to avoid interference due to language or cultural differences, and also to consider the often diverse expectations and desires of the aging patient.

2) Even though rating scales often lack reliability and, to a lesser extent, validity, these scales permit a more objective description of the subjective clinical evaluation of the status of CNS functioning. Rating scales focus mainly on the overt behaviour of the patient and, therefore, provide a somewhat rough measure of the patient's clinical status.

3) Intellectual level and performance scales are most sensitive to reduced validity. This reduction can be accounted for, in part, by the development and validation of these scales for normal adult populations.

4) The self-description scales offer a number of advantages. They can be easily administered and this administration can be repeated to evaluate the clinical status of the patient. However, these scales also present some fundamental problems such as item content comprehension, item relevancy with respect to the aged person's socio-cultural environment, and control for acquiescence tendencies.

5) Projective techniques afford a measure of emotional reactivity but require experienced and complex interpretation which cannot be easily related to other objective sources of information.

Biological measures derived from neurophysiological, hormonal and ANS parameters are interesting when used in the experimental field. Unfortunately, these measures require strict standardized conditions, are difficult to obtain, and, sometimes, are difficult to interpret when they are used in the clinical setting.

This brief survey of the evaluation techniques available for the assessment of aging demonstrates that no one technique alone offers a reliable basis for diagnosis and emphasizes the absence of methods designed to assess the interpersonal and environmental aspects of this process. Clinically oriented techniques focus on the individual and tend to obscure his/her social and environmental characteristics which are very important in shaping the aging process. The aging person develops his/her self-image, in part, from those social cues around him or her, and, often, these external influences can mould the person's actual level of functioning.

When an aged person is in a nonrewarding environment, his or her sensory and motor deterioration as well as memory and intellectual deficits are more likely to be denied or produce a withdrawal reaction. Clearly, a psychological assessment of aging must include measures which evaluate environmental and social influences on this process. Reliable techniques must be developed for this purpose.

PURPOSE

Psychological evaluation of the aged person is of fundamental importance if one is to understand and describe the physiological and pathological changes of this process. The major problem encountered in this field is the identification and differentiation between the normal aspects and pathological aspects of aging.

We have already discussed some of the problems associated with
current assessment methods when they are employed in the clinical
setting.   However, other problems also interfere with any clinical
diagnosis of the aged.   The aged patient's status should be
considered with respect to how much it has changed in relation to
his or her previous characteristics.   For example, a statistically
"normal" IQ score may be obtained from a patient who has suffered
vascular brain damage.   The measure of IQ alone does not permit
assessment of the actual IQ loss.   Accordingly, an aged patient
with a modest sociocultural level may easily obtain a below
"normal" IQ score even when no pathology is present in the aging
process.   Several methods have been proposed to develop IQ
deterioration curves related to aging.   Both the cross-sectional
and longitudinal methods proposed to develop these curves have
practical and theoretical limitations which reduce the ranges of
their application[2].

A possible method for minimizing some of these problems
encountered in the clinical assessment of the aged patient is to
integrate test results with behaviour descriptions and evaluations
from relatives.   These external descriptions add more meaning to
the normative measures obtained from tests.   External evaluation
reflects an "historical" knowledge of the patient from the
relatives' perspectives while test measures reflect a "transversal"
evaluation of the patient.   Hence, it appears important to identify
the relationships between these two diverse descriptions and
evaluations of the aging person when one is concerned with a
comprehensive diagnostic procedure which includes biological,
intellectual, psychological and social elements.

The methodological procedure proposed here has been introduced
at the Institute of Personality Assessment and Research (IPAR) of
the University of California, Berkeley to study personality in
various applied fields, i.e. creativity in architects, success in
professional careers, academic achievement in university under-
graduates and medical students[3].   This procedure consists of
assessment sessions where the subjects are evaluated on the basis
of intellectual level tests, self-description scales, interviews,
and reliable observational procedures.

The procedures employed in this study follow those typically
used at IPAR.   Teams of observers have described each single
subject after observation and interview sessions with subject.
Each observer described the subject with the same instrument to
control reliability among team members and to obtain a combined
description for each subject which minimizes individual observer's
variability.

Several different instruments can fulfill the requirements of
this type of observational description but we have chosen to use

the Adjective Check List (ACL) of H. G. Gough[4,5,6]. The ACL was selected because it has previously been used for this purpose, has demonstrated validity, is easy to use, and permits control for internal concordance between individual observers and groups of observers who have operated under different conditions.

The ACL consists of 300 adjectives or short adjectival descriptions which can be used to describe persons, geographic locations, and historical characters. The observer freely checks the adjectives which he or she considers descriptive of the object of his or her observation. The resultant description is idiographic in the sense that it reflects the individual peculiarities of the subject and not a rank related to other subjects, and is a free choice technique in that checking one item does not influence the checking of others but each observer can check any of the items he or she wants for each case.

The purpose of this research is to explore the possibility of using a complex and polycentric methodology of assessment with the aging person. This methodology is complex because it employs several measurement techniques and polycentric in that its direction is both toward the subject and his or her environment. The aim of this comprehensive evaluation is to provide a more meaningful examination of the process of aging that is not limited only to the clinical and functional evaluation of the subject but also encompasses the social age concept which seems more relevant to the problems confronting the aged in their everyday lives.

METHOD

The participants in this study were 13 male subjects with a mean age of 63.17 (S.D. = 6.48) and a mean scholastic level of 5.0 years (S.D. = 1.65) who voluntarily agreed to take part in the research. All subjects were retired and living in the same geographical zone. Three neuropsychiatrists interviewed the subjects. The Wechsler-Bellevue Intelligence Scale for Adults[7], the Bender-Gestalt Test, and the Tolouse-Pieron Test were administered. Subjects also underwent a neurological examination and a computerized EEG analysis. (Table 1)

Their closest relatives were interviewed by three other clinicians and the 6 interviewers for each subject then completed the ACL to describe the subject. Two types of descriptions were obtained: one based on the direct observation of the subject and the other based on the relatives' picture of him.

The interviewers also completed a questionnaire based on social and interpersonal abilities, ways of coping with his aging process, and an appraisal of his present status with respect to his past.

Table 1. Methodology - Aged Subject (65-70 Years of Age)

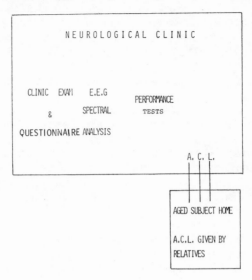

This time, only one questionnaire was filled out for each interview condition.

The following analyses were performed: Kendall's Tau was completed to determine concordance between questionnaire responses obtained from the subjects and the subject's relatives: Chronbach's alpha was computed to determine reliability among observers of each condition; the two alpha series were compared; two combined descriptions for each subject were obtained from each group of three interviewers of both conditions; these two combined descriptions were compared to identify differences; these resultant combined descriptions were then correlated with the results of intellectual level and performance scales; EEG results were compared also to intellectual and performance scales.

RESULTS

Questionnaire

The analysis of concordance demonstrated the following:

a) Subjects describe themselves and are described by relatives as being in good health. Only 35% of the cases reported health problems. However, 45% of the cases complained of unsatisfactory psychological conditions. Clinical examinations of subjects revealed neither physical nor neuropsychiatric pathology.

b) The subjects and their relatives presented discordant pictures of the relationships between subjects and their spouses.   The relatives gave a more favourable description of this relationship. It should be noted that the subjects' wives were included in the relatives group.

c) Both subjects and relatives described his past role in the family as a prominent one.   His relationships with others were not described as changed after retirement.

d) Relatives and subjects disagreed about the presence of cultural, social, and political interests with relatives minimizing their importance.   However, both groups agreed about the presence of a social life outside the family.

e) The relatives expressed a more pessimistic view of the subjects' future.

f) Both groups reported little change with respect to the behaviour of others towards the subjects with advancing age.

g) Eating habits and maintenance of physical appearance were reported as unchanged by both subjects and relatives.

h) The problems of sexuality in the aged were described in a discordant manner with subjects expressing problems in this area and relatives minimizing their importance. (Table 2)

Wechsler-Bellevue, Bender-Gestalt, Toluse-Pieron

The Italian edition of the Wechsler-Bellevue Intelligence Scale for Adults[7] was used and the results of this test are shown in Table 3.   These results suggest that the subjects in our sample were within the normal range of intelligence and that no single subtest was specifically affected by the aging process.   When one takes into account the scholastic level and SES level of our subjects, this sample can be considered as representative of a normal aged post-retirement population even though the sample size is small.

The Bender-Gestalt Test has been scored with two scoring methods: the Pascal-Suttell[8] Scale and the Hutt-Briskin[9] Scale. This double type of scoring was employed to control for the presence of either organic or psychopathological signs.   The results are shown in Tables 4 and 5, and demonstrate an absence of any kind of pathological signs.

The Toluse-Pieron Test was used in this study to evaluate the subject's ability to solve tasks which require attention and which require rapid and precise perceptual discrimination.   The variables measured by this test were: number of correct responses, number of errors, number of missing responses, and response time.   The test is organized in a series of 4 tasks.   The results are shown in

Table 2.   Concordance of Responses by Family Members and the Subject
           to an Interview-Questionnaire (Kendall's Tau)

CONCORDANCE

1° - HOW WOULD YOU DESCRIBE YOUR PHYSICAL CONDITION?

$\tau = 0.79$ P< 0.01
Tau = .79   P< .01

2° - HOW WOULD YOU DESCRIBE YOUR PSYCHOLOGICAL CONDITION?

$\tau = 0.73$ P< 0.01
Tau = .73   P< .01

3° - HAS YOUR RELATIONSHIP WITH YOUR SPOUSE CHANGED OVER TIME?

$\tau = 0.06$ P N.S.
Tau = .06   N.S.

4° - HAVE YOU PLAYED A PROMINENT ROLE IN YOUR FAMILY LIFE?

$\tau = 0.85$ P< 0.001
Tau = .85   P< .001

5° - HAS YOUR RAPPORT WITH OTHERS CHANGED SINCE YOU WERE RETIRED?

$\tau = 0.68$ P< 0.01
Tau = .68   P< .01

6° - DO YOU HAVE CULTURAL, SOCIAL, AND/OR POLITICAL INTERESTS?

$\tau = 0.39$ P N.S.
Tau = .39   N.S.

7° - DO YOU HAVE FRIENDS AND DO YOU ENJOY SEEING THEM OFTEN?

$\tau = 0.64$ P< 0.05
Tau = .64   P< .05

8° - WHAT DO YOU THINK ABOUT YOUR FUTURE?

$\tau = 0.08$ P N.S.
Tau = .08   N.S.

9° - DO YOU THINK THAT THE BEHAVIOR OF OTHERS HAS CHANGED TOWARDS YOU?

$\tau = 0.54$ P< 0.05
Tau = .54   P< .05

10° - DO YOU ATTEND TO YOUR PHYSICAL APPEARANCE?

$\tau = 0.95$ P< 0.001
Tau = .95   P< .001

11° - DO YOU ENJOY GOOD FOOD OR DO YOU HAVE TO BE ENCOURAGED TO EAT?

$\tau = 0.68$ P< 0.01
Tau = .68   P< .01

12° WHAT DO THINK ABOUT SEXUALITY IN THE THIRD AGE?

$\tau = 0.46$ P N.S.
Tau = .46   N.S.

Table 3. Wechsler - Bellevue (Adult Intelligence Scale)

	M̄	SD̄	N
INFORMATION	8.67	1.88	12
COMPREHENSION	9.17	3.19	12
DIGIT SPAN	5.33	2.06	12
ARITHMETIC	7.46	2.16	11
SIMILARITIES	6.92	4.27	12
VOCABULARY	7.75	2.42	12
VERBAL I.Q.	97.25	9.11	12
PICTURE ARRANGEMENT	3.09	1.58	11
PICTURE COMPLETION	6.09	3.42	11
BLOCK DESIGN	6.09	3.05	11
OBJECT ASSEMBLEY	6.40	3.27	10
DIGIT SYMBOL	4.33	2.15	12
PERFORMANCE I.Q.	95.73	11.33	11
FULL SCALE I.Q.	95.36	9.42	11

Table 6 and indicate that all subjects responded within the normal range with few missing responses and no errors.   Execution time was also normal and no negative trend was demonstrated over time.

Adjective Check List

   The alpha coefficients from the reliability analysis demonstrated median values of .52 for both clinical interviewers of the subjects and interviewers of the family members of the subject. The range was .29 to .78 for the subject group and .30 to .75 for the family member group.   These values, although not very high, represent a satisfactory level of reliability when only 3 observers are employed.   This can be considered the minimum number of observers necessary to obtain a valid observation.   The similarity demonstrated between the group reliabilities emphasize that the method that we propose here is not tied to a specific observational

Table 4. Bender-Gestalt Test (Pascal-Suttell Scale)

	$\bar{M}$	$S\bar{D}$
1° FIGURE	3.62	3.75
2° FIGURE	3.77	3.44
3° FIGURE	8.62	5.85
4° FIGURE	12.54	6.92
5° FIGURE	6.31	5.02
6° FIGURE	7.46	4.89
7° FIGURE	9.15	4.58
8° FIGURE	5.85	5.83
CONFIGURATION DESIGN	4.00	4.51
(TOTAL SCORE)	62.08	30.27

condition.   Both the direct observation and the indirect observation exhibited the same degree of reliability.

An examination of the alpha coefficients for the individual descriptions shows that the larger differences in reliability between the subject and family member conditions were found for those subjects whose performance and intellectual scores or whose personal history had elements which presented difficulty in interpretation even though they were not pathological.   This illustrates how eventual problems in the aged are perceived in diverse manners when the conditions of observation are different.

The two combined ACL descriptions, one from the three interviewers of the subject and one from the three interviewers of the family members, were derived according to the following criterion: only items checked by at least 2 of the 3 observers were included in each combined description.   These two descriptions were then compared and the results are presented in Table 7.

Family members describe the aged in a more critical and depreciative way than clinicians.   Family members are less likely to describe their aged relatives as logical, leisurely, painstaking,

Table 5.   Bender-Gestalt Test (Hutt-Briskin Scale)

	$\bar{M}$	$\bar{SD}$
SEQUENCE	3.08	2.57
POSITION, 1ST DRAWING	1.18	0.64
USE OF SPACE	1.00	0.00
COLLISION	3.89	2.84
SHIFT OF PAPER	1.00	0.00
CLOSURE DIFFICULTY	1.28	0.94
CROSSING DIFFICULTY	3.46	2.50
CURVATURE DIFFICULTY	5.39	2.63
CHANGE IN ANGULATION	2.23	1.54
PERCEPTUAL ROTATION	3.54	3.64
RETROGRESSION	1.00	0.00
SIMPLIFICATION	1.92	2.57
FRAGMENTATION	1.46	1.13
OVERLAPPING DIFFICULTY	4.00	3.35
ELABORATION	2.39	2.90
PERSEVERATION	1.69	1.32
REDRAWING, TOTAL FIGURE	1.87	1.73
TOTAL SCORE	40.05	12.58

Table 6.   Test Di Toulouse-Pieron

PROVE	1		2		3		4	
(TRIALS)	$\overline{M}$	$\overline{SD}$	$\overline{M}$	$\overline{SD}$	$\overline{M}$	$\overline{SD}$	$\overline{M}$	$\overline{SD}$
RIGHT CHECKS	10.69	5.14	11.31	3.68	12.31	4.03	12.15	4.43
MISSING CHECKS	2.46	3.89	1.92	3.15	1.85	3.83	1.62	3.28
WRONG CHECKS	0.00	0.00	0.00	0.00	0.00	0.00	0.00	0.00
TIME SEC.	64.15	43.21	52.39	29.07	52.69	32.97	53.85	31.32

Table 7.   Characteristic and Noncharacteristic Adjectives
Employed to Describe a Group of Aging Male Subjects
By Family Members and Clinical Interviewers

	FAMILY MEMBERS	CLINICAL INTERVIEWERS			FAMILY MEMBERS	CLINICAL INTERVIEWERS
CLEVER	+	+		∧ PAINSTAKING	−	
WARM		+		HONEST	+	+
FRIENDLY	+	+		∧ PESSIMISTIC		−
ACTIVE	+	+		∧ PRACTICAL	−	
∧ LEISURELY	−			SINCERE	+	+
CIVILIZED		+		SPONTANEOUS		+
∧ LOGICAL	−	+		∧ SUGGESTIBLE		−
COOPERATIVE	+	+		∧ MASCULINE	−	

" + " INDICATES THAT THE ADJECTIVE HAS BEEN CHECKED FOR MORE THAN 50 % OF THE SUBJECTS.

" − " INDICATES THAT THE ADJECTIVE HAS NOT BEEN CHECKED FOR ANY OF THE SUBJECTS

NO SIGN INDICATES THAT THE ADJECTIVE HAS BENN CHECKED FOR LESS THAN 33 % OF THE SUBJECTS AND THERE WAS A
    DIFFERENCE BETWEEN OBSERVER GROUPS.

" ∧ " DISCORDANCE IN JUDGMENT.

practical, and masculine. These adjectives were never used by family members in their descriptions of subjects but clinicians did use them. One can interpret these differences in descriptions as due to the different perspectives, i.e. the clinical and familial. The clinician notes the lack of pathology while the family member stress the changes in behaviour related to the aging process and subject's role modification after retirement.

These combined descriptions were subsequently correlated with the Wechsler-Bellevue scores of verbal, performance and total IQ and with the Bender-Gestalt scores of the Pascal-Suttell and Hutt-Briskin scales. The results from both groups of descriptions can be found in Tables 8, 9 and 10.

The correlated patterns are different for each group of descriptions. Clinical evaluation of the aged person is strongly influenced by the observational context and the amount of available information. In fact, all of the variables considered were correlated with different descriptions with respect to the group of descriptors. Clinicians described the subjects in a less precise and less detailed way with a different perspective when compared to family members. Family members, instead, evaluated their aged relatives mainly on the basis of efficiency and abilities and gave less emphasis to verbal aspects of behaviour. Verbal ability, in fact, may be less important for this SES level where the elderly person is considered mainly on his ability to be autonomous, his ability to control his own behaviour, and his ability to have satisfactory interpersonal relationships. Clinicians, instead, are less sensitive to practical abilities relevant to everyday life and tend to emphasize the presence of macroscopic pathology of behaviour and the higher functions of the CNS. Hence, when gross pathology is absent (as in the case of our normal subjects) clinicians evidence less precise and less focused descriptions and evaluations of the aged.

EEG Analysis

Eight of our subjects agreed to undergo an EEG examination. All of them presented normal histograms of frequencies for their biological age. The fundamental activity of their alpha rhythm had a mean of 8.6 c/s while their beta rhythm's mean was 14.5c/s and their theta and delta rhythms had respective means of 5.7c/s and 2.2c/s. (Tables 11, 12, 13 and 14)

Significant correlations were found between delta and theta rhythms and verbal I.Q. with an increment of the activity being associated with a lower verbal ability.

Table 8. Family Members' and Clinicians' ACL Descriptions
          Significantly (P < 0.05) Correlated with Age, I.Q.
          Measures and Bender-Gestalt Test Scores

A G E		V E R B A L I.Q.	
CLINICIANS	FAMILY MEMBERS	CLINICIANS	FAMILY MEMBERS
CHARACTERISTIC ADJECTIVES		CHARACTERISTIC ADJECTIVES	
SINCERE	ACTIVE		
MODEST	MODERATE		
DIGNIFIED	WARM		
UNSELFISH	CHEERFUL		
APPRECIATIVE	UNSELFISH		
ACTIVE	CIVILIZED		
UNCHARACTERISTIC ADJECTIVES		UNCHARACTERISTIC ADJECTIVES	
UNDERSTANDING	PESSIMISTIC	UNSELFISH	CHEERFUL
FRANK	MODEST	APPRECIATIVE	UNSELFISH
	IMPULSIVE	MODEST	MODERATE
	SOCIABLE	CALM	SIMPLE
	SUGGESTIBLE		CAPABLE
			WARM

CONCLUSIONS

The traditional approach to studying the effects of neuro-
pharmacological substances reputed to alter the higher functions
of the CNS (such as memory, attention span, logical thinking, etc.)
has always encountered problems with the objectivity, both
quantitative and qualitative, of therapeutic results. Also, as
clinicians, we have frequently noted that our clinical judgements
concerning the treatment outcome do not coincide with either the
patient's, family's or nurses' judgements. Furthermore, the
same methods of evaluation and same therapies often obtain diverse
results when applied to inpatients and outpatients. These
discrepancies are also reported in the literature. We are,
therefore, confronted with the problem of identifying a valid
methodology which permits a more precise assessment of the stages
of aging. The methodology employed in this research not only
uses the more traditional psychological and psychometric parameters

Table 9.   Family Members' and Clinicians' ACL Descriptions
           Significantly (P < 0.05) Correlated with Age, I.Q.
           Measures and Bender-Gestalt Test Scores

PERFORMANCE I.Q.		FULL SCALE I.Q.	
CLINICIANS	FAMILY MEMBERS	CLINICIANS	FAMILY MEMBERS
CHARACTERISTIC ADJECTIVES		CHARACTERISTIC ADJECTIVES	
SYMPATHETIC	INTELLIGENT		INTELLIGENT
LOYAL	FRANK		FRANK
	PRECISE		
	SPONTANEOUS		
	SELF-CONTROLLED		
	JOLLY		
	TOLERANT		
	OPTIMISTIC		
UNCHARACTERISTIC ADJECTIVES		UNCHARACTERISTIC ADJECTIVES	
SPONTANEOUS		SPONTANEOUS	CHEERFUL
		UNSELFISH	UNSELFISH
		APPRECIATIVE	
		SELF-CONTROLLED	
		FRANK	

(interview, questionnaire, I.Q. measures, and performance measures)
together with with the biological parameters (EEG frequency
analysis) but also includes measures related to the social age
which is mainly determined by social and environmental influences.

Our study has been limited to normal aged male subjects with
the intention of identifying aspects associated with the normal
process of aging.   However, a natural extension of this research
would be the application of the same methodology to a group with
pathological ageing processes.

Despite the limited size of our sample, the subjects were
homogeneous with respect to sex, age, SES, and absence of pathology.
Our results clearly demonstrate the differences in descriptions
between clinicians and family members when related to the same aged
subjects.   These differences are related to the different

Table 10. Family Members' and Clinicians' ACL Descriptions
          Significantly (P < 0.05) Correlated with Age, I.Q.
          Measures and Bender-Gestalt Test Scores

BENDER-GESTALT PASCAL-SUTTELL SCALE		BENDER-GESTALT HUTT-BRISKIN SCALE	
CLINICIANS	FAMILY MEMBERS	CLINICIANS	FAMILY MEMBERS
CHARACTERISTIC ADJECTIVES		CHARACTERISTIC ADJECTIVES	
SPONTANEOUS	SIMPLE	INTERESTS NARROW	SIMPLE
SELF-CONTROL	MODERATE	SPONTANEOUS	
UNEXCITABLE	CHEERFUL	WHOLESOME	
UNSELFISH	UNSELFISH	UNDERSTANDING	
APPRECIATIVE	SERIOUS		
INTERESTS NARROW	HUMOROUS		
FRANK			
UNCHARACTERISTIC ADJECTIVES		UNCHARACTERISTIC ADJECTIVES	
	LOYAL	HEALTHY	RESERVED
	ASSERTIVE	RIGID	CONVENTIONAL
	PRECISE	RELIABLE	APPRECIATIVE
	INTELLIGENT	DETERMINED	SUPERSTITIOUS

perspectives assumed by the clinicians and the relatives with
respect to the aging process. The clinician's views are
focused on signs of pathology while the family members consider the
aged subject on the basis of usefulness within the nuclear family.

The comprehensive approach we have employed here permits one
to obtain a more global picture of the aging process by including
methods to assess biological, chronological, psychological, as well
as social age.

This complex and polycentric methodology when validated also
with pathological subjects can permit a useful evaluation of the
interaction between pharmacological therapies and the behaviour as
well as the higher functions of the CNS in the aging patient.

Table 11. E.E.G. Analysis in the Aging Person (N = 8)

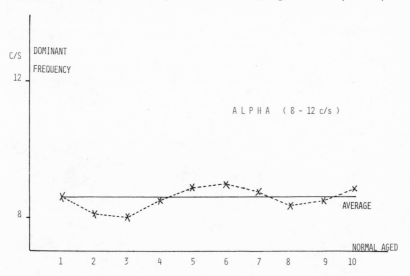

Table 12. E.E.G. Analysis in the Aging Person (N = 8)

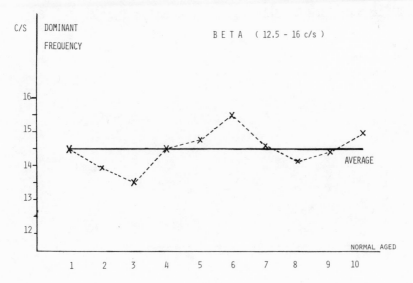

Table 13.   E.E.G. Analysis in the Aging Person (N = 8)

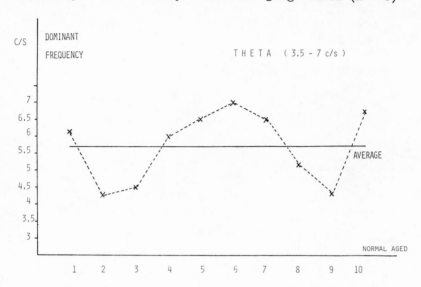

Table 14.   E.E.G. Analysis in the Aging Person (N = 8)

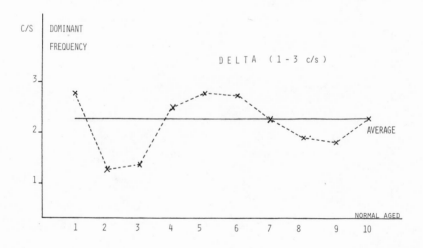

REFERENCES

1. J. E. Birren and V. J. Renner, Research on the psychology of
    aging: research and experimentation, in: "Handbook of the
    Psychology of Aging", J. E. Birren and K. W. Schaie, eds.,
    Van Nostrand Reinhold Co., New York (1977).
2. J. D. Matarazzo, "Wechsler's Measurement and Appraisal of Adult
    Intelligence", The Williams & Wilkins Co., Baltimore (1979).
3. J. S. Wiggins, "Personality and prediction", Addison-Wesley
    Pub. Co. Mass (1973).
4. H. G. Gough, Un'edizione italiana del The Adjective Check List,
    Bollettino di Psicologia Applicata, 79/82:1 (1967).
5. H. G. Gough, Lo sviluppo di una taratura italiana per
    l'Adjective Check List, Bollettino di Psicologia Applicata,
    88/90:23 (1968).
6. H. G. Gough and A. B. Heilbrun, The Adjective Check List Manual,
    Consulting Psychologists Press, Palo Alto Ca.(1965).
7. D. Wechsler, 'W-B I; Manuale", Organizzazioni Speciali,
    Firenze (1955).
8. G. R. Pascal and B. J. Suttell, "The Bender-Gestalt Test"
    Grune & Stratton, New York (1951).
9. M. L. Hutt, "The Hutt Adaptation of the Bender Gestalt Test",
    Grune & Stratton, New York (1977).

PAIN AND THE AGING PATIENT: A CRITICAL APPROACH

A. Moricca* and E. Arcuri[+]

* Head of Dept. of Anesthesiology, Resuscitation and
  Pain Therapy
  Regina Elena Cancer Institute, Rome

[+] Anesthesiologist, Dept. of Anesthesiology
  San Giuseppe Provincial General Hospital
  Marino-Ciampino (Rome)

To the researcher, aging and pain have several aspects in common. These aspects do not concern so much the complexity of the subject or the insufficiency of specific knowledge (the technical problems), but also and above all the method of approach and study (the ontologic problem).

The technical problems are linked to the high number of scientific disciplines that are involved in the study of the individual components of the global phenomena: pain and aging (e.g. macro- and micro-anatomy, biochemistry, biophysics, physiology, neuropathology, gerontology, etc.).

The ontologic problems arise from the need for the evaluation of single clinical facts not so much "per se" as in the context of biological and philosophical reality.

Thus, the most recent and modern trend for research on aging and pain is orientated in two main directions:

a) the neurophysiopathologic approach;
b) the behaviouristic approach.

Of these two types of approach, the second is the indispensable key for the understanding of the facts of the first, and for attributing significance to both. In other words, a biochemical or neurophysiopathologic phenomenon cannot be given any absolute

value unless it is related to the actual psychological and socio-
biological context.   This is because the latter can, in fact,
modify it extensively.

This concept of "relativity" can be illustrated by means of
two examples having to do with pain threshold variability and the
very concept of age itself which we shall examine here.

PAIN SENSITIVITY IN OLD AGE

It is known that the pain threshold varies significantly and
sometimes surprisingly between different human beings and under
different conditions, and that - even more surprisingly - variatio
may occur in the same human being.

Even age is not so much a numerical fact as a biological
condition because we know just as well that extraordinary
differences can be found between chronological and biological age.
Both age and youth are, in fact, little more than opinions in the
conventional measurement: everyone here has certainly met "youthf
old men" and "old youths", and all of us, during the trials of li
have felt young or old within a period of a few days, even a few
hours.

We therefore feel that the relatively commonplace belief tha
the elderly possess a lower sensitivity to pain (that is, a highe
pain threshold) - although it is possible that this is "in itself
a simplistic concept because it is based on a quantitative
evaluation of two conditions (aging and pain) about which one of
the few indisputable facts refers to this extraordinary variabili
between individuals.

Moreover, there are a multitude of questions that undermine
the certainty of that belief, and the most obvious are the
following: what type of pain and what type of aging are we
actually referring to? psychic pain or physical pain? so-called
rheumatic pain which is chronic but moderate, and which the patie
carries with him for decades? or that equally chronic but severe
pain of advanced coxarthrosis? Acute pain of abdominal perforatio
or the unbearable, so-called intractable pain of lung cancer that
has invaded the brachial plexus?

Then we must ask: who is the aged patient? Is he an old
hypocondriac with advanced arteriosclerosis or a lively, critical
seventy-year-old whose sharp eye can still analyse the world arou
him and his emotions? Is he the patriarch surrounded by the
affection of several generations of children and grandchildren or
that classical old man increasingly isolated by a society which i
becoming continually less human and which has no space for the

elderly?

It is clear that there cannot be only one response to this
virtually infinite series of possible combinations of clinical and
psycho-sociological aspects which frequently make up the everyday
reality of the physician.

However, the major difficulty in establishing whether the aged
patient is actually less sensitive to pain lies in the fact that
progress in the study of the neurophysiopathological and biochemical
aspects of pain have, in regard to ageing, provided only highly
fascinating but incomplete results.   We are referring to a whole
set of sometimes contradictory and still poorly defined information about the biological phenomena of aging[1].

In fact, some phenomena - quite independently of their role in
the process of aging - appear to confirm the presence of a minor
sensitivity to pain among the aged in general, whereas others tend
to prove the contrary.

INVESTIGATION OF PAIN SENSITIVITY

It must be pointed out that the greater part of the existing
literature supports the statement to the effect that sensitivity to
pain decreases with the advance of age.   The data that are provided
by way of proof can be grouped under two main headings as follows:

a) biohumoral and structural data;
b) experimental and clinical data.

The first category of data refer primarily to alterations in
the nerve ultrastructures, to biochemical alterations, and to links
between them and possible repercussions on pain perception.   In
this regard, the studies of the modifications of hypothalamic
factors which condition hypophyseal hypofunctioning are highly
important.   The most careful research has indeed been focussed
specifically on the study of the alterations of the dopaminergic
system, for it is believed that the latter may condition hypothalamic activity (directly) and hypophyseal activity (indirectly).

Significant variations have also been described regarding the
serotoninergic and gabaergic neurons; the resulting changes in the
synaptic links may be the true mediators of certain alterations in
the sensitive and motor responses among the aged.

On the other hand, far less is known about the modifications
that occur in the endorphin-enkaphalin system.   Since we are now
aware of the role that these neuropeptides play in the genesis of

pain and in pain relief[2], it would be interesting to find out how
the processes of their synthesis and inactivation may vary during
aging.   Apparent reductions in the synthesis and secretion of
endorphins and enkephalins that have been recently observed[3]
(although these data are still tentative) are in line with the
general knowledge that the biological processes of synthesis and
secretion are greatly reduced in the elderly.

Even before their discovery, however, many authors had
attempted to link this presumed minor sensitivity to pain of the
elderly to alterations in their responses to pain-relieving drugs.
Among the most interesting studies in this respect was that of
Farner and Verzar[4], who observed that the elderly generally showed
an exalted response to depressants of the central nervous system
(CNS) whereas the effects of stimulants appeared to be reduced.
This behavioural response to drugs was attributed (in a very
obvious but also rational manner) to the continuous and irreversible
loss of cells that occurs in the CNS with advancing age.   A
practical deduction in this respect was drawn in a subsequent study
by Belleville and his colleagues[5] who observed that the aged patient
shows higher response to analgesic drugs but, quite unexpectedly,
that this response is not associated with a higher incidence of
side effects.   As a result, although it may be valid practice to
establish drug administration on the basis of body weight, height
and surface for younger patients, these parameters lose a great
deal of their importance in the aged patient.

Another way to verify the degree of pain sensitivity in the
elderly is based on the extremely vast range of pain measuring
methods that exist today, all of which are more or less complicated
and artificial: thermal or pressure stimulations, provoked ischemia
etc.   They all have the same defect.   That is, besides being
fundamentally subjective, they all refer to experimental pain and
not to clinical pain which we all know is something entirely
different, especially when that pain is chronic.

It may be that the recording of cortical potentials evoked by
pain stimulation will, in the future, provide a more objective
method of measuring pain, but it will certainly not provide the
final word on the problem of clinical pain.   In this respect, it
is also significant that alterations in electroencephalographic
responses have not yet been clearly and completely related to the
encephalic alterations that are typical of aging.

Although isolated, it would appear that Pontoppidan's[6]
observations concerning the progressive and age-related loss of
protective reflexes in the airways may be of greater significance.

On the whole, however, we feel that all the above approaches
suffer from excessive analytical technicality and that they are

fragmentary: there is no true interest in synthesis. Their
conclusions tell us that pain is generally perceived to a minor
degree by the aged patient, but clinical practice, although less
well documented, points to a completely different conclusion.
In fact, pain that is perceived is not pain that is lived.

Many elderly patients actually appear to suffer more intensely
than younger ones from pain that might appear "less serious", and
it is in the aged patient that we observe most often the appearance
of "sine materia" or chronic pain. This lays a heavy burden on
both the patient and his family.

The loneliness and depression of the elderly are actually
primary causes of certain pain symptoms which the physician may
consider to be doubtful or even non-existant but which are nonetheless severe in the perception of the patient. To the latter, such
symptoms may be one of the few opportunities of obtaining sincere
human contact. It is a paradox that solicitous courtesy on the
part of his doctor or the attention and sympathy of family members
may provide the patient with an unconscious motivation for perpet-
uating his pain symptomatology. He in fact refuses to be denied
a potentially gratifying relationship.

THE PROCESSES OF AGING AND PAIN

Apart from such considerations of psychological order, it is
necessary to remember that a whole series of inevitable deterior-
ations of body tissues (especially cardiovascular, osteoarticular
and nerve tissues) end up by provoking more-or-less intense and
continuous pain in the aged. The main features of this deterioration are:

a) demyelinization of nerve fibres as a consequence of "age";
b) modifications in the ionic environment (Ca^{++}, Na^+, K^+)
   leading to differences among neurophysiologic responses
   of the mechanoreceptors;
c) increased exposure, with the advance of age, of the
   so-called "pain spots" (free nerve endings and inter-
   alia pacinocorpuscles) which are particularly relevant
   along the muscle structures and vertebral and limb joints.

These features indicate that the elderly, independently of
their degree of sensitivity to pain, are "more exposed" to pain.
Therefore, if it is possible (and it is) that a "mistake" in the
perception process can raise the threshold for one or more pain
stimulations, it is also possible that a massive and continuous
input of pain stimulations can amplify normal integration and
reaction. This is what we have observed in certain types of
chronic pain, especially in the elderly.

The many contradictions between experimental data and clinical practice lead us to conclude that the problem must be examined in a different light.   The key for this somewhat new approach might, in our opinion, be found in the more organic theories concerning the process of aging.   By way of example, there are the theories of Still and Vernadakis[7,8], according to which the CNS and its modifications over time might be the "pace-maker" of the process of aging at each stage of the biological organisation.   It is clear that the painful inputs <u>cannot fail</u> to vary in parallel with the variations of such a system which is characterised by a cybernetic structure.   However, it must be just as clear that:

a) pain is a complex phenomenon that is generated by an indivisible set of perceptions (input) and reactions (output); i.e.that is regulated by a feedback mechanism;

b) input and output may vary in parallel but not necessarily in proportion, making it always difficult to foresee the feedback responses;

c) such responses, however, in the aged patient are always the expression of a whole series of pre-existing factors that are not part of the process of aging but which can be modified by it (ethnic, cultural, religious, sociological and biological factors, previous pain experience,history of drug use, etc.).

The approach briefly outlined above is, to our mind, a promising one, and it is in line with the most recent discoveries regarding the mechanisms of pain.   Furthermore, it has important practical consequences in terms of analgesic therapy.

The geriatric condition is, in fact, characterised by progressive disturbances of consciousness and responsiveness.   The therapy of geriatric pain should therefore follow this <u>conditio sine qua non</u>: <u>avoid interfering negatively with the mental functions, be they normal or naturally (or pathologically) hypofunctional</u>.

This principle is supported above all by the results of over thirty years of experience in the relief of pain through nerve blocks.   In aged patients, these blocks have been performed to relieve pain (whether linked or not with ageing) of all types and intensities, generally with satisfactory to very good results. It suffices to mention sympathetic block which may resolve, in some cases with surprising rapidity, both the symptoms (pain-claudication) and the evolution (trophic disturbances - gangrene) of peripheral vascular diseases; Gasserian ganglion block which may also resolve selectively, using Ecker's technique*, those trigeminal neuralgias which are proper to advanced age; chemical rhyzotomies and/or sympathetic blocks for the relief of unbearable post-herpetic neuralgias; subdural, epidural and trigger-point injections for the relief of myofascial and low-back pain.

Finally, antalgic blocks for the treatment of cancer pain have allowed us to give hope and confidence to patients who had lost all will to live due to both age and pain.

Where cancer pain is diffuse and cannot be tracked down individually, an improvement over nerve blocks is represented by pituitary neuroadenolysis (NAL), a technique which is now widely practised and which offers many advantages in respect of neuro-surgical pain-relieving methods. The advantages that are of decisive importance in the treatment of the aged patient include minor risk (at lower cost).

We repeat that our experience has consistently shown that the disappearance - even partial - of pain and the abandonment of "heavy" drugs have a decisive role in improving the quality of life of the aged patient, even those whose state of health is seriously deteriorated.

In conclusion, the approach described above is not intended to present an established truth regarding the problem of pain in the aged patient. Nor is it our aim to make any overall synthesis because we feel that this is improbable. The prevailing concept according to which sensitivity to pain diminishes with the advance of age may actually hide our perhaps unconscious lack of sensitivity, our uncomfortable sense of responsibility. Most certainly, it hides our degree of ignorance regarding a category of the population and a type of pain that have wrongly been considered "second class".

----

* In line with the recent trend towards abandoning ethanol injection (except perhaps for the Ecker technique) in favour of other methods (percutaneous thermorhyzotomy, Giannetta-Loeser neurosurgical operation, etc.), we have been using cryoprobes since 1978. The results are promising but the relatively short period involved does not as yet allow us to draw valid conclusions.

REFERENCES

1. P. F. Spano and M. Trabucchi, "Aspetti biologici e farmacotera-
    peutici dell'invecchiamento cerebrale", ed., ESAM, Rome
    (1978).
2. E. Costa and M. Trabucchi, "Endorphines", Raven Press, New York
    (1978).
3. G. Quintarelli, Ageing and Pain, Paper presented at the
    International Symposium on Pain, Sorrento (June 1979)
    (unpublished).
4. D. Farner and F. Verzar, The age parameters of pharmacological
    activity, Experientia 17:421 (1961).
5. J. W. Belleville, W. H. Forrest et al, Influence of age on pain
    relief from analgesics - A study of postoperative patients,
    J.Amer.Med.Assoc. 217:1835 (1971).
6. H. Pontoppidan, H. K. Beecher, Progressive loss of protective
    reflexes in the airway with the advance of age, J.Amer.Med.
    Assoc. 174:2209 (1960).
7. J. W. Still (Quoted by Spano) The cybernetic theory of aging,
    J.Am.Geriat.Soc. 17:637 (1969).
8. A. Vernadakis, E. A. Petropoulus, C. V. Clark, (Quoted by Spano)
    Progr.Brain Res. 40:231 (1973).

AGING AND EXTRAPYRAMIDAL SYNDROMES (PARKINSONISM)

A. Agnoli, M. Baldassarre, E. Ceci, S. Del Roscio
and A. Cerasoli

Dept. of Neurology
School of Medicine
University of L'Aquila, Italy

INTRODUCTION

The cerebral catecholaminergic (CA) system, and in particular the dopaminergic (DA) system, is involved in the process of aging. The activity of tyrosine-hydroxylase, key-enzyme of the CA metabolism, is diminished in the caudate nucleus and putamen, especially in relation to a numerical decrease of the cells of the substantia nigra[1]. Dopa-decarboxylase, an enzyme limiting the metabolism of dopamine (DA), has an analogous behaviour[2].

The monoamineoxidases show, on the other hand, an increase in activity and, consequently, can be responsible for lower DA levels[3]. Furthermore it has been recently demonstrated in various DA areas of the SNC the presence of a DA-dependent adenyl-cyclase.

The activity of this enzyme was measured both in young and old rats: the response to the DA stimulus at the level of the striatum, nucleus accumbens and substantia nigra was lower (about 50%) in the older rats while in the retinal DA pathway the opposite reaction occurred. This indicates that a reduction of adenyl-cyclase in the brain is not a generalized phenomenon.

Nevertheless, it can still be hypothesized that the increase of the response of the adenyl-cyclase to the DA stimulus is a consequence of a reduced functioning of the DA neurons, and, therefore, of a receptor hypersensitivity due to a functional denervation.

Recently specific methods of binding have been identified; they are able to measure the capacity of the neurotransmitters to tie themselves to their own receptors.   This method was also used for dopamine and haloperidol (DA antagonist).   It has been demonstrated that the binding activity for these two substances is notably diminished in the striatum of older rat, in comparison to the normal one.

It has also been observed, in these areas of the brain, that a reduction in the turnover of dopamine occurred, which demonstrates a reduced conversion of tyrosine into dopamine.   An analogous behaviour was noted in the hypothalamus for the synthesis of noradrenaline.

When the pathology of aging is studied analogous data to that of physiological aging can be observed.   The disappearance of noradrenergic fibre in the cortex of Alzheimer type dementia has been demonstrated, while the DA axons were strongly reduced[4].

The main metabolite of DA, homovanillic acid (HVA), is notably reduced in striatum of the patients with Alzheimer type dementia in comparison to the older subjects in whom it is already normally reduced[5].   DA is reduced in the thalamus and pons as well as in the frontal cortex and hippocampus, while noradrenaline is reduced only in the putamen and frontal cortex in strict proportion to the worsening of the dementia symptomatology[6].   In vivo studies indicate that the acid metabolites HVA for dopamine and 5HIAA for serotonine are greatly reduced in the CSF; such reduction can be observed both under basal conditions or after accumultation with the Probenecid test.   Moreover, clinical and pharmacological data exist which demonstrate the involvement of the extrapyramidal system in dementia and this is confirmed by the positive result obtained with pharmacological stimulation of the DA system[7].

In fact, Alzheimer type dementia can present some extrapyramidal symptoms and dementia symptoms can be found in Parkinson's disease; an improvement of the dementia picture was noted after L-dopa therapy or DA agonists[8].

It is well known that DA is involved as a neurotransmitter in the control mechanisms of the selective attention and short-term memory; it follows that DA modulates some superior cortical functions[9].

It has been demonstrated, furthermore, that there are DA pathways which originate in the ventral tegment of mesencephalon (A 10) and project to neocortical areas (frontal cortex, anterior cingulum and enthorinal cortex of the rat)[4].   In the monkey these projections are typically prefrontal, also in man they project at prefrontal levels[10].

EXTRAPYRAMIDAL DISTURBANCES AND AGING

In accordance with Critchley[11] it can be affirmed that among the commonest neurological manifestations of old age are those usually labelled as extrapyramidal in type which, when occurring in complete form, constitute a variety of parkinsonism without tremor. So common indeed are these minor extrapyramidal signs in the aged that they are considered to be almost a normal feature of aging.

A more detailed analysis permits the identification of diverse aspects:

1. an attitude of general flexion and in marked cases the presence of the so-called "psychical pillow",
2. a rigidity which in typical cases assumes the aspects of the Foerster's arteriosclerotic rigidity[12] and consists of a uniform resistance to passive manipulation, more obvious in the extremities than in the trunk, and in the proximal rather than the distal segments of the limbs.
3. a general reduction of movement, specially of so-called "associated" movements as the secondary movements, synergias and cooperation movements
4. bradykinesis or slowness of movements so that both voluntary and automatic acts are reduced.

A modification of this picture, which when present in minor degree is a feature of normal old age, can be found in:-

a) arteriosclerotic parkinsonism when phenomena of vascular disorders determine the pathological variation of the clinical picture,
b) the senile rigidity of Jakob[13] with aspects of complete parkinsonism associated with symptoms of "senile dementia" type.

The occurrence of these physiological and physiopathological modifications, typical of old age and expressing a diffuse involvement of DA systems, does not actually surprise us since the pathogenesis can be rationally explained.   In fact, the presence of extrapyramidal symptoms of the parkinsonian type is explained by disorders of the nigrostriatal system; symptoms of dementia type by disorders of the mesocortical system;   psychotic symptoms by disorders of the mesolimbic system and neuroendocrine disturbances by involvement of the tuberoinfundibular system[14].

ANATOMOPATHOLOGICAL CORRELATIONS BETWEEN NORMAL AGING AND PARKINSONISM

One of the principal problems is to determine if particular anatomo-pathological alterations exist in the aged which explain the diverse extrapyramidal symptoms and if in the diverse types of parkinsonism, with or without dementia, particular anatomo-pathological patterns exist.

A) Observations on the Presence of Specific Changes of Parkinsonism at the Level of the Substantia Nigra and Locus Coeruleus

It has been observed that in normal older people there may be present specific changes, such as the Lewy bodies, Alzheimer's neurofibrillary degeneration and neuron loss at the level of substantia nigra and locus coeruleus.   In general, such phenomena increase with advancing age, in particular the Alzheimer's neurofibrillary degeneration is present already at age 30 and increases markedly with age.

Comparing these data with those obtained with parkinsonian patients it is found that:

1. 56% of patients with Parkinson's disease present a serious loss of cells in the substantia nigra, while such a phenomenon is present only in 12% of normal elderly subjects.
2. Many parkinsonian patients present numerous Lewy bodies at level of the substantia nigra and Alzheimer's neurofibrillary degeneration while these occur in very few normal elderly subjects[15].

In a recent study[16] it was found that in parkinsonian patients, again at level of substantia nigra and locus coeruleus, there are increased senile plaques, Alzheimer's neurofibrillary degeneration, granulovacuolar degeneration and neuron loss greater than that observed in normal controls.

B) Observations on the Modifications of the Striatum in Older Subjects and Parkinsonian Patients

The necessity for a study of the modifications of anatomo-pathological type of the striatum is justified by the fact that the neurons of the nigrostriatal DA pathways have a synapse at the striatal level.

Two cellular populations exist at this level: small cells, presumably GABAergic with an output toward the pallidum and substantia nigra, and large cells, acetylcholinergic in type on which the DA nigrostriatal pathways synapse.   A recent study[17] has shown that:

a) in the aged, compared to normal controls, a notable reduction of the two cellular components exist and in particular in the small cells,

b) in parkinsonian patients there is a reduction of 35% of the small
cells and of 50% of the large cells with a variation (69%) of
the ratio small/large cells, compared to the normal.

These data not only demonstrate a determination in the function of
the striatum in senility and Parkinson's disease, but can also
explain particular modifications of the therapeutic response in the
course of the evolution of the disease and even in the more aged
patients.

THE PROBLEM OF ARTERIOSCLEROTIC PARKINSONISM

In the realm of so-called cerebral arteriosclerosis some forms
are considered which we feel useful to describe, before we consider
the problem, much debated, of an aetiological connection between
arteriosclerosis and Parkinson's disease.

In literature these forms are described:

1. Paraplegia of the old of cerebral origin[18] with a progressive
   spastic paralysis and dysbasia. This form is different from
   syndrome of Yakovlev or pyramido-striatal syndrome of the aged
   with flexor contractures.
2. Malaise's arteriosclerotic dysbasia with worsening of the
   characteristic aspects of senility associated with signs of a
   slowing of gait; pyramidal, extrapyramidal and cerebral signs
   can also be present.
3. Pierre-Marie's lacunar disintegration, with pseudobulbar palsy,
   striatal and pallidal syndrome (Wilson type)[19].
4. Arteriosclerotic rigidity syndrome of Foerster[12], Parkinson's
   disease without tremor predominantly hypertonic with atypical
   symptomatology of Parkinsonism, such as catalepsy.

Other than these forms of historical interest, more codified
descriptions exist of arteriosclerotic Parkinsonism that are
principally inspired by the classic work of MacDonald Critchley of
1929. In the group of arteriosclerotic Parkinsonism 5 types are
distinguished:

1. the first type with rigidity, amimia and short-stepping gait

2. a second type which includes the signs described for the first
   type plus pseudo-bulbar signs

3. a third with rigidity, amimia, short-stepping gait and dementia
   with incontinence of urine and faeces

4. a fourth with the signs of the first but also with pyramidal
   signs leading to a mixed pyramido-pallidal syndrome

5. a fifth type which consists of the signs of the first and
   cerebellar symptoms causing a mixed pallido-cerebellar syndrome.

These forms can have in general a gradual onset, or, occasionally, an acute onset as a result of a little stroke or following a stroke and therefore manifesting a unilateral Parkinsonism.

Schneider and his colleagues[20] tried to correlate arteriosclerosis cerebral atrophy and Parkinson's disease, observing that the physical signs indicative of arteriosclerosis (risk factors, calcification of the internal carotid artery and aorta) increase with advancing of age, but they are not related to either Parkinsonism or to cerebral atrophy. It was concluded that atrophy and parkinsonism are not correlated with arteriosclerosis and that both cerebral atrophy and calcification in the carotid siphon are considered as a symptom of old age. Eadie and Sutherland[21] discussing arteriosclerosis and parkinsonism evaluated the signs of arteriosclerosis by palpation of the carotid and brachial arteries, the measurement of arterial blood pressure, the appearance of the fundus oculi and the tolerance to physical exercise. It was observed that there was no difference between Parkinsonian patients and normal controls and they concluded that there is no association between Parkinsonism and arteriosclerosis.

These data are open to criticism since cerebral arteriosclerosis can be present without peripheral symptomatology and therefore an evaluation using only physical signs is at risk of underestimating the problem.

Pollock and Hornabrook[22] furnished criteria for the definition of Parkinsonism with associated ateriosclerosis:

1. Parkinsonism with acute onset or progressive with successive episodes
2. Presence of slowly evolving dementia,
3. Presence of pseudobulbar or pyramidal signs,
4. Marked degree of rigidity predominating in the flexor muscles of the arms and extensor muscles of the legs, and
5. Short-stepping gait.

It was observed that out of 131 patients, 83 (63.3%) presented an idiopathic form, 37 (28.3%) an arteriosclerotic and 11 (8.39%) a postencephalitic form. Of the arteriosclerotic group, 16 (43.2%) had severe hypertension and, of these, 11 had signs of arteriosclerosis.

VARIATIONS OF THE RESPONSE TO THERAPY ACCORDING TO THE AGE OF ONSET AND DURATION OF PARKINSONISM

There are data in the literature[23] which tend to demonstrate that while youthful Parkinsonism can be considered a mononeuronal form principally involving the DA system, in the elderly one finds rather a "multineuronal" form, involving several neurotransmitter systems.

The loss of cells and the diminution of the neurotransmitters with ageing[24] tend to suggest that in addition to normal aging which causes the already described modest pyramidal-extrapyramidal anomalies, there exists an accelerated neuronal aging which is manifested by signs of dementia.

According to Granerus et al[25] there are two types of Parkinson's disease accounting for differences in therapeutic response:-

1. a pure form with onset at an early age and more rapid development presenting greater risks of a syndrome of decreasing therapeutic efficacy (On-Off)
2. a second form with onset at a later age with a weaker response to L-dopa but with minimum risk of on-off syndrome and dyskinesia, but with a tendency to dementia.

The response to L-dopa therapy and the possibility of side-effects in relationship to the age at which Parkinson's disease appears can be summarised:-

1. the therapeutic response is modified by age,
2. the therapeutic effect is better in young patients rather than older, in whom a positive therapeutic response is only obtained in predominantly hyperkinetic syndromes
3. the on-off syndrome is decidedly more frequent and more probable the younger the patient or the younger the patient was at the onset of Parkinson's syndrome and the onset of L-dopa therapy.
4. lastly, patients with the on-off syndrome require elevated dosages of L-dopa and have a greater duration of therapy than any other group.

ANALYSIS OF DIFFERENT CASE-MATERIAL IN RELATION TO THE AGE OF ONSET, PATHOGENESIS AND POSSIBILITY OF DEVELOPMENT OF DEMENTIA

Relationship between Age and Parkinsonism

From the analysis of results obtained from different surveys regarding the period 1955 - 1977 and in different geographic areas (Table 1), it has been shown that the incidence of parkinsonism in relationship to the total population is maximum at ages between 50 and 70 years, when the rates are between 0.19 and 1.91% (mean 0.8).

Furthermore, the examination of four different surveys (Table 2) shows a substantial correspondence between the age of onset and the

Table 1. Prevalence and Incidence Rates for Parkinson's Disease
          According to Various Studies (from Marttila & Rinne '76
          mod.)

COUNTRY	YEAR	PREVALENCE PER 100 POPULATION AGE GROUPS				TOTAL
		49	50-59	60-69	70	
ROCHESTER USA	1955	0.005	0.24	0.76	1.53	0.19
GOTHENBURG SWEDEN	1960	0.02	0.09	0.19	0.50	0.07
CARLISLE ENGLAND	1961	0.02	0.16	0.31	0.52	0.11
WELLINGTON NEW ZELAND	1962	___	___	___	___	0.10
ICELAND	1963	0.009	0.16	0.93	1.51	0.17
VICTORIA AUSTRALIA	1965	0.006	0.16	0.29	0.92	0.08
ROCHESTER USA	1965	0.008	0.02	0.66	1.91	0.15
TURKU FINLAND	1971	0.004	0.11	0.50	0.75	0.12

clinical form.   In fact, against an early onset (from 28 to 36.2
years) of the postencephalitic form, the idiopathic form shows a
later onset (between 55.3 and 61.6 years) and much later is the
onset of arteriosclerotic parkinsonism.   According to Marttila-
Rinne[26]post-encephalitic parkinsonism has a maximum incidence (64%)
under 39 years of age and idiopathic parkinsonism shows an age-
related progressive increase with a maximum (36.6%) between 60 and
69 years.

Pathogenesis

     At present the factors determining the onset of Parkinson's
disease are unknown.   Borromei and Parmeggiani[27] (Fig.1), have
drawn attention to the importance of encephalitis lethargica among
the infective pathologies. They have demonstrated (Table 3) a
greater global incidence of the complete type, and a relative
scarcity of the hyperkinetic pictures in the post-encephalitic form.
By contrast, there is a lower incidence of akinetic-hypertonic
pictures in the idiopathic form.   Pollock and Hornabrook[23]
utilizing a semiquantitative scale, have shown a greater gravity in
the clinical picture of the arteriosclerotic form compared to the
idiopathic type (Table 4).

Table 2.   A Survey on Different Case-Material in Relation with Age
on Onset and Clinical Type

CLINICAL TYPE	HOEHN - YAHR '67		POLLOCK & HORNABROOK '62		MARTTILA RINNE '76		KURLAND '55		TOTAL	
	N° CASES	AGE AT ONSET (MEAN)	N° CASES	AGE AT ONSET (MEAN)	N° CASES	AGE AT ONSET (MEAN)	N° CASES	AGE AT ONSET (MEAN)	N° CASES	AGE AT ONSET (MEAN)
IDIOPATHIC	672	55.3	80	59	459	61.6	39	59	1250	58.7
POSTENCEPHAL.	96	28.2	11	28	25	36.2	7	30	136	30.6
ARTERIOSCLER.	—	—	37	70	—	—	12	70	49	70

Parkinsonism and Dementia

Another aspect, and certainly one of the most important, is the
relationship existing between parkinsonism and dementia.

Using the Wechsler-Bellevue test it has been shown (Table 5)
that the more severe cases of Parkinson's disease (stages III and

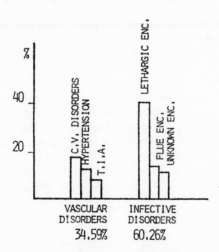

Fig. 1.   Diseases responsible for the onset of parkinsonism.
(n=total patients 402) (from Borromei & Parmeggiani
'76 mod.)

Table 3.   Clinical Type and Pathogenesis of Parkinsonian in
           Percent. (n=number of patients). From Borromei and
           Parmeggiani '76 mod.)

CLINICAL TYPE	POSTENCEPHALITIC	IDIOPATIC	ARTERIOSCLEROTIC
HYPERKINETIC	13.1	38.3	32.0
	(12)	(72)	(39)
AKINETIC - HYPERTONIC	21.7	13.8	22.9
	(20)	(26)	(28)
COMPLETE	65.2	47.9	45.1
	(60)	(90)	(55)

IV in the Hoen-Yahr scale[28]) have a significant reduction in
performance in the sub-tests "digit-span", "recording" and "puzzle".
The sub-test "reordering" shows a corresponding deterioration for
those over 60 with more than 4 years duration of the illness and
with the arteriosclerotic form.   Moreover, impairment of performance
is noted in the oldest patients with the greatest duration and
severity of disease and particularly in the arteriosclerotic form.
Thus among the Wechsler-Bellevue sub-tests the most important seems
to be "reordering"; it can be efficiently used to differentiate the
aspects of parkinsonian syndromes.

    These data, even if they can be criticised on methodological
grounds (for example, lack of comparison with controls of the same
age group) confirm what has been known for a considerable time, that
in Parkinson's disease there is a deterioration in cognitive
functions in addition to impairment of motor functions.   Similar
findings have been reported by Martilla-Rinne[29] derived from a study
of a series of patients suffering from Parkinson's disease with or
without dementia (Table 6).

    A first observation concerning the frequency of dementia in the
total population of parkinsonian patients shows a proportional
increase to the age of incidence of dementia (Fig. 2).

    From the examination of the characteristics of the two groups,
confronted in the various classes of age, emerges a highly significant greater gravity in the demential pictures, especially for the
groups up till age 74 (Table 7).

    The correlation affected between cases of Parkinson's disease
with or without dementia, with respect to the duration of the
illness, age at moment of onset, (Fig. 3) does not evidence significant variations in the two groups.

Table 4.   Disability of Patients with Parkinson's Disease by
           Clinical Type.   (n=percent) (from Pollack and Hornabrook
           '62 mod.)

| | DISABILITY GRADE | | | | TOTAL | % |
	1	2	3	4		
IDIOPATIC	56	11	16	___	83	63.3
	(67.46)	(13.25)	(19.27)			
POSTENCEPHALITIC	6	1	3	1	11	8.4
	(54.54)	(9.09)	(27.27)	(9.09)		
ARTERIOSCLEROTIC	8	11	13	5	37	28.3
	(21.62)	(29.72)	(35.13)	(13.51)		

A further distinction within the group of those suffering from
dementia between the subjects with clinical signs of arteriosclerosis
and without, in relationship to the seriousness of Parkinson's
disease, evaluated by Hoehn-Yahr scale (Fig.4), shows a substantial
homogeneity among the two groups.   Analogous homogeneity is
evidenced (Fig. 5) in the confrontation of the clinical picture:
the symptoms tremor, rigidity and hypokinesis are substantially the

Table 5.   Assessment of intelligence (Wechsler-Bellevue R.S.) in
           parkinsonian patients. $p < 0.005 = {}^{\circ}$ $p < 0.01 = {}^{\circ\circ}$
           (from Roccatagliata et al '77 mod.)

| ITEMS | TOTAL SCORE | AGE | | DURATION | | SEVERITY HOEHN - YAHR S. | | CLINICAL TYPE | |
	(MEAN)	60 Y 11 CASES	60 Y 11 CASES	4 Y 11 CASES	4 Y 11 CASES	I - II	III - IV	IDIOPAT. 12 CASES	ARTERIOS. 10 CASES
VERBAL S.	91	91	91	92.60	90.30	94.80	88.20	91.5	90.5
PERFORMANCE	79	79.5	78.4	83.45	72.55	81.50	77.50	81	77
I.Q.	86	85.6	85.5	87.70	83.40	89.50	81.60	87.5	84
% DETER.	44%	43%	46%	41%	48.5%	40.	48%	43%	45%
DIGIT SPAN	3.45	3.45	3.63	3.80	2.90	5	1.9 °°	3.4	3.5
REORDERING	3.68	4.63	2.72 °	4.6	2.7 °	4.6	2.7 °	4.83	2.3 °°
PUZZLE	5.18	4.81	5.54	4.8	5.55	6.7	3.6 °	5.41	4.9

Table 6. Overall Occurrence of Dementia in Idiopathic (n=421) and
          Postencephalitic (n=23) Parkinsonian Patients
          (n=percentage) (from Marttila & Rinne '76 mod.)

	DEMENTED			TOTAL	NON DEMENTED
	STAGE				
	1	2	3		
TOTAL	67	37	24	12,8	316
	(15,1)	(8,3)	(5,4)	(28,8)	(71,2)
IDIOPHATIC	64	35	23	122	299
	(15,2)	(8,3)	(5,5)	(29)	(71)
POSTENCEPHALITIC	3	2	1	6	17
	(13,0)	(8,7)	(4,4)	(26,1)	(73,9)

same in the two groups, while there seems a greater incidence of
total gravity, even if not statistically significant, in the group
of parkinsonian patients without clinical signs of arteriosclerosis.

RESULTS OF OUR CASE-MATERIAL ON 154 PATIENTS WITH PARKINSON'S
DISEASE (L'AQUILA STUDY)

     A series of 154 cases of Parkinson's disease, (77 men and 77
women whose ages ranged from 33 to 83) was studied between 1975
and 1979.   Subdividing the patients in four age groups (under 50,
50-60, 60-70 and over 70 years) and relating them to the clinical
and diagnostic type it was found (Table 8) that there was a low
incidence of the arteriosclerotic form (8.5%) and in the age group
under 50 years a high incidence of post-encephalitic forms.   The
largest proportion of patients was in the age group 60 and over
years.

     The examination of risk factors (Table 9), divided into
vascular and metabolic, showed, apart from an increase of the
presence of each with advancing age, a relatively higher proportion
of vascular factors compared with metabolic.   A similar analysis,
based on neurological signs present in the Parkinson's group
(Table 10) does not show substantial modifications of the different
symptoms in the different age groups.

     In order to evaluate the extent of impairment of cognitive
functions the Hachinski Scale[30]was utilized.   This consists of three
sub-scales whose items permit the quantification of a deficit with
respect to intelligence, information-memory-concentration

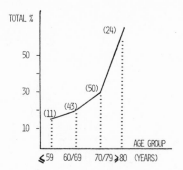

Fig. 2. Frequency of dementia in various age groups in
         parkinsonian patients (n=444)
         (n) = number of demented patients in the age group
         indicated. (from Marttila & Rinne '76 A, mod.)

(sub-scales 1 and 2) and, lastly, the incidence of vascular factors
(sub-scale 3).   Dementia is present when the scores exceed 50% in
sub-scales 1 and 2 and vascular damage when the score is equal to
or greater than 6 in sub-scale 3.

When the scores of the scale are compared in the various age
groups (Table 11) it is found that there is a uniform distribution
of vascular damage at different ages.   Vascular damage (score > 6)
was present in only 20.7% of cases (Table 12).   The occurrence of
vascular damage shows a high correlation with the clinical diagnosis

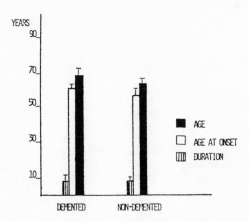

Fig. 3. Age and duration (mean ± S.E.) of the disease in demented
         (n=128) and non-demented (n=316) parkinsonian patients.
         (from Marttila & Rinne '76 A, mod.)

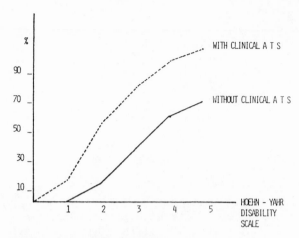

Fig. 4. Percentage of demented patients with or without
        clinical arteriosclerosis and stage of disability.
        (from Marttila & Rinne '76, mod.)

of arteriosclerotic parkinsonism (10 cases out of 14) and conversely
the absence of vascular damage correlates with the diagnosis of
Idiopathic Parkinsonism (108 cases of 130). Examination of the
clinical symptoms shows a distinct relation between parameters of
diffuse cerebral damage and high scores on the vascular scale.

CONCLUSIONS

    From the literature and the examination of our case material
it can be concluded:-

1. A high incidence of parkinsonism is observed between ages 60
    and 70 and, in particular, over 70; the arteriosclerotic form
    occurs most frequently betwen 60 and 70, the idiopathic between
    55 and 60 and post-encephalitic under the age of 50. Ageing
    clearly worsens the disability making the akinetic-hypertonic
    and hyperkinetic forms with respect to the complete form more
    frequent.

2. Signs of vascular damage (about 80%) and of damage to the
    superior nervous activity (about 85%) are present in Parkinsonian
    patients and they are equally distributed in the age groups.

3. Reduced performance in the Wechsler-Bellevue test is found
    particularly in patients aged over 60 years and is related to
    the duration of Parkinson's disease and to the seriousness of
    the clinical picture (according to Hoehn-Yahr Scale) especially
    when clinical signs of arteriosclerosis are present. The most
    serious forms of Parkinsonism associated with dementia are of

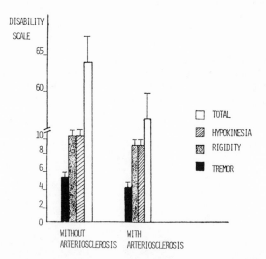

Fig. 5. Severity of disease in the demented patients with or
without clinical arteriosclerosis (mean $\pm$ S.E.)
(from Marttila & Rinne '76, mod.)

the akinetic-rigid type and are aged between 60 and 70 years.

4. Arteriosclerotic Parkinsonism accounts for 9 to 28% of cases
in the series reported in the literature.   The characteristic
clinical type consists in a greater prevalence of akinetic-
hypertonic forms and a lower frequency of complete forms;
the arteriosclerotic form is the one causing most disability.

5. The use of the Hachinski scale for assessing vascular risk-
factors is of value in the diagnosis of arteriosclerotic
Parkinsonism.   A high score ($\geqslant$ 6) is found in 32% of clinically
diagnosed arteriosclerotic parkinsonism compared with only
3% of non-parkinsonian patients.

6. When vascular damage is assessed on the Hachinski scale in
arteriosclerotic Parkinsonism there is found to be a high
frequency of pyramido-extrapyramidal signs and of the
primitive reflexes (palmo-mental and snout-reflex) and to a
lesser extent of supranuclear ophthalmoplegia.

Table 7. Disability of the Patients and Severity of the Diseases,
          by Age in Demented and Non-demented Patients.
          $p < 0.05 = {}^{\circ}$ $p < 0.01 = {}^{\circ\circ}$ $p < 0.001 = {}^{\circ\circ\circ}$
          (from Marttila & Rinne '76 A, mod.)

AGE	MEAN SCORES							
GROUPS	TREMOR		RIGIDITY		HYPOKINESIA		TOTAL SEVERITY	
	DEMEN.	NON DEMEN.	DEMEN.	NON DEMEN.	DEMEN.	NON DEMEN.	DEMEN.	NON DEMEN.
59 (N = 63)	4.1	3.1	8.6°	6.6	9.0°°	5.6	54.6°	38.0
60/64 (N = 91)	3.6	3.8	9.5°°	6.8	9.6°°°	6.2	58.6°°	42.1
65/69 (N = 105)	5.2	4.3	10.6°°°	7.2	10.3°°°	6.6	67.6°°°	45.2
70/74 (N = 96)	5.3°	4.1	9.9°°	7.8	9.1°	7.3	62.2°°	48.2
75/79 (N = 55)	5.0	5.1	8.7	7.5	9.2°	6.9	58.2	49.0
80 (N = 34)	4.4	3.8	9.1	7.5	9.2	7.0	59.8	45.5

Table 8. Correlation between Age, Clinical Type and Diagnosis
          in Parkinsonian Patients. L'Aquila Study.
          (n): Number of Patients.

AGE GROUPS	CLINICAL TYPE			DIAGNOSIS			TOTAL
	HYPERCINETIC	AKINETIC HYPERTONIC	COMPLETE	IDIOPATHIC	POSTENCEPHAL.	ARTERIOSCLER.	
50 (N=10)	-----	20 (2)	80 (8)	40 (4)	60 (6)	-----	6.6 (10)
50/60 (N=25)	8 (2)	20 (5)	72 (18)	84 (21)	12 (3)	4 (1)	16.2 (25)
60/70 (N=58)	10.4 (6)	12 (7)	77.6 (45)	86.2 (50)	3.4 (2)	10.4 (6)	37.6 (58)
70 (N=61)	14.7 (9)	9.8 (6)	75.5 (46)	90.1 (55)	-----	9.9 (6)	39.6 (61)
TOTAL (N=154)	11.0 (17)	13.0 (20)	76.0 (117)	84.4 (130)	7.1 (11)	8.5 (13)	100.0 (154)

Table 9. Correlation between Risk-factors and Age Groups in
          Parkinsonian Patients. L'Aquila Study.
          (n): Number of Patients

RISK - FACTORS

AGE GROUPS	VASCULAR DISORDERS				METABOLIC DISORDERS		
	CEREBROVASCULAR DISORDERS	HYPERTENSION	CORONARIC DISORDERS	VASCULAR FAMILIARITY	DIABETES	DYSLIPIDEMIA	HYPER-URICEMIA
50 (n=10)	—	10.0 (1)	10.0 (1)	50.0 (5)	—	10.0 (1)	—
50/60 (n=25)	—	16.0 (4)	12.0 (3)	56.0 (14)	—	24.0 (6)	—
60/70 (n=58)	3.4 (5)	31.0 (18)	22.4 (13)	44.8 (26)	15.5 (9)	29.3 (17)	12.1 (7)
70 (n=61)	4.9 (3)	36.0 (22)	32.8 (20)	32.8 (20)	16.4 (10)	9.8 (6)	6.6 (4)
TOTAL (n=154)	3.2 (5)	29.2 (45)	24.0 (37)	42.2 (65)	12.3 (19)	19.5 (30)	7.1 (11)

Table 10. Correlation between Clinical Symptoms and Age Groups in
           Parkinsonian Patients. L'Aquila Study.
           (n): Number of Patients

AGE GROUPS	FOCAL SYMPTOMS	EXTRAPIRAMIDAL SYMPTOMS			DIFFUSE SYMPTOMS										
		BLADDER DYSFUNCTION	CONIUGATE VERTICAL MOV. PALSY	PYRAMIDAL EXTRAPYRAM. GAIT	SNOUT REFLEX	PALMOMENITAL REFLEX	GRASP REFLEX	ROSSOLIMO'S REFLEX	PYRAMIDAL EXTRAPYRAM. SYMPTOMS	PSEUDOBUL. SYMPTOMS	DYSPHAGIA	DYSPHONIA	DYSARTHRIA	INVOLUNTARY LAUGHING AND CRYING	BILATERAL PYRAMIDAL SYMPTOMS
50 (n=10)	30.0 (3)	10.0 (1)	70.0 (7)	20.0 (2)	20.0 (2)	10.0 (1)	–	–	40.0 (4)	–	10.0 (1)	50.0 (5)	50.0 (5)	–	30.0 (3)
50/60 (n=25)	16.0 (4)	4.0 (1)	48.0 (12)	20.0 (5)	40.0 (10)	12.0 (3)	–	12.0 (3)	16.0 (4)	4.0 (1)	20.0 (5)	36.0 (9)	60.0 (15)	–	20.0 (5)
60/70 (n=58)	29.3 (17)	13.8 (8)	46.5 (27)	17.2 (10)	43.1 (25)	22.4 (13)	3.4 (2)	1.7 (1)	32.8 (19)	5.1 (3)	10.3 (6)	32.7 (19)	63.8 (37)	3.4 (2)	27.6 (16)
70 (n=61)	29.8 (17)	11.5 (7)	34.4 (21)	24.6 (15)	34.4 (21)	22.9 (14)	3.3 (2)	1.6 (1)	24.6 (15)	3.3 (2)	14.7 (9)	37.7 (23)	54.1 (33)	8.2 (5)	24.6 (15)
TOTAL (n=154)	26.6 (41)	11.0 (17)	73.4 (67)	20.8 (32)	37.7 (58)	20.1 (31)	2.6 (4)	3.2 (5)	27.3 (42)	3.9 (6)	13.6 (21)	36.4 (56)	58.4 (90)	4.5 (7)	25.3 (39)

Table 11. Correlation between Hachinski Scale and Age Groups in
          Parkinsonian Patients. L'Aquila Study.
          (n): Number of Patients

HACHINSKI SCALE

AGE GROUPS	DEMENTIA SCORE AND INFORMATION MEMORY - CONCENTRATION TEST		ISCHEMIC SCORE	
	< 50%	≥ 50%	< 6	≥ 6
50 (N=10)	80.0 (8)	20.0 (2)	80.0 (8)	20.0 (2)
50/60 (N=25)	100.0 (25)	——	88.0 (22)	12.0 (3)
60/70 (N=58)	82.8 (48)	17.2 (10)	69.0 (40)	31.0 (18)
70 (N=61)	86.9 (53)	13.1 (8)	82.0 (50)	18.0 (11)
TOTAL (N=154)	87.0 (134)	13.0 (20)	77.9 (120)	22.1 (34)

Table 12. Correlation between Diagnosis, Clinical Type and
          Ischemic Damage of Hachinski Scale (Ischemic Damage < 6
          Non-ischemic Damage ≥ 6)   L'Aquila Study.

	DIAGNOSIS			CLINICAL TYPE		
	IDIOPATHIC	ARTERIOSCLER.	POSTENCEPHAL.	COMPLETE	HYPERCINETIC	AKINETIC HYPERTON.
ISCHEMIC DAMAGE (N=32)	68.7 (22)	31.2 (10)	——	68.7 (22)	9.4 (3)	21.9 (7)
NO ISCHEMIC DAMAGE (N=122)	88.5 (108)	3.2 (4)	8.2 (10)	76.2 (93)	12.3 (15)	11.5 (14)

REFERENCES

1. E. G. McGeer, Aging and neurotransmitter metabolism in the
    human brain, in: "Aging" Vol.7, R. Katzman, R. D. Terry,
    K. L. Bick, eds., Raven Press, New York (1978).
2. E. G. McGeer and P. L. McGeer, Neurotransmitter metabolism in
    the aging brain, in: "Aging" Vol. 3, R. D. Terry, S. Gerhon,
    eds., Raven Press, New York (1976).
3. D. S. Robinson, Changes in monoamine oxidase and monoamines
    with human development and aging, Fed.Proced. 34:103 (1975).
4. E. Scarnati, C. Forchetti, S. Ruggieri and A. Agnoli,
    Dopamine and dementia. An animal model with destruction of
    the mesocortical dopaminergic pathway: a preliminary study,
    in:"Aging of the brain and dementia", Raven Press,
    New York, (1980) In Press.
5. C. G. Gottfries, I. Gottfries and B. E. Roos, Homovanillic
    acid and 5-hydroxyindolacetic acid in cerebrospinal fluid
    related to mental and motor impairment in senile and
    presenile dementia, Acta Psychiatr.Scand. 46:99 (1970).
6. R. Adolfsson, C. G. Gottfries, L. Oreland, B. E. Ross and
    B. Winblad, Reduced levels of catecholamines in the brain
    and increased activity of monoamine oxidase in platelets in
    Alzheimer's disease: therapeutic implications, in:"Aging",
    Vol.7, R. Katzman, R. D. Terry and K. L. Bick, eds.,
    Raven Press, New York, (1978).
7. J. Pearce, Letter: Mental changes in parkinsonism,Br.Med.J.,
    2:445 (1971).
8. A. Agnoli and A. Polleri, Prolactin circadian rhythms in
    elderly. Modifications of the pattern after treatment with
    dopamine agonists (in press).
9. A. Agnoli and G. Squitieri, A study upon short-term memory
    interaction with nooanaleptic and nootropic drugs, in:
    Proc. Symposium Nooanaleptic and Nootropic Drugs",
    A. Agnoli, ed., 3rd Congr.Intern.Coll.Psychosom.Med.,
    Rome (1975).
10. M. C. Leonard, The prefrontal cortex of the rat: I) Cortical
    projection of the mediodorsal nucleus. II) Efferent
    connection, Brain Res. 88:190 (1969).
11. M. Critchely, The neurology of old age, Lancet 1:1119 (1931).
12. O. Foerster, Die Arteriosklerotische Multestarre, Allg.Zeit.
    f.Psych. 66:902 (1909).
13. A. Jakob, "Die extrapyramidalen Erkrankungen", Springer,
    Berlin (1923).
14. A. Agnoli, N. Martucci, M. Baldassarre and A. Polleri,
    Correlation du spectre EEG et de la prolactine plasmatique
    chez le sujet agé, Rev.Franc.EEG et Neurophisiol., In press.
15. L. S. Forno and E. C. Alvord, Some new observations and
    correlation, in: "Recent advances in Parkinson's disease",
    F. H. McDowell and C. H. Markhan, Blackwell Scientific
    Publ., Oxford (1973).

16. A. M. Hakim and G. Mathieson, Basis of dementia in Parkinson's disease, _Lancet_, 2:729 (1978).

17. O. Bugiani, F. Perdelli, S. Salvarani, A. Leonardi and L. Mancardi, Loss of striatal neurons in Parkinson's disease. A cytometric study, _Europ.Neurol._, (1979) In press.

18. J. Lhermitte, Etude sur les paraplegies des vicillards, These de Paris (1907).

19. H. Claude and J. Cuel, Le meiopragie cerebrale par angioscler-ose precoce sans ischaemie en foyer (forme de demence presenile arteriosclereuse), _L'Encephale_ 22:151 (1927).

20. E. Schneider, P. A. Fischer, H. Becker, H. Hacker, A. Penc, and P. Jacobi, Relationship between arteriosclerosis and cerebral atrophy in Parkinson's disease, _J.Neurol._ 217:11 (1977).

21. M. S. Eadie and J. M. Sutherland, Arteriosclerosis in Parkinsonism, _J.Neurol.Neurosurg.Psychiat._ 27:237 (1964)

22. H. Pollock and R. W. Hornabrook, The prevalence, natural history and dementia of Parkinson's disease, _Brain_, 89:429 (1962).

23. A. K. Granerus, G. Steg and A. Svanborg, Clinical analysis of factors influencing L-dopa treatment of Parkinson's syndrome, _Acta Med.Scand._ 192:1 (1972).

24. A. Carlsson, The impact of catacholamine research on medical science and practice, presented at IV International Catecholamine Symposium, Asilomar,Cal., (1978). In press.

25. A. K. Granerus, A. Carlsson and A. Svanborg, The aging neuron. Influence on symptomatology and therapeutic response in Parkinson's syndrome, _in_: "Adv. in Neurology", Vol. 24, L. J. Poirier, T. L. Sourkes and P. J. Bedard, eds., Raven Press, New York (1979).

26. R. J. Marttila and U. K. Rinne, Epidemiology of Parkinson's disease in Finland, _Acta Neurol.Scandinav._ 53:81 (1976).

27. A. Borromei, A. Parmeggiani, Esperienze cliniche e rilievi statistici su un campione di 425 parkinsoniani, _in_: "Atti III Riunione Lega Ital. Morbo di Parkinson e Malattie Extrapiramidali", A. Agnoli, G. Bertolani, eds., Roma (1976).

28. M. M. Hoehn and M. D. Yahr, Parkinsonism: onset, progression and mortality, _Neurology_ 5:427 (1967).

29. R. J. Marttila and U. K. Rinne, Dementia in Parkinson's disease, _Acta Neurol.Scandinav._ 54:431 (1976 A).

30. V. C. Hachinski, L. D. Iliff, E. Zilkha, G. H. Du Boulay, V. L. McAllister, J. Marshall, R. W. Ross Russell and L. Symon, Cerebral blood flow in dementia, _Arch.Neurol._ 32:632 (1976).

MANAGEMENT OF PARKINSON'S DISEASE IN THE ELDERLY

M. Hildick-Smith

Consultant Geriatrician
Nunnery Fields Hospital
Canterbury

What is the chance that any one of us will develop Parkinson's
disease at some time in our lives?   This chance is now assessed
at 1 or 2 in a hundred - the prevalence of the disease for those
who are now over 60 years old being 1%.   I would maintain,
therefore, that management of Parkinsonism in the elderly forms an
important element of our medical practice.

This disease has been responsible for much long-continued
handicap in the past.   In Britain in the early 1970s there were
22,000 patients handicapped by the disease living at home[1] and
15,000 in institutions[2] such as hospitals and nursing homes.
Since the advent of levodopa treatment patients stay in hospital
for less time, and survive longer with the condition[1].   This
means we are less occupied with managing patients with long-term
and incurable handicap in hospitals, and more occupied with trying
to maintain the patients as mobile and independent as possible
outside hospitals until the terminal phase of the disease. At the
same time, because patients live longer with the condition now,
we see more advanced manifestations of the disease, which has time
to develop further.

Now that the generation who suffered the encephalitis epidemic
of the 1920s is dying out, we see less and less post-encephalitic
Parkinsonism.   The main three groups I see are those with idiopathic disease (occasionally with the reported familial tendency[4,5]),
those with drug-induced disease, and the difficult group whose
Parkinsonism is accompanied by evidence of arteriosclerosis and/or
dementia[6,7].

DIAGNOSIS

How do the well-known signs and symptoms of Parkinsonism differ in elderly patients?   I find the disease more difficult to recognise in an elderly person especially if, on that vital first glance as he enters the consulting room, he needs support in walking or is using a Zimmer Frame.   This may "blur" the clinical impression - and as you all know, it is difficult to think of this particular diagnosis after you have seen the patient several times.   Family doctors, who may see an elderly patient at frequent intervals with complaints of slowness or tiredness or stiffness may miss the characteristic facial and postural changes or dismiss them as "just old age".   Similarly the diagnosis may be made more difficult in the elderly because the characteristic tremor is more frequently absent than in younger patients.   The characteristic leadpipe or cogwheel rigidity of Parkinsonism may begin unilaterally and then the family doctor may mistake it for the spasticity of a stroke. Alternatively he may be unsure whether the patient's resistance to movement may be due to apprehension or to fear of movement of stiff or painful joints and not to Parkinsonism - further clouding the diagnosis in the handicapped patient.

Patients and their relatives are particularly likely to accept without complaint the increasing disability due to bradykinesia - they may not realise that slowness in doing up buttons, difficulty in cutting up food or slowness in writing is anything other than the result of ageing.   Turning over in bed is another characteristic difficulty which may not be volunteered by an elderly patient.

If I suspect a patient of having this disease I watch him walk at least 15 to 20 feet, turn, and if possible negotiate an obstacle such as a doorway or the edge of a carpet.   The characteristic early loss of arm-swing, short length of steps, turning all in one piece and hesitation or "stuttering" steps at an obstacle often reveal the diagnosis.   I watch for tremor at the same time - even if absent at rest it may be present under the stress of negotiating an obstacle while being observed.

Another test which I do regularly is to time the patient writing his name and address in the clinical notes.   It often takes quite a time to ensure spectacles, hearing-aid and pen are all in working order, and the writing surface is at a comfortable height before one can start!   An old person who is used to writing will take $\frac{1}{2}$ - $1\frac{1}{2}$ minutes to complete his name and address, whereas a Parkinsonian patient may take 2 - 3 minutes or more, and the writing may be characteristically small or tremulous.   The written and timed record then stays in the notes as a useful baseline for future comparison whether treatment is started or not.

Another problem in the elderly is the non-specific presentation of Parkinsonism.   "Slowing down" has already been mentioned,
and another non-specific symptom in which Parkinsonism should be
suspected is in repeated falls.   Unfortunately there are many
other causes of falling and Parkinsonian signs may not be sought.
Again incontinence could be due to Parkinsonian slowness in undoing
buttons or hesitation in negotiating a toilet doorway - and the
part played by the Parkinsonism may be underestimated or missed.

Hypersalivation may be worsened in the elderly Parkinsonian
patient by pronounced flexed posture of the head.   The indignity
of dribbling saliva is particularly distressing to sensible patients.
There is, not infrequently, a problem with swallowing in
Parkinsonism, possibly due to disorder of lower oesophageal movement
which is under vagal control[8,9].   Sometimes this problem, and
associated weight loss, may be severe enough to raise suspicion of
a neoplasm of the oesophagus.

Occasionally in elderly patients as in younger ones the
autonomic disturbances of Parkinsonism are severe, including undue
sweating or seborrhoea or poor regulation of blood pressure or
temperature control.   I have recently seen a patient with motor
overactivity (akathisia) who also showed excessive autonomic
involvement (wide-eyed facies, seborrhoea and heat intolerance)
together with weight loss and who looked very like classical hyper-
thyroidism but who in fact responded to anti-Parkinsonian drugs.

In practice, I find that the main differential diagnoses which
must be considered in the elderly include depression leading to
retardation and withdrawal, and myxoedema leading to slowness,
confusion or repeated falls.   Occasionally one has to decide
between different causes of tremor.   Sometimes the disease pursues
a relentless downhill course and proves to be a more severe and
widespread neurological condition such as nigrostriatal degeneration
or Shy-Drager syndrome.

As multiple pathology can be expected in elderly patients with
Parkinsonism, it is obviously of importance to note those conditions
likely to affect the choice of treatment.   A patient who has
glaucoma or prostatism may be worsened by anticholinergic drugs.
Pre-existing postural hypotension must be noted though now that we
use 5 fold smaller doses of levadopa in the combined tablets the
postural change may not be affected by treatment.   The same may
apply to recent myocardial infarction, or the presence of an
arrhythmia neither of which now need prevent essential treatment.

A full review of recent medication is always essential in the
elderly.   The patient may have been taking prochlorperazine
(stemetil) or other phenothiazines such as chlorpromazine (largactil),
or a drug of the butyrophenone group such as haloperidol (serenace).

Any of these can produce drug-induced Parkinsonism by blocking the post-synaptic dopamine receptors and should be withdrawn. As the dopamine receptors are blocked, it follows that dopamine treatment will not help, though anticholinergic drugs can be of benefit.

I seldom find elderly patients treated by mono-amine oxidase inhibitors for depression, but these drugs must be discontinued for a month before starting a levodopa preparation.

PARKINSONISM AND CONFUSION

One of the most important aspects of assessment in the elderly Parkinsonian patient is to judge whether the patient is confused or not. Many patients who have a good social facade, may be able to conceal their confused state and it is essential to have some objective measurement. The questionnaire I use[10] is one which takes about $2\frac{1}{2}$ minutes to run through, and covers aspects of orientation and awareness of current affairs, as well as containing a question which tests short-term memory loss. (See Table 1) There are some problems in using the questionnaire if the patient is deaf or if there is any element of dysphasia. Similarly the drowsy or oversedated patient may not bother to answer all the questions. A depressed or poorly motivated patient may refuse to answer. The questions must be introduced tactfully, otherwise an intelligent patient may be antagonised by their simplicity.

Despite the problems I find the objective score obtainable from this questionnaire is much better than the usual vague assessment, such as "confused" or "rational" or "disorientated in time". A further value of the questionnaire is that a patient can act as his own control - so that later improvement or deterioration of his score can be viewed in the context of drug treatment, intercurrent illness and so on.

What proportion of patients with Parkinsonism have mental symptoms? This question has proved difficult to answer, particularly in the elderly, where it is hard to disentangle the effects of ageing from those of the disease. In a good study by Loranger[11] in 1972 comparing elderly Parkinsonian patients with others of the same average age, intellectual impairment was present in 36%. The researchers concluded that this impairment could not be explained by depression, motor slowing nor the anticholinergic drug treatment available then. They found no relationship between mental impairment and age or length of illness. Another study has suggested that 40% may be confused[12] so that the problem is considerable.

Table 1

Abbreviated Mental Test[*]

1. How old are you?

2. What time is it? (to nearest hour)

3. Address for recall at end of test - this should be repeated to
   ensure it has been heard correctly: 42 West Street.

4. What year is it?

5. What place is this?   Name of institution or address.

6. Recognition of two people (doctor, nurse, etc.)

7. Day and month of birth.

8. Year of start of the First World War.

9. Name of present monarch.

10. Count backward 20 to 1.

Total number of correct answers = mental score.

*Make sure patient is not dysphasic, depressed, deaf or drugged.

Not infrequently we all see patients whose confusion, hallucinations, etc. are directly related to inappropriate drugs given for their Parkinsonism.   In our experience benzhexol (artane) is the worst offender, being unsuitable for use in the elderly, while amantadine (symmetrel) is sometimes implicated.   Withdrawal of the offending drug may be sufficient to restore the patient's previous intellectual functions but on other occasions there is true underlying dementia or poor cerebral reserve.   Other, and normally more harmless drugs may also precipitate confusion in these patients as may episodes of intercurrent illness, so the drug management becomes more complex and difficult.   In one large series Pollock and Hornbrook[13] found that 25% of patients were more trouble to their families because of their dementia than because of their physical disability.

Levadopa treatment for elderly Parkinsonian patients had a bad press at first.   Two papers in the Lancet in 1970[14,15] showed the mental ill-effects of large doses of L-dopa in older patients, and quoted 3 severe reactions to L-dopa in demented patients. The general impression at first was that L-dopa was unsuitable for the elderly, especially if any confusion or any evidence of arteriosclerosis complicated the Parkinsonism.   The suggestion was made that anticholinergic or amantadine treatment was more suitable for the elderly.   Later work, particularly by Sutcliffe[16] and by Vignalou[17] showed that in suitable smaller dosage levadopa is effective in the elderly and gives response in 75% of patients. Anticholinergics and amantadine are not only relatively ineffective, but have confusional side-effects of their own[18].   Levodopa became the treatment of first choice from 1970 or so, but care was needed with confusional side-effects as these occurred twice as often in the elderly as in younger cases[17] (30% compared with 10 - 15%).

With the advent of combined tablets many peripheral toxic effects of L-dopa, such as nausea and vomiting became much less common.   Unfortunately the mental confusion is a central, not a peripheral effect, and its incidence is just the same on combined tablets - so the problem continues.

When the patient is confused, or shows signs of arterio-sclerotic brain disease (perhaps with pyramidal tract signs, etc.), individual judgement is needed whether to attempt treatment. Many physicians would favour a short trial of sinemet or madopar in small doses since the response can be good.   Where confusion and aggression must be treated together with Parkinsonism thioridazine (melleril) seems a valuable drug.   This phenothiazine has a low affinity for dopamine receptor sites and hence little tendency to worsen Parkinsonism.

TREATMENT OF NON-DISABLING PARKINSONISM

As previously discussed, Parkinsonism in the elderly patient is
more often missed than diagnosed at an early stage.   However, there
are some patients with, for example, a characteristic tremor, who may
be diagnosed long before they have significant other disability.
Some of these may be cases of so-called "benign Parkinsonism" where
the level of disability may remain mild for a decade or so[19].
There is a tendency now to treat these patients without drugs, if
possible.   They would be advised to walk a mile a day, and take
deep breathing exercises.   The physiotherapist e.g. in the day
hospital can show them musical exercises to practise at home,
including rolling, trunk and head movements.   Judith Davis, a
therapist in Baltimore has described her regime in detail;[20] and I
am sure we will hear more about physical treatment from Dr. Morosini
later today.   We must all emphasize the necessity to keep active
in this condition - a few days' bedrest for a cold may necessitate
weeks of effort to get the Parkinsonian patient back to his previous
independence.   We must not forget the contribution of the occupational therapist.   Raising the height of bed or chair, or changing from
buttons to zip fastenings or velcro may be all that is needed to
make a slight disability manageable.

If a drug is employed at this stage it might be orphenadrine
(disapal) 100 - 300 mgms a day as this drug has a mildly alerting
and euphoriant action in addition to helping with the rigidity.

TREATMENT OF EARLY (DISABLING) PARKINSONISM

Treatment with dopa plus an inhibitor will probably be needed
once disability is present, and has replaced levodopa as the treatment of first choice now since 1973 or so[21,22].   Most patients will
gain benefit from these drugs for about 5 years, but the disease
process continues and eventually the benefit of the drugs will begin
to fade.   The benefit period may be related to the total levodopa
dosage over the years, so there is a tendency to put off using these
most effective drugs until they are needed, and also to use the
smallest effective doses that can be achieved.   My practice, like
that of many physicians working with the elderly, is to start with
sinemet 110 using either half or one tablet twice daily and building
up slowly until an adequate relief of disabling symptoms is
achieved.   On average, elderly patients can often be stabilised on
500 mgms levodopa plus inhibitor, divided into 4 or 5 daily doses
and may remain satisfactorily controlled on this dosage for several
years.

If the patient has already tried this particular combined tablet
and has suffered nausea or dyskinetic movements, I use the alternative madopar 125 (or even the new madopar 62.5), again building up

the dosage cautiously to similar final levels (about 500 mgms
levodopa and inhibitor).   Family doctors often push up the sinemet
or madopar dosage too fast or too high and do not recognise the
characteristic dyskinetic movements of the face, tongue, limbs and
trunk as a sign of overdosage.   The remedy is to reduce dosage by
half and start building up again cautiously.

Many patients find the Parkinson's Disease Society helpful at
this stage - to meet and talk over problems with other sufferers or
to hear talks or read one of the Society's helpful booklets on this
condition[23].

TREATMENT OF LATER PARKINSONISM

After some years of successful levodopa treatment, patients beg
to experience side-effects of the drugs such as dyskinesias at
dosages which were previously satisfactory.   Other patients find
the tablets effective for only 3 hours instead of 4 or 5.   Yet
others experience the distressing "on-off" syndrome - abrupt onset
of akinesia followed by equally sudden return of therapeutic respons
to their levodopa dosage[24].   This symptom may be less frequent in
the elderly because of the lower levodopa doses commonly used.

For some time these symptoms can be kept at bay with increasing
difficulty by altering the timing of the tablets or their strength,
or by adding a supplementary drug such as anticholinergic.   Howeve
sooner or later these measures will fail.   The patient should
perhaps then be tried on bromocriptine (parlodel)since improvement
shows rapidly if it is going to occur and if not the trial can be
abandoned.   The main indication for this relatively expensive drug
is in late levodopa failure.   It can be effective in the "on-off"
phenomenon and in end-of-dose deterioration as it is long-acting
and may smooth out the response[25].   The dosage is not yet clear
and daily totals of 50 mgms or so have been advised.   For the
elderly I start with 2.5 mgms once or twice daily (probably the
equivalent of $\frac{1}{2}$ tablet sinemet 110) and build up to an average daily
dose of 10 - 20 mgms while the levodopa dosage can be decreased[26].
Although bromocriptine may be effective, its side-effects of
hallucinations and postural hypotension may limit its usage and
dosage in the old.   The new agent deprenyl is being evaluated in
research centres[27] as a means of enhancing levodopa but its
position and that of other new drugs is not yet clear.

Apart from the dyskinesias and "on-off" phenomena, there are two
other-aspects of long-term Parkinsonism which are now appearing as
the patients survive longer.   Postural instability leading to
frequent falls may be a sign that benefit from treatment is being
lost.   Confusion and dementia are increasingly seen in the later
stages of the disease.

PARKINSONISM IN THE TERMINAL STAGES

   Though patients with Parkinson's disease now remain out of
hospital and relatively mobile and independent for some years,
the disease process continues to advance in the nigral cells despite
drug treatment.   The mortality in Parkinson's disease is 1.9 times
normal despite levodopa treatment[29].

   The final stages of the disease may still require hospital
in-patient treatment.   There is often physical wasting obvious in
the face, difficulty with swallowing and with phonation, and poor
or varying response, often accompanied by mental confusion.   The
patient may be on numerous drugs which should be slowly withdrawn
as some of them (phenothiazines and tricyclic antidepressants for
example) may be causing the confusion.   After assessing the patient
again a minimal drug regime may be evolved.   In my view,
investigation of the terminal dysphagia or antibiotic treatment for
an aspiration pneumonia is not indicated and the patient should be
allowed to die with as much peace and dignity as possible.

   Although patients still die from their Parkinson's disease, the
last 10 - 15 years have shown remarkable advances in the treatment
and management of Parkinsonism.   The elderly have shared in these
benefits, and there is every indication of further new advances in
the decade to come.

REFERENCES

1.   A. I. Harris, "Handicapped and Impaired in Great Britain",
        H.M.S.O. London (1971).
2.   "Parkinson's Disease", Studies of Current Health Problems No. 51.
        Office of Health Economics, London (1974).
3.   R. D. Sweet and F. H. McDowell, Five years treatment of
        Parkinson's disease with levodopa: therapeutic results and
        survival of 100 patients, Ann.Intern.Med. 83:456 (1975).
4.   H. G. Garland, Parkinsonism, Br.Med.J. 1:153 (1952).
5.   L. T. Kurland, in "Pathogenesis and Treatment of Parkinsonism,"
        W. S. Fields, ed., C. C. Thomas, Springfield, Illinois (1958).
6.   M. Critchley, Arteriosclerotic Parkinsonism, Brain 52:23 (1929).
7.   J. D. Parkes, C. D. Marsden, J. E. Rees, G. Curzon, B. D.
        Kantamaneni, R. Knill-Jones, A. Akbar, S. Das and M. Kataria,
        Parkinson's disease, cerebral arteriosclerosis and senile
        dementia, Q.J.Med. 43:49 (1974).
8.   W. J. Nowack, J. M. Hatelid, R. S. Sohn, Dysphagia in
        Parkinsonism, Arch.Neurol. 34:320 (1977).
9.   J. A. Logemann, E. R. Blonsky and B. Boshes, Dysphagia in
        Parkinsonism, J.Amer.Med.Assoc. 231:69 (1970).
10. H. M. Hodkinson, Evaluation of a mental test score for assessment
        of mental impairment in the elderly, Age & Ageing 1:233 (1972).

11. A. W. Loranger, H. Goodell, F. H. McDowell, J. E. Lee and
     R. W. Sweet, Intellectual impairment in Parkinson's
     syndrome, Brain 95:405 (1972).

12. G. C. Celesia and W. M. Wanamaker, Psychiatric disturbances in
     Parkinson's disease, Dis.Nerv.Syst. 33:577 (1972).

13. M. Pollock and R. W. Hornabrook, The prevalence, natural
     history and dementia of Parkinsonism, Brain 89:429 (1966).

14. R. B. Jenkins and R. H. Groh, Mental symptoms in Parkinsonian
     patients treated with L-dopa, Lancet 2:177 (1970).

15. O. W. Sacks, C. Messeloff, W. Schartz, A. Goldfarb and M. Kohl,
     Effects of L-dopa in patients with dementia, Lancet 1:1231
     (1970).

16. R. L. G. Sutcliffe, L-dopa therapy in elderly patients with
     Parkinsonism, Age & Ageing 2:34 (1973).

17. J. Vignalou and H. Beck, La L-dopa chez 122 Parkinsoniens de
     plus de 70 ans, Gerontol.Clin. 15:20 (1973).

18. F. I. Caird and J. Williamson, Drugs for Parkinson's disease,
     Lancet, 1:986 (1978).

19. M. M. Hoehn and M. D. Yahr, Parkinsonism: onset, progression
     and mortality, Neurology 17:427 (1967).

20. J. C. Davis, Team management of Parkinson's disease,
     Am.J.Occup.Ther. 31(no.5):300 (1977).

21. C. D. Marsden, J. D. Parkes and J. E. Rees, A year's comparison
     of treatment of patients with Parkinson's disease with
     levodopa combined with carbidopa versus treatment with
     levodopa alone, Lancet 2:1459 (1973).

22. A. Barbeau, H. Mars, M. I. Botez and M. Joubert, Levodopa
     combined with peripheral decarboxylase inhibition in
     Parkinson's disease, Canad.Med.Assoc.J. 106:1169 (1972).

23. R. B. Godwin-Austen, Parkinson's disease. A booklet for
     patients and their families, published and distributed by
     the Parkinson's Disease Society, 81 Queen's Road, London.
     (1971).

24. C. D. Marsden, J. D. Parkes, "On-off" effects in patients with
     Parkinson's disease on chronic levodopa therapy. Lancet
     1:292 (1976).

25. J. D. Parkes, A. G. Debono, and C.D. Marsden, Bromocriptine in
     Parkinsonism: long term treatment, dose response and
     comparison with levodopa, J.Neurol.Neurosurg.Psychiatry
     39:1101 (1976).

26. D. B. Calne, C. Plotkin, A. C. Williams, J. G. Nutt,
     A. Neophytides and P. F. Teychenne, Long term treatment of
     Parkinsonism with bromocriptine, Lancet 1:735 (1978).

27. W. Birkmayer, Medical treatment of Parkinson's disease:
     General review, past and present, in "Advances in Parkin-
     sonism", W. Birkmayer and O. Hornekiewicz, eds., Roche,
     Basle (1976).

28. W. Birkmayer, L. Ambrozi, E. Neumayer and P. Riederer,
     Longevity in Parkinson's disease treated with L-dopa,
     Clin.Neurol.Neurosurg. 1:15 (1974).

29. U. K. Rinne, Recent advances in research on Parkinsonism,
     Acta.Neurol.Scand. 67(Suppl) 57:77 (1978).

NEUROMUSCULAR REHABILITATION IN AGED HEMIPLEGIC PATIENTS

A. Baroni and P. Kelly

Centro di Riabilitazione
I.N.R.C.A.
Ospedale "I Fraticini", Firenze

The common course of recovery of motor function following hemiplegia shows a regular sequence of reflex changes, each of which is associated with a corresponding increase in ability for voluntary movement. The initial phase of flaccidity is sooner or later replaced by that of spasticity. The resurgence of this increased tone does not constitute, however, a simple entity, for spasticity is modified and conditioned by other factors, such as stretch of associated muscles, the position of the head in relation to the body, the position of the body in relation to the supporting surface, and various other stimuli, producing a marked influence on motor responses[1]. Recovery, which may become arrested at any stage in the sequence of reflex changes, depends to a large extent on the degree of spasticity, its rate of evolution, and the type and extent of associated sensory disturbances[2].

For a long time it has been said that spasticity depends merely upon the exaggeration of the myotatic reflex which impedes normal movements being carried out by voluntary control. But many different signs are included under the term spasticity more than just the exaggerated myotatic reflex of Hoefer-Putman. We refer to the hyperactivity of other reflexes, such as the cutaneous plantar response (Babinski), the defence reflex of Marie-Foix, and the postural reactions of righting and equilibrium, as well as to the coactivation of agonists and antagonists, primitive spinal cord patterns of movement (synergies) and to its particular distribution to antigravity musculature.

From neurophysiological studies we know that following decerebration there is an increased activity of static as well as dynamic fusal activity. This supported the concept that spasticity

355

is strictly related to enhancement of a peripheral mechanism (the servo loop system) and, consequently, the main aim of therapy is to reduce tone by proprioceptive input.

THE AIM OF REHABILITATION

More globally, the goal of rehabilitation in hemiplegia is to restore and improve function and to promote a continuum of motor learning for skilled activity. Rehabilitation is, in fact, an education process; the therapist teaches by applying appropriate facilitatory or inhibitory stimuli to activate a response and by giving adequate feedback to a correct response in order to enhance motor learning.

Two well-known treatment approaches are those of Brunnstrom and Bobath. According to Brunnstrom[3], the hemiplegic patient should be aided and encouraged to gain control of the basic limb synergies that appear during the early spastic phase of recovery.

The gaining of control of the synergies from involuntary to voluntary motor responses seems to constitute a necessary intermediate stage for further recovery of movement, combinations which deviate from the synergies. In contrast, Bobath[2] states that the aim of treatment should be to suppress these abnormal patterns of movement which are the result of spasticity and to introduce more normal ones. Reflex-inhibiting movement patterns are used; these not only inhibit abnormal postural reactions, but, at the same time facilitate active and automatic movements. Two other approaches in the rehabilitation of neuromuscular dysfunction that are also used in the treatment of hemiplegia are those of Kabat (Knott and Voss) and Rood.

The techniques of Kabat aim more directly at proprioceptive facilitation of those muscles groups which act in movement patterns opposing those which are spastic[4]. The techniques of Rood, which are based on normal sensorimotor development of coordinated motor activity, seek to restore the absent or reduced components of normal skilled movement in the sequence and manner in which it is developmentally acquired; abnormal movements, thus, are not to be reinforced[5].

THE CONTROL OF SPASTICITY

Despite the differences in these approaches, common to all of them is the transformation of neurophysiological mechanisms of movement into techniques and the use of sensory input to enhance motor learning. As stated previously, the therapist re-educates neuromuscular function primarily by use of stimuli which can be

enteroceptive, proprioceptive and interoceptive, as well as those
of the special senses (vision, hearing, etc.).   Sensory inputs,
however, frequently have opposite effects depending upon their
method of application.   Some general parameters which can guide
the selection of application include the rate at which the
stimulus is given, its duration and frequency.   Let us consider
how some stimuli affect spasticity, a major problem in the
treatment of hemiplegia.

Slow, rhythmical motion, a proprioceptive stimulus, act to
inhibit arousal from the reticular activating system (RAS) as well
as to avoid stimulation of the dynamic component of the muscle
spindle; thus, slow, rhythmical movements can decrease spasticity.
By contrast, quick stretch, which is a brief,rapid movement, acts
to facilitate arousal from the RAS and to stimulate the dynamic
component of the spindle; therefore, quick stretch is a proprioceptive stimulus that can increase spasticity.

Some other commonly used stimuli which can decrease spasticity
include maintained manual contacts, prolonged icing, mild heating
or neutral warmth, slow, steady joint compression or traction,
maintained positioning and slow spinning; while tapping, quick
icing, intermittent joint compression or traction and fast spinning
are stimuli that tend to increase spasticity.   In general, then,
slow rhythmical stimuli are inhibitory and fast irregular stimuli
are facilitatory[6].

TECHNIQUES OF PHYSIOTHERAPY

These mechanisms of sensory input, currently used in physiotherapy to activate motor responses, ought to be reviewed,
in the light of new developments in the neurophysiology of motor
learning.   To regulate movement an appropriate set of muscles
must be activated in proper temporal relationship while the
antagonists are inhibited.   In the peripheral control theory, the
value of sensory information in movement is recognized: the motor
output is built up from smaller, discrete phases linked together by
chain reflexes with sensory feedback as a trigger for the subsequent
one[7].   In the emerging central control theory, claims are made that
feedback from movement is unnecessary for motor output[8] because the
brain already possesses all the information necessary to specify
the temporal and quantitative aspects of movement (the feed-forward
versus the feedback concept of motor control).   The central control
theory tends to support the idea that subjects can monitor their
behaviour internally, being able to correct movements prior to the
arrival of peripheral input when a discrepancy between the intended
and actual command arises.   Clearly, the higher centres of the
nervous system are using a different form of information according
to the type of movement requested; these sources have been

summarized by Kelso and Stelmach as follows:
1) proprioceptive feedback of muscular contraction,
2) external environmental feedback as an indirect consequence of
   muscular contraction usually in reference to a goal, namely,
   the knowledge of results, and
3) internal feedback or information generated prior to the
   response from structures within the nervous system.
Presumably, man's early effort at movement control would be
predominantly dependant on peripheral feedback, while as learning
progresses the large and slower external loop becomes less
necessary with feed-forward control assuming a major role. This
shift helps to explain the acquisition of highly skilled movements
through a process of prepropgramming. Proposing a model of the
nervous system, Nashner and Woolcott[10] suggest that in postural
adjustment the processes which are organised during the latent
period of a triggered reaction (e.g. rapid sway of the standing
basement) are programmed within the peripheral element of the
sensorimotor system and are activated primarily by intersegmental
somatosensory inputs. In contrast, organisational and adaptive
changes, which require several trials, involve the integrative
functions of the central nervous system necessitating more complex
combinations of somatosensory, vestibular and visual inputs.

   Recently, Houk[11] has also suggested an organisational scheme
involving two rather different types of neural processes:
one analogous to a servomechanism and another based on logical
operation. The first is a continuous processor which amplifies
and combines signals from muscle proprioceptors to regulate
"stiffness", i.e. a combined property constituted by the ratio of
force to length change, rather than controlling length and force
separately as in previous gamma loop theories. The second
processor is a stimulus-response (S-R) processor, a subsystem of
sensorimotor regulation, that produces the central motor command
for triggered movements. This classic reaction-time mechanism,
whose latencies change according to the variability of choices
introduced in the task, uses a simple decision-making process
that selects from already available preprogrammed responses.
It does not use peripheral feedback but, rather uses afferent
input only to detect environmental stimuli (open-loop motor control)
Unlike the servo-motor mechanism, S-R processor has an adaptive
capability, either altering the sensory cues required to trigger
a response or modulating the quality and quantity of the response
to any given sensory cue using olivary and cerebellar neurons as
the base for this "detection-read out" type of processing. We do no
know how far such neurophysiological advances will affect the
management of hemiplegia, but it does seem that a new step in
rehabilitation should be to give further consideration to the
internal as well as the external feedback mechanisms. Many authors
are beginning to direct more and more of their effort to manipulate
the learning processes. We refer to the recent proposals, from

the simplest ones of goal-orientated programmes, to those of
biofeedback, up to the more complex ones as the method of Peto[12]
which combines visual and verbal stimulation with timing of any
activity of daily living.

No doubt however, it is a very difficult job to transfer these
neurophysiological advances into treatment techniques because at
the base of the problem lies the fact that we do not have a
specification of the end-product of the behaviour we seek to
understand since we do not yet know how to relate the behaviour of
the components to that of the whole[13]. We should look to the
future of our work keeping in mind the evolutionary landmarks
recapitulated by Sherrington[14] on primates.

> "The parallelism of the ocular axes and the overlapping
> of the uniocular field....together with promotion of the
> forelimbs... to a delicate explorer of space,... together
> also with the organization of mimetic movements to express
> thoughts by sounds"

have been the outstanding feature of man's dominion of environment,
"the universal goal of animal behaviour".

REFERENCES

1. T. E. Twitchell, The restoration of motor function following
   hemiplegia in man, in: "Neurophysiologic approaches to
   therapeutic exercise", O. D. Payton, S. Hirt and R.A. Newton,
   eds., F. A. Davis Company, Philadelphia (1977).
2. B. Bobath, "Adult Hemiplegia:Evaluation and Treatment",
   William Heinemann Medical Books Limited, London (1978).
3. S. Brunnstrom, "Movement Therapy in Hemiplegia", Harper and
   Row Publishers, New York (1970).
4. K. E. Hagbarth, G. Eklund, The muscle vibrator a useful tool
   in neurological therapeutic work, in: "Neurophysiologic
   approaches to therapeutic exercise", O. D. Payton, S. Hirt
   and R. A. Newton, eds., F. A. Davis Company, Philadelphia
   (1977).
5. S. A. Stockmeyer, An interpretation of the approach of Rood to
   the treatment of neuromuscular dysfunction, Am.J.Phys.Med.
   46(1):900 (1967).
6. S. D. Farber, A. J. Huss, "Sensorimotor evaluation and treatment procedures", Indianapolis, Indiana University - Purdue
   University at Indianapolis Medical Center, (1974).
7. J. A. Scott Kelso, G. E. Stelmach, Central and Peripheral
   Mechanism in motor control, in: "Motor Control: Issues and
   Trends", Academic Press, New York - London (1976).
8. J. R. Higgins, R. W. Angel, J.Exp.Psychol. 84:412 (1970)
   quoted by J. A. Scott Kelso,G. E. Stelmach (see above)

9. R. A. Schmidt, <u>Psychol.Rev</u>.82:225 (1975)
        quoted by J. A. Scott Kelso, G. E. Stelmach (see above)
10. L. M. Nashner, M. Woollacott, The organization of rapid postural
        adjustments of standing humans: An experimental conceptual
        model,<u>in</u>: "Posture and Movement", Raven Press, New York
        (1979).
11. J. C. Houk, Motor control processes: New data concerning
        motoservo mechanism and a tentative model for stimulus
        response processing, <u>in</u>: "Posture and Movement", Raven Press
        New York (1979).
12. M. Hari, K. Akos, "Konduktiv Pedagogia TankonyuKiado", Budapest,
        (1971).
13. J. M. Brookhart, Convergence on an understanding of motor
        control, <u>in</u>: "Posture and Movement", Raven Press, New York
        (1979).
14. C. S. Sherrington, "The integrative action of the nervous
        system", Scribuer, New York, 2nd Ed. Cambridge University
        Press (1947).

ASSESSMENT AND TREATMENT FOR REHABILITATION OF PARKINSON'S DISEASES

C. Morosini*, F. P. Franchignoni**, and G. Grioni***

* Cattedra di Terapia Fisica e Riabilitazione, Milano
** Centro Medico di Riabilitazione di Veruno, Pavia
***Servizio di Terapia Fisica e Riabilitazione, Milano

PREMISE : DIAGNOSIS FOR REHABILITATION

When a subject exhibits any alteration of the neuromuscular functions and, generally, any loss of psychosomatic integrity, in order for him to achieve the maximum independence and social integration, we feel compelled to overcome the assessed pathology of organs.   Over and above the identification of the damaged system, of apparatus or structure, nature or amount of damage, qualitative and quantitative manifestations arising from it, we have to consider the repercussions on the whole psychosomatic patrimony and the individual's own capacity to defend it.

The diagnosis will take into account all damage causing disability and particularly the alterations in the neuromuscular function, the assessment of factors causing lack of adaptability to the environment and therefore the study of disorganization and re-organisation of movement in three aspects: execution, organization and motivation.

The diagnosis for rehabilitation will then be based on the assessment of:

1) primary damage and its direct consequences, i.e. the study of the nature, area and extent of the anatomical damage caused by any lesion and also a careful and detailed analysis of the systems connected with the damaged structure, both for proximity and dependence (functional damages that in the long run become structural).   In terms of movement disorganisation it is necessary to assess and where possible prevent the pathological patterns of compensation.

2) secondary damage i.e. of the functional disorder affecting
   systems, also far from the damaged one and not directly
   connected; a disorder that, if not prevented, may lead to a
   total psychosomatic dissolution.

2) tertiary damage or of the extent of disability: determined by
   a pathology of immobility or by fixed patterns of posture and
   movement and/or by a psychic pathology, due to lack of
   communication between the chronic patient and the environment.

4) pathological potential or of premorbid condition, that a
   disease can demonstrate if the intrinsic defensive mechanisms
   are inadequate.

5) health potential which indicates the degree of autonomy of a
   patient and may totally differ, both in positive or negative
   sense, from the seriousness of the disease as well as the degree
   of disability.

These five aspects of the diagnosis for rehabilitation can easily
lead to an equally specific prognosis: to a correct diagnosis for
rehabilitation, summarized in the light of the degree of autonomy
and disability, we have to associate the study of the premorbid
personality and the repercussions the disease might have on it
(strength of ego, cultural and social background, emotional and
working life, degree of motivation, on the one side, and self-
derogatory image, depersonalization, catastrophic reactions and so
on, on the other side) and equally important we have to associate
the study of the dynamic relationship between the patient and his
family environment, social and health service and interaction among
them; in this way we can set up and modify the treatment in the
successive check-up; qualify the health potential of each individual
all the time and eventually analyse the influence that medical and
psycho-social diagnosis, health potential and treatment play as a
whole.   To the established prognosis related to the diagnosis by
the physician, a similar prognosis related to treatment is carried
out by the physiotherapist.

     The scope of rehabilitation can be re-defined as: selection,
guidance, improvement of the neurophysical potential of every handicapped individual, and then re-learning such an individual to a
functional standard beyond his physical deficiency.   In order to
do so it is essential to comprehend the intellectual overall
condition of the patient (psycho-motor, intellectual and emotional
levels, his deficiencies, his capabilities and the mode of utilization, correct or not, or non-utilization of such capabilities);
environmental interrelations (the reaction of the family and the
local attitude towards handicap) and also the presence of doctors
and therapists in the team, well experienced in diagnosis and
treatment of a particular pathology, who should know how to convey

their knowledge to the rest of the team as well as the interdisciplinary exchange of ideas between the team and other
specialists.

THE DIAGNOSIS FOR REHABILITATION OF PARKINSONISM

Therefore the scope of the physician, in the specific case of
Parkinsonian pathology is to establish not only the extent of the
primary damage (i.e. major damage in the nigric area added to a
dopaminergic deficiency, with direct consequences in the proximity
and connected to motor, vegetative and psychological areas); or of
the secondary damage (consisting of the damage of apparatus not
directly connected to the damaged system, mainly at osteoarticular
and cardio-respiratory level); or of the tertiary damage (caused by
pathology of immobility i.e. deformity and related psychic pathology
i.e. lack of autonomy) but also to correlate to these damages the
concept of health potential, intended as the result of neurological
deficiency, the overall functional disability and the reaction of
the patient and his surrounding environment, of the pathological
potential, embracing all psychosomatic effects caused by the
vicious circle of disease and protracted pharmacological therapy.

The importance of the rehabilitation, not succeeding like the
pharmacological and neurosurgical therapy in contrasting the abio-
trophic degenerative pathology, lies in the association, from the
initial stage, with other therapies, so as to reduce the use of them
and contrast the effects of the chronic therapy (and besides restrain
the primary damage, prevent, if it is possible, the secondary damage,
withstand and correct the tertiary damage).

The rehabilitation of the Parkinson patient demands an assessment at neurological, orthopaedic, psychiatric and psychological
levels, which gives an overall picture of the patient in his family
and social environment. Besides, it requires a precise coding of
the re-education techniques, deduced from the already established
principle and applicable according to the semeiologic, neurophysiologic and psycho-dynamic condition.

1) Neurological Assessment

To assess the functional inability deriving from neurological
symptomatology we have to examine in depth the specific semeiotic,
stressing on the symptoms that cause such deficiency, with particular attention to rigidity, tremor and akinesia.

We would like to underline the importance of the initial
symptoms that in the Parkinson's disease are not often recognized
as such, but most of the times they seem to be the signs of aging.

In fact, this process is manifested in general tiredness, slowing down in the execution of certain movements, an overall reduction of motor initiative a tendency to remain in a certain posture for sometime. Often the patient feels rheumatic pain in the arms, headache, flushing and intestinal colic pain. Other minor signs appearing less frequently are: decreased winking, sometimes with the presence of the Stellwag sign (enlargement of the palpebral fissure); a preference for keeping the fingers in flexed and adducted position, the presence of a tremor, not visible but felt by the patient, who explains it as an internal vibration. The specific neurological test then will not only be limited to the assessment of the tone and associated (increase in the tone of posture, attitude, support and strength, phenomenon of serrated trochlea, plastic contracture) reflexes, analysing both extra-pyramidal reflexes and pyramidal and extrapyramidal reflexes in conjunction, of the sensitivity objective but mainly subjective (paraesthesia, cramps localized in lower limbs appearing mainly at night) of neurovegetative function (sialorrhea, hyper-sweating, sebaceous hypercrinia, disorder in the respiratory and cardiac rhythm, daytime lethargy, psychomotor agitation at night) but it will have to take into account certain tests of semeiology, significant from the functional point of view:

- head dropping test, which shows a slow and reluctant descent in a Parkinson patient (sign of the psychical pillow);
- swing test of the upper limbs in which particularly the anterior swing amplitude diminishes rapidly;
- ballottement test of the shoulders, that exhibits reduced swinging on the affected side of the arms;
- test of the falling arm (in Parkinson's disease the fall of the abducted arms is delayed);
- test of swinging leg, which in the Parkinson patient is of very short duration;
- pushing test, which demonstrates the lack of rear flexion of the toes and feet, maintaining balance;
- test of deviation of the upper limbs or test of convergence, which reveals the tendency of slow deviation inwards and downwards of the upper limbs when extended horizontally in front of the patients;
- Lerry's eyebrow sign (looking up, while the eyelid moves, the eyebrow and the frontal muscle fail to follow it);
- Lerry's first sign (in flexion of the fingers, the extensor muscles of the hand do not contract synergically);
- Hunt's sign: while sitting the patient drops on the chair. Often he fails in his attempt to get up because he cannot draw in his feet properly;
- Bastroem's sign: the execution of two simultaneous movements is impossible;
- speech test: speech is monotonous, without inflexions, "saccadee" or trembling; sometimes the tone decreases to a whisper or to an

inaudible murmur. Repetitiion of a word or a syllable can often
be noticed (palylalia) sometimes pronounced in a diminished tone
(aphonic palylalia), or to the phenomenon of festination of the
word (paroxistic tachyphemia). It is important to underline
how Parkinsonian dysphonia affects the mode of speech such as
volume, tone and intonation leaving intact the symbolic
implications;
- writing, which becomes precociously small, very slow, difficult,
  irregular and tends to get smaller at the end of the lines,
  (micrographia), also leaving the symbolic implications intact;
- walking test which demonstrates how the Parkinsian patient finds
  difficulty in walking on even or uneven grounds, through narrow
  passages and doors, when he is asked to stop suddenly or to turn
  backwards;
- test of the use of the hand: the patient looks extremely uneasy
  in using his hands in comparatively easy tasks and particularly
  in deft use of fingers like turning the pages of a book, knotting
  a tie, shaving etc.
- standing upright test: the Parkinsonian patient suffers from lack
  of equilibrium with a tendency to fall forwards, backwards or
  sideways, owing to involuntary movements, such as of the arms.
  Sometimes the phenomenon of hypokinesis is evident; often in the
  Romberg's position one can see a massive intentional hypertonia
  (Lhermitte);
- test of changes in posture: in Parkinsonian patients, the change
  from supine to sitting position, from supine to prone, from
  sitting to standing, from the floor to kneeling and then on
  standing, always show some deficiencies in the Normal Postural
  Reflex Mechanism, above all in the reduction of quickness and
  stability.

The Normal Postural Reflex Mechanism is a Bobath concept which sums
up the evolved automatic reactions and constitutes the basis of
voluntary movement; it is formed by all the postural changes, both
unconscious and automatic, that precede the voluntary movements and
that consist of invisible changes or widespread fluctuations in the
muscular tone or in compensatory movements. During development
the functional movements require the formation of co-ordinated
postural schemes for all persons, even if the modes of expression
vary for different individuals. To the above schemes belong the
reaction of righting, equilibrium and parachute, as well as the
"placing" or capacity to arrest the movement, either automatic or
voluntary, at any stage, without resisting or relaxing too much and
the "grading" i.e. the capacity to control every step of the
movement in the right direction.

2) Orthopaedic Assessment

We assess the muscular strength and the joint flexibility by
an exacting balance method.   Particular emphasis should be put on
the evolution of malposition manifested in: the flexion of the neck,
kyphosis and kyphoscoliosis of the dorsal spine, flexion of the
elbows and wrists, interosteal hand or in the shape of a claw or
a cup, flexion of the hips and knees, equinovarism of the feet.
All these malfunctions affect respiration, gait and balance, these
two latter functions are influenced by both static and tonic
disorder maintained by labyrinth integration.   Besides we emphasize
the importance of assessment and treatment of the deformity of hands
and fingers, which have a peculiar analogy with the rheumatoid
arthritis and which hinder self-sufficiency and the skill of
manipulation.

3) Psychological and Psychiatric Assessment

In a Parkinson patient the akathisia phenomenon is shown as a
continuous restlessness and inability to keep still together with
attacks of paradoxical hyperkinesia, akairia (remarkable querulity
and insistent importunity), echopraxia, echolalia (tendency to
repeat the same actions and same words), bradipsychism (neothic
activity, slow and monotonous, reduction of attentiveness,
concentration, in one word lack of initiative, which in the worst
cases can lead to cessation of all psychomotor activities).

Besides, one has to remember the relationship between motor
symptoms and psycho-pathological processes, that are shown in the
neurosis and psychosis, as well as in the extrapyramidal illnesses
such as: twitches, spasms, cramps and convulsions, catatonia and
catalexia.   Some manifestations of both disorders are similar.
Therefore it is not inappropriate to draw a parallel connection
between instinctive motor activity and basic psyche, and it is not
out of context to interrelate disturbed motor processes regulated
by the extrapyramidal system and psychic processes.   In this way
the extreme feebleness of control and the weak modulation and
elaboration of the instinctive emotional responses, in the
Parkinsonian patient (marked slowing in reaction, lack of memory,
and insight etc.) could also be linked to an alteration of the
connections between extrapyramidal, rhinencephalon and diencephalon
structures.   On the other hand without trying to find evidence in
the pathology, in order to find out the correlation between the
emotional condition and automatic gestures, it will suffice to
remember that before starting a voluntary movement, the soma assumes
a preparatory attitude (pre-pattern) caused by a tonic postural
motor activity influenced either by attention, correlated to
emotional condition, or by an anti-gravitational pattern that

prepares and modulates every precise teleokinesis. This is because
the emotional condition that accompanies the idea conceived through
the extrapyramidal system has obtained an early and precocious
facilitation and inhibition, alerting the structures that will
receive the voluntary message and those who would enable to carry
out the guiding movement assuming synergic support, according to a
particular style, reflecting the motility of every patient. Such
style, which has its anatomical base in this balance between
inhibition and facilitation of tonic postural motor activity, leaves
an imprint on the particular individual, i.e. a certain pattern of
psycho-motor activity: the capacity for self-expression, to interact
with the world, to decide, to attain a purpose, to use and live the
interior somatic experience possibly through more complex cortical-
subcortical integration, determined by historical and cultural
experiences.

One can assume then that the optimum motor possibilities,
depending on the integrity of the extra-pyramidal system, will
produce an element favouring a good image of oneself; on the contrary
the loss of automatisms will cause an overall change of personality,
aggravating the feeling of being imprisoned inside walls of rigidity
and akinesia.

To all this we add a research for the pre-morbid personality
of the patient and the interrelations with the social and family
environment, before and after the disease and in the context of
rehabilitation, deduced from questionnaires of the anamnestic,
functional and psychological types.

The physician will combine all these aspects with an assessment
of the degree of disability, for which many scales exist, (North
Western University Disability Scale, Webster Rating Scale, Hoen and
Yahr, etc.) both neurological and functional, which we think are
incomplete because in our opinion the problem lies in comparing
immediately the degree of disability with autonomy, which partly
depends on the neurological and functional deficiency, but can also
differ in the positive or negative sense. For instance, in the
initial stage of Parkinson's disease, in patients with high socio-
cultural level and a minimum functional deficiency one can notice
limited autonomy, as a result of intolerance to frustration for
the reduced image of oneself, that will lead them to refuse a social
role which could still be sustained by their psycho-motor resources;
and viceversa, in a patient who for years has been affected by
Parkinsonism, treated perhaps with psychotherapy, one can notice the
capacity to accept with serenity his own disability, being able to
be sufficiently autonomous.

0) the patient can still work and is completely self-sufficient in
   everyday life, though he feels some discomfort, or tremor,

usually unilateral, rheumatic pain, general tiredness, slow
movements etc;

1) the patient is unable to perform some fine and discriminating
   activities.   His gestures are slow and he cannot perform two
   complex movements simultaneously;

2) the patient cannot use freely one of the two upper limbs in the
   manipulating activities.   When he is still, normally a tremor
   is evident.   The working capacity is decreased.   He may walk
   more slowly.   His mimic get poorer;

3) the patient acquires the typical flexed attitude of Parkinson
   cases.   He is still able to look after himself, though slowly
   and awkwardly.   He can no longer devote himself to his usual
   work.   Akinesia, tremor and rigidity are evident in various
   degrees.   Problems of speech, writing, neurovegetative etc.
   may also appear;

4) the patient shows lack of balance, above all facing obstacles,
   turning backwards, or in any abrupt movement.   He walks with
   short steps or festination.   Rigidity, akinesia and tremor will
   hinder him from achieving autonomy in dressing, washing and
   feeding.   The speech and writing are at risk.   The amimia is
   evident;

5) the patient can no longer walk by himself.   His hands acquire
   the typical form and their use is minimum.   The flexed positions
   of the head, trunk and limbs become permanent;

6) the patient can no longer get up and sit by himself. He has to
   be dressed, washed and fed;

7) the patient has lost his will to move.   He could still be
   partially autonomous but his motor initiatives will function
   only on insistence;

8) the patient walks only for short distances at home, supported by
   others.   By now the abnormal posture, with the typical deformities, has become permanent.   If tremor exists, while making a
   voluntary effort it becomes more evident;

9) the patient no longer gets up, neither sits, nor walks, even
   when supported.   In bed he shows the typical deformities in
   flexion-adduction of upper limbs; flexion of the lower limbs
   (the knees do not touch the bed), flexion of the head (psychic
   pillow).   All other symptoms are clearly visible: tremor,
   sialorrheoa, psychic disturbances, amimia, etc.;

10) the patient does not even know how to turn in bed. He often
    feels acathisic pain.   This state of marasmus will normally
    bring death by multiple diseases.

    We can also grade autonomy simply as: minimum autonomy when
the patient is bedridden or just succeeds in sitting or standing;
average autonomy, when the patient is self-sufficient for daily
needs and is able to move around; maximum autonomy when the patient
goes out on his own, drives a car, has some work activity.   This
functional test, associated with neurological testing, enables us
to realize in time when the medical re-education programme is no
longer effective, and hence requires modification and regulation.

    Also the treatment, like the assessment, cannot be standard-
ized because the patient's main problem needs to be analysed, both
in relation to his multiple symptoms and to his reactions to disease
and treatment, and in relation to his environment.   As for all
assessments, it is essential to consider not only the patient's
incapability but above all his capability, how he performs, which
pathological compensations are there and how they can be modified.
For this purpose the co-operation of the physiotherapist is
essential.

    For many years we have carried out rehabilitation therapy
which deals with the apparent symptoms on the one hand e.g.
hypertonia, akinesia, tremor and orthopaedic deformities (according
to the methods described by Licht[1], Rusk[2], Kiernander[3], Fasio and
Soriani[4], Doshay[5], Ribera[6], Farmhouse[7], Umbach[8], Barie[9] and others),
and on the other is based on those more general premises that take
into account the overall situation of the patient and his immediate
environment and also of the neuropsychological consequences of
neurological and orthopaedic symptoms [10].

1) The psycho-motor deficiency, poor capacity of learning,
    fickleness of will and activity, of interests and aspirations,
    the sense of imprisonment in rigidity, the loss of cogenital
    and acquired automatisms and the automatic components of
    voluntary activity (with consequent disorder of patterns,
    preparatory and complementary to the movement itself), the
    impossibility of performing two movements simultaneously, the
    emotional stress increasing the tremor to the extent of slowing
    down and altering the skilled movements, the lack of initiative,
    physical participation and interest in the movement due to
    akinesia, that affects posture, gait, mime and speech, are all
    neurophysiological and psychological reasons that justify the
    application of a rehabilitation therapy for the whole life,
    characterized by order, rhythm, slow but steady tasks, patience
    and firmness in continuous physiotherapy and introduce the
    necessity of a psychotherapeutic component aiming at strengthen-
    ing the functions of the ego of Parkinson patients.

It is necessary to plan the number of exercises to be taught
one after another, increasing progressively both the effort and
the concentration of the patient and plan precise tasks throughout the day.

It is necessary to impress a rhythm in the life of a Parkinson
patient both by a schematic plan of the exercises, by being
constantly urged by the physiotherapist during the session and
finally by group therapy, as early as possible, increasing
gradually.

At the end of the re-education session the pronounced verbal
rhythm should have influenced the patient enough to cope with the
difficulties of daily life.

2) We would like to add that the physician's task is also to guide
a team of therapists to advise the patient and his family in the
proper choice of modes of life: from diet, to house and table
decoration, type of clothes, a rational programme of the day's
activities, continuation of working life, as long as possible,
especially if it does not entail time keeping or working
under dangerous conditions; method of self-relaxation (especially exercises for heavy segmental concentration and control of
breathing, which have a positive effect on the psyche, tremor
and rigidity).

3) Another essential element for a correct rehabilitation programme
lies in the exact assessment of the general conditions of the
patient, especially of cardio-vascular and respiratory system,
that could deteriorate due to orthopaedic deformities and since
patients, especially the elderly ones, are likely to get a
respiratory disorder it has to be fought with proper exercises
(relaxation, manipulation of stiff joints, training of diaghragm
respiration, evacuation of stagnating secretions etc.)

4) The need for an orderly and varied activity throughout the day,
if possible in the context of a community life. It counteracts
akinesia and improves the functional activities and co-ordination
in daily life. It can stimulate creative processes, making the
patient aware of his capabilities, reducing his apprehension and
anxiety, lifting his psychic tone, restoring his self-confidence
and making him feel useful to others. It is necessary to teach
the patient how to use his hands in more complex and fine
activities, starting with exercises of ability and skill with
games and tools in the gymnasium, taking care of his own person,
including taking food, to teach him writing, handicrafts, etc.
It is very important that the choice of ergotherapy should be
suitable to influence the attention and concentration of the
patient and stimulate the psychic resources, in relation to age,
socio-cultural background and pre-morbid personality.

5) We have stated that in the late stages the main purpose of re-
education is the maximum self-sufficiency permitted by the
deformities and by the severity of the neurovegetative symptoms.
In such cases great importance lies in the role played by the
family (or by the personnel if he is in hospital) on whom the
Parkinson patient depends, because they can help in the passive
segmented or global mobilization and modify items of clothes,
personal hygiene, cutlery etc. to enable the patient to make
use of his affected limbs.

During the initial stages, in addition to the above general
rules, we have to set up a programme in order to attack the symptom
or the predominant symptoms:

1) First of all head and trunk exercises are required against the
hypertonia: in fact the flexed position of the patient greatly
affects his balance, not only because the incorrect posture projects
the centre of gravity of the body axis forward, but also distorts the
global postural tone, maintained by the various righting reactions,
mostly of labyrinthine cervical origin.

These exercises are performed in supine, prone, sitting and
standing positions (the last two positions facing a mirror). To
begin with, the head will be moved passively in all directions,
slowly and at full amplitude then guided active movements, active
against resistance, free but regulated with rhythm and amplitude.
The mirror has a fundamental importance in the re-education
treatment. We can say it fulfils three essential purposes: the
awareness and acceptance of reality on the part of the patient; his
active participation to the recovery with a constant check on the
modifications necessary to his postural tone schemes, especially
the correct alignment of head-trunk-limbs and regulation of mimic
expressions, of phonation and respiration; the dialogue with his
own "impressive psychomotricity".

For the trunk the exercises will be kept more or less in the
same progression as for the head: flexion-extension, rotation,
even better in combination, and with the limbs moving following
Kabat's techniques, freeing the upper and lower trunk, extension
from prone position, rotation on the major axis, on the mattress,
then followed by movements in all directions and at full amplitude,
with the patient sitting facing the mirror, and the last check will
be the amplitude and rhythm of such movements. When the patient
has learned with sufficient freedom to move head and trunk, he
should keep on doing such movements in front of the mirror stopping
when asked and doing them slowly or fast. Useful is the combination of movements with respiratory and phonation exercises
(the latter will be more and more complex, from the rhythmical to
vocal tone, to a shout, to a phrase, to singing etc.).

For the upper limbs it will be necessary, first of all, to relax the shoulder's adductors and the elbow's flexor muscles, the agonist antagonist passive movements, (slow and at full amplitude), to break the anomalous patterns through inhibitory postures, according to Bobath or overall movements in diagonal directions, according to Kabat.

Lapidari's wheel could be useful to perform movements of circumduction of the shoulder with increasing resistance.   The segmented mobilization of the hand and fingers is very important.

For the lower limbs, the first thing to be treated will be the mobilization of the hip, completing the relaxation on the bed and exercises on the mattress; very useful for the relaxation of trunk, head and limbs have proved the patterns of passive mobilization, according to Doman and active mobilization, according to Temple Fay.

In the very serious cases of rigidity, in which contractures are very painful, careful and segmented overall massage is of some benefit.

2) For the akinesia and the overall psycho-motor slowness we utilize the semeiotics techniques already described. Most of the time the re-education sessions will be devoted to exercises of shifting positions: from sitting to standing, utilizing Hunt's test to teach the patient to get up, drawing his feet backwards and outwards. In the upright position the pushing technique will be utilized together with the fluctuating platform to reinforce the equilibrium reactions.   The exercises for walking will be performed on outlined distances, with long and wide steps, or with goose-step (Doshay[5]). Then there will be associated dangling movements of the upper limbs, firstly at the parallel bars, then with the help of two sticks and of the physiotherapist who will impress and stimulate the crossed movement of the lower and the upper limbs (Farmhouse[7]).

It can also be useful to employ the method of walking with two sticks placed diagonally behind the back of the patient, who will catch them with the upper limbs lifted (while the physiotherapist pushes the lower limbs forward).   Subsequently the patient will be trained to walk overcoming various obstacles in shape and amplitude, to turn, to walk on non-rectilineal paths, uphill and downhill, walking backwards, on uneven ground, and to climb up the stairs.

Against the amimia, performed facing the mirror, exercises of the mobilization of the eyes in all directions, and then faster and faster, with the patient looking at an object moved at increasing speed and keeping his head still, will be very useful. Subsequently, these exercises may be combined with those already mentioned, i.e. mobilization of the head, so to balance all the righting reactions, either of labyrinth, cervical or of visual origin.

The patients have to learn how to wrinkle up the forehead, to move the tongue in the oral cavity, to protrude it in all directions and space, to blow, to whistle etc.   It is suggested that the mimic exercises and mobilization of facial and lingual muscles should be combined with exercises of phonetics and respiration.

If it is possible to organize a group who has learned to feign the most varied expressions in front of a mirror and to check the voice and breathing, everyone can be assigned a task: teaching dynamic co-ordination and motor ability, acting and singing in a psycho-dramatic activity.

3) The tremor is perhaps less affected by the re-education techniques than the akinesia.   Anyhow with the aid of electromyography suitable exercises can be chosen to lessen it and slowly become a harmonious voluntary activity.   In the initial stage, with the patient at rest, rhythmical stabilization i.e. isometric exercises alternating agonist and antagonist movements can prove helpful; in the cases of serious hypertonia, where tremor also appears in voluntary movement, we can use isotonic contractions at full amplitude and against resistance.

4) The alterations in postural tone associated with peripheral damage (retractions and muscle-tendon fibrosis, joint rigidity, arthritis etc.) can be evident in all segments with equal intensity in the same patient or some can predominate on others.   Suitable for the joint deformities are passive mobilization in gradual extension of head-trunk with the patient sitting, and active mobilization in overall extension head-trunk-limbs with the patient in a prone position.

Both for the limbs and the trunk, it will be necessary to improve overall alignment.   It can be started on the statics table, then pass on to a bicycle for tetraplegics, on which the Parkinson patient must lean on a high and hard back and move the upper and lower limbs very slowly, without using any resistance, organizing mentally the motive harmony of the four limbs at every turn of the pedal.

For the deformities of the hands, all the mechanical devices applied to polyarthritic cases can be adopted, with the purpose of correcting contractures and joint rigidity; above all the skilled exercises already described are useful with the possible adaptation of instruments and tools for daily and working life.

We quote here some physio-patho-genetic considerations that have led, in the past years, some authors to suggest other specific exercises for the Parkinson patient.

Vannini and his colleagues[11] write that:
1) In Parkinson's disease only the more recent automatisms are selectively damaged,
2) The automatisms are built on through the repetition of a gesture or a voluntary posture, until they are received by the extra-pyramidal archive,
3) The automatic motor patrimony of old acquisition remains unchanged.

According to them the re-education treatment will have to:
1) Utilise the voluntary gesture for the reconstruction of motor schemes in relation to the environment and then in centrifugal directions, and repeat them constantly, leading to a future facilitation and automatic recruitment;
2) the schemes evoked by the voluntary gesture must strictly run over again the directions and purposes of lost automatic schemes, limited to the essential motor schemes, aiming at the patient's autonomy.
3) The Parkinson patient in compensating for the lack of motor activity with the stereotypia of residual automatism, seeks for compensation that will end in damaging more his level of autonomy Then it appears necessary that the therapeutic gesture is aggravated in a direction diametrically opposite to the residual spontaneous gesture.   Such a phenomenon serves to fix more the new gesture in the central neuron maps, underlining it through apparent proprioceptive information and through the image of vision.

Borromei[12] also suggests exercises with coloured obstacles, very near to the physiological movement of the step, but capable of compelling the patient to overcome the spontaneous amplitude of the movement, making it difficult for him to discriminate the task in advance and therefore improving the voluntary component of the motor choice.   In addition to that, it compels the patient to control the amplitude of the step, measuring it on horizontal lines, outlined on a transparent sheet of sicoglas, which he will hold in front of him.

Perfetti and Grimaldi[13] maintain that in the treatment of Parkinson patients it is necessary to adopt functional systems different from the damaged ones.   Neither the repetition of the same motor scheme, nor the activation of absolute reflexes, evoked by vestibular or proprioceptive afferents, nor the adaption of the so-called voluntary movements will help the patient in the re-acquisition and activation of the lost automatism.   According to Perfetti and Grimaldi such failures are due to the fact that these therapeutic movements are directed to deficient structures and their intervention at an elaboration level is inadequate.

Their treatment is based upon:

1) Exercises aiming at maintaining the articulation and the muscular trophism, in which it is advisable that the control of the excursion of the movement is entrusted, above all, to tactile and visual afferents, for which a kind of automatic utilization emerges much less frequently than for the proprioceptive ones;

2) Exercises of relaxation following methods based on tactile perceptions and on kinesthesic awareness;

3) Exercises for akinesia and for postural reflex deficiencies, and the purpose must be to: a) encourage on the patient's part, the conscious replacement of kinesthesic and proprioceptive inputs with the non automatic inputs, such as vision, which permit the completion of the afferent synthesis; b) activate the selector of action, no longer on the base of peripheral parameters (internal or external) but referring to the higher level.

CONCLUSIONS

In the light of what has been said, the techniques of rehabilitation of the Parkinsonian patient still appears to be susceptible to modification by better scientific understanding. On one side we have to rely on the neurophysiological and neuro-pharmacological considerations of physio-patho-genesis of Parkinson's diseases and on the other we must not forget the psychic component of posture and movement, related through the extrapyramidal system.

Only in this way we will obtain a treatment resuming balance between facilitating and inhibiting scales, supported by the balance of pharmacological systems, an equilibrium that can be achieved either utilizing the carticalization of voluntary movement or building up again the learning of the automatic type of movement on a recomposed attentive and postural tone regulation.

Anyhow from now on we think that the treatment for rehabilitation ought to be carried out as early as possible: then associated with drug treatment and not following it, so as to prevent rather than curing; continuous, in order to limit the functional damage and avoid, as far as possible the tertiary pathology; personalized and not standardized, with the physician's and physiotherapist's guidance, both at home and in hospital; modifiable depending on the kind of pathology, of the time of evolution, on the treatment results and on the reaction of the patient and his family.

To conclude we want to stress the importance of a multi-disciplinary approach to study means of fighting efficaciously the overall evolution of such pathology.

REFERENCES

1. S. Licht, "Therapeutic exercise", Waverly Press, Baltimore (1961)
2. A. H. Rusk, "Rehabilitation Medicine", The C.V. Mosby Co.,
   Saint Louis (1964).
3. B. Kiernander, "Physical medicine and rehabilitation", Blackwell
   Scientific Publications, Oxford (1953).
4. C. Fazio, S. Soriani, Il problema della riabilitazione nelle
   malattie neurologiche, La Ginnastica Medica 8   (1960).
5. L. J. Doshay, Parkinson's disease, J.Amer.Med.Assoc. 174:10
   (1960).
6. V. A. Ribera, "Rehabilitative measures in parkinsonism".
   Comptes Rendus di IV Congr. Int. de Medicine Physique,
   Paris 6-11/9/1964 Internat.Congress Series n.107-Excerpta
   Medica Foundation Amsterdam, New York, London, Milan, Tokio,
   Buenos Aires (1964).
7. M. Farmhouse, "Reéducation dans la maladie de Parkinson",
   Comptes Rendus de IV cong.Int.de Med.Physique. Paris 6-11/9/
   1964. Intern.Congress Series n.107 Excerpta Medica
   Foundation. Amsterdam, New York, London, Milan, Tokio,
   Buenos Aires (1966).
8. W. Umbach, H. Leube Teirich, T. Riechert, "ABC fur Parkinson-
   kranke", Georg Thieme Verlag, Stuttgart (1967).
9. M. L. Barrié, "La kinésitherapie de la maladie de Parkinson",
   Libr.Maloine S.A. Paris (1970).
10. C. Morosini, "Il problema delle sindromi parkinsoniane e 10
    studio delle metodiche riabilitative ad esse applicabili",
    Tamburini Ed. Milano (1967).
11. A. Vannini, L. Buscaroli, F. Servadei, G. Faccani, La rieduca-
    zione del parkinson, Eur.Medicophysica 13:115 (1977).
12. A. Borromei, "Rieducazione neurofisiomotoria della deambul-
    uzione nella malattia di Parkinson", Atti della IV Riunione
    della Lega Italiana contro il morbo di Parkinson et le
    malattie extrapiramidali, A. Agnoli and G. Bertolani, eds.,
    D. Guanella, Roma (1978).
13. C. Perfetti, L. Grimaldi, Aspetti neurodinamici della rieducaz-
    ione motoria del parkinsoniano, La Riabilitazione 122 (1975).

FEATURES OF PSYCHOGERIATRIC ASSISTANCE FOR HOSPITALIZED AND NON-HOSPITALIZED PATIENTS[*]

I. Simeone[+]

Médecin-Directeur Adjoint
Centre de Gériatrie de Genève

INTRODUCTION

The medical exigencies linked with aging increase rapidly each year and more doctors are called upon or will be called upon to concern themselves with aged subjects. But every medical discipline or doctrine can only be constructed from knowledge resulting from research and transmitted by teaching. Thus, there is the necessity to create first a model of geriatric medicine for the future doctor, whether he be the general practitioner or the specialist of tomorrow[1]. The Geriatric Institutions of Geneva represent one of the first responses to this new demand.

The concept of care practised in Geneva is, in certain aspects, a little different from traditional concepts. The history of medicine is perhaps the history of an eternal gap between the psyche and the soma. The incomprehension or the reciprocal distrust between various specialists of psychic illness and those of somatic illness, both of which may be present at the bedside of the patient, has become paradoxical, if not dangerous.

Geriatrics thus offers a privileged field of a new science which can avoid the major pitfalls of modern medicine that favour a one dimensional technique to the detriment of a relational dimension;

* Work performed in the Geneva Institutions of Geriatrics (Director Prof. J.P. Junod) Departments of Psychiatry and Medicine, University of Geneva.

+ Psychiatrist - Associate Director.

for if such medicine is not suited to man, it is even less suited
to aged man. Everyone is aware that in certain hospital
establishments the relationship of care-giver to patient has become
so inadequate in recent years due to the technicological development
of care, that it has been necessary to create a group of studies
for the "humanization of hospitals".

The method of care practised in the Geriatric Institutions of
Geneva was, from the very outset, conceived as a multidisciplinary
care animated by a medical doctrine described as "integrated
medicine".

As for this "multidisciplinary" aspect, let me say briefly that
all the units of hospital and extra-mural care group together
doctors, nurses, social workers, occupational therapists, physio-
therapists etc. in coherent and homogeneous working teams. On the
other hand, we define the term "integrated medicine" as the very
close collaboration between somatic medicine and psychological
medicine in our daily practice between internal medicine and
psychiatry[2].

This second aspect of geriatrics in Geneva is the object of
this presentation and we make reference here particularly to the
contribution and the presence of psychogeriatrics in these
institutions not only as an integral part of the medical action but
also as an instrument of analysis of the psychological interreactions between the patient and his illness, the patient and his
care-givers, the patient and his family and social environment.

WHY PSYCHIATRY IN A GERIATRIC INSTITUTION?

This is the question we must ask ourselves before continuing
further with this presentation. The answer may be that after
serious consideration and after many years of clinical experience,
we are convinced that the aged patient can only be understood
within the context of his personal history and that it is by the
decoding of the subjective world within which the patient lives
that the illness has a sense and a significance for the care-giver.

We know that the state of the well-being and that of the illness
of a given subject depends always upon the equilibrium and the
results of innumerable interreactions existing between his biology
and his psyche, these two variables which are linked to the
characteristics of his living space. For these reasons, aging is
an astonishing differential process, for it is the culmination of
one's past life and personal experience at the same time and with
the same significance as the biological evolution. The ontogenetic
future of man consists of the convergence of all these factors,
including aging. From this point on we are far from the old

interpretations of senescence exclusively organically based. In order to really understand the evolution of man (for it is a question of evolution, not involution), it is indispensable to see it as a bio-psychodynamic process with multiple existential components, whether they be normal or pathological.

Modern psychiatry (by this we mean psychiatry with psycho-analytical orientation) possesses the necessary tools with which to understand the psychodynamics of the patient and to investigate, if necessary, his conflicts in the face of illness, aging, and death. This can clarify the often ambiguous and difficult relationships between the care-giver and his patient, the patient and his family.

However, the existence of psychological conflicts and psychiatric troubles of the patient require a good clinical knowledge in order to establish an exact diagnosis and adequate therapy. Clinically, if we attempt to define the nosographic scope of present-day psychogeriatrics, we find that the mental troubles we see regularly during the course of aging are less the major schizophrenic syndromes or delusions, but rather that which we call the "minor psychopathology of aging".

The important chapters of psychiatry weaken and fade with age. Follow-up studies of psychoses and classic neuroses demonstrate this. Recalling the studies in Lausanne (L. Ciompi and C. Muller[3,4]) we see that half of the schizophrenic patients hospitalized at a young age or in adult life, have, in their old age, evolved favourably towards a cure or with a slight residual condition; only a fifth presented a less favourable evolution. Among the neurotic patients (same method of research) the clinical pictures of three classic neuroses (hysteric, phobic and obsessional) had disappeared or had greatly diminished in at least half of the cases. A very small percentage suffered a deterioration. Among a great number of these patients the old neurotic symptomology was replaced by a new physical symptomology with indistinct and hypochondriacal contours.

The scope of these clinical pictures, appearing for the first time during the course of senescence, forms the field of study of psychogeriatrics today more than a late evolution of the old pathology.

Some new forms are appearing such as: atypical depressions; hidden depressions where the somatic symptoms, the only symptom present, hides the psychogenic origin of the syndrome; states of acute or chronic regression (often taken for senile dementia); states of confusion (often of iatrogenic origin); senile delusions; chronic hypochondria and numerous somatic symptoms. For these last syndromes geriatrics offers again new aspects to psychosomatic medicine although the syndromes typical to a younger age group (asthma, colitis, stomach ulcer, etc.) also decrease in number

(Oules, J.[5]).   It should be noted, however, as has recently been
said by certain German authors[6], that the pathology of senescence
can also offer the model for a new somatopsychic medicine where the
organic syndromes bring about a forced alteration of some psychological defences, by the appearance of a new medical approach of
mixed components, with a biological beginning and a psychic
consequence, contrary to the classic psychosomatic doctrines of
the intrapsychic conflict as generator of the organic lesion.

The ensemble of these pathologies represents a large percentage
of the aged population, up to 25 - 30% for Zimmerman[7], and from
30 - 35% for Roth[8].   The senile-dementia syndromes represent only
3.4% of this same percentage but they cover a great number of
hospitalizations and efforts on the part of the geriatric care-
givers[9].

What are the present views on this syndrome? Is it that perhaps
we are witnessing today the transformation of the traditional concept
of senile-dementia? With the years senile-dementia has become a
"mythical" illness, as has cancer; that is to say, an illness where
"man has involuntarily placed all his secular fear in the presence
of death" and madness.   But only a small number of specialists
know that cancer and senile-dementia are not necessarily synonymous
with death and institutionalized insanity.   Similarly, for the
neoplasms and the schizophrenias, where the etiopathogenesis is still
obscure, one is aware of several degrees of evolution and of
malignancy, from the slightest to the most serious.   For senile-
dementia as well, our comprehension must be even more on the alert
and our hope as great as possible, and in all event abandoning the
false belief in ineluctable illness or hereditary destiny.   Such
false ideas and beliefs which blame senile-dementia for everything
such as, for example, the character problems which accompany it,
or which place upon the same level of illness, the delirious
interpretations which stem rather from the personality of the subject
than from his psycho-organic deficiency.

On the same theme of old prejudices, today we also know what
must be done or what should not be done in our method of treatment
of senile-dementia so that we do not plunge these patients more
deeply into illness.   As in schizophrenia where, independent of
the etiopathogenic causes which still remain rather obscure, there
exists a group of therapeutic norms and attitudes to observe which
are already actively used with good results. It is by means of
these methods that we see psychogeriatrics not just limiting itself
to clinical investigation, but watch its action broaden also in
psychotherapy.   Psychotherapy, individual or in groups, supportive
or psychoanalytical, family, or by physical relaxation, offers
many possibilities to the specialists as well as to the general
practitioners.

The same rules and the same principles which animate the psychotherapeutic field with a patient can and should be applied in the comprehension and the analysis of the emotional attitude of the care-giver.   Nowhere, more than in a geriatrics service, is the identification and the emotional response of personnel put to such a difficult test.   Frustrations and failures are many and can generate sentiments of impotence, of guilt, and thus, of rejection.

In psychoanalysis the term "counter-transfer" is given to these reactions and the help of such psychological knowledge is extremely important within a team in order that these sentiments may be freely exteriorized in discussions.

We should also bear in mind that, at times, the attitude of the care-givers, on the contrary perhaps, may be so imbued with too much of a maternal attitude which may worsen the regressive behaviour of the patient by increasing his dependence.

This is medical psychology, the relational area between the care-giver and the patient, and where psychiatry has a particular place in the sensitizing of the doctors, the nurses and the para-medical professions to these problems.

It is in this way that the role of the teaching of psychiatry in an institution presents itself to medical students, interns, nurses and other paramedical professions.   It should be an instruction which teaches not only the clinical aspects of mental illness but especially, as one can easily imagine, includes also the psycho-affective reactions of the care-givers to the problems of aging, death, social deterioration, and psychic annihilation[10].

If, to begin with, it is very difficult for the younger care-giver to identify with the aged, it is even more difficult to identify with the aged sick.   Thus it is necessary to make a continued and sustained effort towards the training and the sensitizing to these gerontological aspects.

The following instruction is presently being given in under-graduate and postgraduate courses,and continued in the case of medical psychology and psychiatry: a weekly multidisciplinary seminar, this academic year centered upon depression; a seminar for doctors only on the psychotherapy of the aged; another seminar on relaxation; weekly instruction in clinical psychiatry; various other clinical aspects and aspects of research discussed when reports are made or upon the presentation of cases;   finally, additional instruction is planned particularly for medical students, the nursing care services in the hospital, and in the extra-mural services, the nurses aides and the volunteer workers, in the form of traditional instructions or in small groups for sensitizing and discussion.

A few words finally concerning the research in progress on
this subject recalling that in the last three years about a hundred
papers have been published concerning the Geriatric Institutions of
Geneva dealing with subjects in all medical domains, and other are
areas.   There are: 1) Studies of the psychological phenomena
associated with senescence; 2) Geriatric studies (for example -
therapeutic groups, short psychotherapy, relaxation, family groups
etc); 3) Studies of models of geriatric action, both extra-mural
and institutional; 4) Medical-social gerontology; 5) Medical ethic

WHAT IS THE PRACTICAL ACTION OF PSYCHOGERIATRICS?

The geriatric Institutions of Geneva, directed by Professor
J. P. Junod, consists of a hospital for short and medium length
stays (Hospital de Geriatrie), a hospital for long stays (Centre
de Soins Continus de Collonge-Bellerive), an ambulatory polyclinic
and home care treatment (Centre de Geriatrie), a specialized
service involving psychogeriatrics at the General Hospital
(Service de Consultations de Psychogeriatrie), and one Day Hospital
which completes the range of possibilities offered.   At the
Geriatrics Hospital (Hopital de Geriatrie), three qualified
psychiatrists work full-time and as Chiefs of Service they super-
vise the interns and all other psychiatric activities, both
clinical and didactic.

In the Consultation Service of the General Hospital there are
two psychiatrists working full-time.   At the Geriatric Center
(Centre de Geriatrie) of the seven doctors working full-time, four
have psychiatric training.   In all, as consultants, there are two
psychogeriatricians and one psychoanalyst.

How does the psychiatrist intervene in the procedure of treatment and instruction in these institutions?   At the hospital the
patient may, upon his admission, present a psychiatric symptomology
and require a first intervention of the care indicated, urgent or
not.   A preliminary diagnosis is established and a first plan of
action put into effect.   Then, in the different care units a
thorough and complete psychic assessment is made along with routine
or particular somatic investigations.

The composition of the dossier is the reflection of this double
orientation.   The complete case history contains not only any
possible surgical intervention or the first signs of hypertension,
but will also take note of important emotional experiences or
conflicts, past and present, as well as the most striking psychological traits and significant relational aspects.   Following this
the somatic status and the psychic status will be established, both
becoming part of the dossier. For example, concerning the mental
status, the intern will observe particularly the semiology of

behaviour such as the patient's appearance, his reactions to the examiner, his behaviour in daily life (as described by himself and those close to him), the semiology of basic psychic activity with a description of possible pathological symptoms present such as illustions, hallucinations; his morale (for example, a depressive or manic state); ideation (problems with thinking process, delirious or melancholy ideas), etc.

A psychometric examination will complete this data and the investigation is then directed toward the memory, the disintegration of language, the praxis and the gnosis, according to the psychometric technique developed in Geneva by the school of Professor Ajuriaguerra which, in turn, derives from Piaget's research with children.   It should be clearly understood that all these examinations are routine and done for each patient in the same manner as the clinical examinations.   The aim is not to "psychiatrize" the patient, but to establish the most complete psychic assessment possible, in the same manner as a cardiac assessment or electrolytes or blood assessments.   If nothing pathological is found this status at admission will be very important for a good understanding of the future evolution with the patient's return to his home or placement in a Retirement Home. In the case of deterioration (one has only to think of the regressive or confused states which occur during hospitalization) the assessment made upon admission and which was considered normal, will determine the diagnosis and future treatment.

In the case of a manifest psychiatric pathology the intervention will be made in rapport with the clinical condition.   The presence of a depressive state, a delirious condition, or a state of agitation, will be given the necessary care and appropriate treatment, and the care-giving team will be alerted to the problems of the treatment to be followed.   In connection with this last action the contribution of the psychiatrist even becomes indispensable in the multidisciplinary teams where, in their common meetings there is frequent discussion of problem cases.   One has only to think of the patients who have behaviour problems, character problems, delirious or suicidal ideas which, at times, are associated with the aggressive reactions of the family or the rejection of the other patients.

For some patients it may well be the dosage of psychotropic medications which pose some problems.   In such cases also, the advice of the psychiatrist can be very important if one bears in mind the difficulties of choosing an antidepressive for such an aged person.   For example, there are the risks of too much sedation, asthenia, fatigue, somnolence, tremors, insomnia, agitation, confusion, difficulties of adjustment, urinary retention, etc.   These are only some examples, and the same holds true for the neuroleptic medications which can provoke a series of even

greater accidents, not to mention the tranquillizers which also
provoke their quota of observable secondary effects.   It is
unnecessary to mention that in daily medical practice, these
complications of medication often cause the non-specialist to rejec
such treatment or to lower the dosage to such a low level that it
is no longer effective.

Prudence in the use of psychotropic drugs in no way implies
that they should be avoided in geriatrics.

Continuing our description of the present structures, we must
point out the presence of psychologists who are available to
conduct the psychometric or projective tests and verbal and
corporal psychotherapy (relaxation). They also animate the
discussion groups as well as the therapeutic groups.   A propos of
this last-mentioned group, many are held with the patients; with
the families of the patients there are also the Balint-type groups
for the medical and paramedical personnel.

The above summary description of the psychogeriatric
organisation applies also to the extra-mural responsibilities of
the Geriatric Center with the slight difference that the Center has
an occupation more specifically psychiatric.

The five regional and pluridisciplinary teams of the Center
are composed of a doctor, whose training is in general psychiatry,
a social worker, an occupational therapist, a secretary, a
psychologist, and, if needed, a physiotherapist.   These teams are
all supervised by a psychiatrist, Chief of Service.   They have the
entire responsibility for aged patients in their homes, in about
forty regularly supervised homes for the aged, in three Day Centers
and in six sheltered housing establishments.

About 24% of the patients have only psychiatric problems; 5%
have only purely somatic difficulties, and about 71% suffer from
both kinds of problems.

The Geriatric Center provides consultant psychiatric advice
for a certain number of large Nursing Homes which have medical
services, and it also has a consultation service for psychogeriatric
cases in the Cantonal Hospital which contains 1800 beds and in which
the number of aged persons exceeds 50%.

With the years the Geriatric Center has seen the number of
consultations requested by private doctors in the city augment
considerably.   However, this is always a parallel treatment where
the Center does not substitute for the practising doctor but
contributes to the treatment with his specific psychiatric training.
This specific psychiatric training has been acquired by the Center's
fourteen years of working within the medical structures existing in
Geneva.

In concluding we will not dwell further upon the multiple aspects of the Geriatric Center, for they are identical to those described for the Geriatric Hospital of Geneva, with the exception that the Center has a character and service more particularly psychiatric. We have preferred to speak rather about the Geriatric Hospital of Geneva and its "integrated medicine" in order to demonstrate an unusual aspect of hospital medicine and the model of an original concept of medical service.

If this type of medical approach is the best that can be offered to a patient, it is not accomplished without institutional conflicts and problems, apparent and latent.   Actually, medicine of this nature requires the psychological apprenticeship of the patient and the institution by all the care-givers. New rules must follow which are not those learned in the Medical Faculty or in the Schools of Nursing.   Nor are they the simplest.

For it is not by accident that throughout the centuries madness has always been relegated to the psychiatric ghettos in the same way as the millenary separation of the body and soul of a patient.

SUMMARY

Since they began, the Geriatric Institutions of Geneva have practiced an "integrated medicine", namely, the integration into institutional practice of somatic medicine and psychological medicine, in this case internal medicine and psychiatry.

After many years of experience of hospital and extra-mural geriatrics this integration has shown itself to be very beneficial in the method of treatment of the aged patient and here, psycho-geriatrics fulfills two roles.   The first role is a clinical one, that is, investigation, diagnosis and treatment, always bearing in mind the interreactions between the patient and his illness, the patient and his family and social environment.   The second role is that of sensitizing the doctors, the nurses and the other care-givers to the multiple psycho-affective aspects of the relationship care-giver to patient, which is especially difficult in geriatrics.

REFERENCES

1. J. -P. Junod, I. Simeone, La geriatria. Pubblicazioni mediche ticinesi, 33:9 (1975).
2. K. Fortini, La médecine intégrée en Gériatrie. Le rôle de la psychiatrie en gériatrie (à paraître).
3. L. Ciompi, C. Muller, Lebensweg und Alter der Schizophrenen. Springer, Berlin - Heidelberg - New York (1976).
4. C. Muller, Influence de l'âge sur les maladies mentales préexistantes. Schweiz. med. Wschr. 95: (1965).

5. J. Oules, Les névroses du troisième âge. _Confrontations psychiatriques_, 5:83 (1970).

6. U. Lehr, R. Schmitz-Scherrer, L'état de bonne santé et le processus psychique du vieillissement, _Gérontologie_ 75:19 (1975).

7. R. E. Zimmermann, Alter und Hilfsbedürftigkeit, Enke, Stuttgard (1977).

8. M. Roth, The principles of providing a service for psycho-geriatric patients, _in_ "Roots of Evaluation", J. K. King and H. Häfner, eds., Oxford University Press, London (1973).

9. A. Jablensky, La sante mentale du troisième âge. Division de la Santé mentale. Organisation Mondiale de la Santé, Genève.

10. G. Goda, I. Simeone, Rôle de la psychiatrie en pratique gériatrique, _Médecine et Hygiène_ 35:3702 (1977).

11. P. Tavernier, Analyse de certaines activités gériatriques extra-hospitalières à Genève, Thèse. Genève (1977).

SUBJECT INDEX

Acetylcholine, 28, 29, 35, 89

Acetylcholine esterase, 28, 89, 136, 276

ACTH, 43

Acute brain syndrome
  see: Confusional states

Adenylyl cyclase, 16, 17, 39

Adjective check list, 299, 303

Adrenergic receptor, 20, 135

Adrenal, 42, 43

Affective disorder, 74, 145, 147
  see: Depression

Aging, 1, 11, 15, 25, 36, 88, 133-135, 217, 229, 244, 276, 295, 316, 323

Agnosia, 75, 82

Alcoholism, 90, 158, 211

Aluminium, 27

Alzheimer's disease, 26, 27, 28, 80, 89, 324
  see: Dementia

Amantadine, 348

Amimia, 83

Amitriptylline, 212

Angiography, 165, 225

Anomia, 202

Anticholinergic drugs, 147, 158, 213, 345

Antidepressants, 92, 134, 145, 212

Antiparkinson drugs, 144

Antiplatelet drugs, 126

Anxiety, 67, 72, 75

Aphasia, 75, 82, 108
  Broca's, 198
  Wernicke's 196
  global, 199

387